Perspectives in Cognitive Neuroscience

Stephen M. Kosslyn, *General Editor*

Hemispheric Asymmetry: What's Right and What's Left

Joseph B. Hellige

HARVARD UNIVERSITY PRESS
Cambridge, Massachusetts
London, England
1993

Printed in the United States of America
10 9 8 7 6 5 4 3 2 1

This book is printed on acid-free paper, and its binding materials have been chosen for strength and durability.

Library of Congress Cataloging in Publication Data
Hellige, Joseph B.
Hemispheric asymmetry : what's right and what's left / Joseph B. Hellige.
p. cm.—(Perspectives in cognitive neuroscience)
Includes bibliographical references and index.
ISBN 0-674-38730-9 (alk. paper)
1. Cerebral dominance. 2. Dominance, Cerebral. I. Title. II. Series.
[DNLM: 1. Brain—physiology. WL 335 H477h]
QP385.5.H45 1993
612.8′25—dc20
DNLM/DLC
for Library of Congress 92-49175
CIP

To the memory of my parents,
Bernard Francis Hellige and Mary Margaret Moffitt

Contents

Preface

The goal of cognitive neuroscience is to understand how the brain produces the range of information-processing abilities and propensities that characterize our species. This is a very large goal and it has many components. This book focuses on just one aspect of the project: cerebral hemispheric asymmetry. It has been known for literally hundreds of years that the two sides of the human brain are different in their processing abilities and preferences and that some of the differences are quite dramatic. During the last thirty years or so there has been a nearly geometric increase in the number of published articles dealing with hemispheric asymmetry. The popularity of this topic can be attributed to the realization that aspects of hemispheric asymmetry have important implications for a wide range of topics in cognitive neuroscience. *Hemispheric Asymmetry* reviews recent research on the differences between the left and right cerebral hemispheres and examines many of the implications of these differences.

The very large number of research studies makes it impossible for me to provide anything even remotely close to an exhaustive review. Instead, I concentrate on research published within the last decade or so, with the intent of being illustrative rather than exhaustive. I have tried to show how this recent research is related to earlier theory and data and to include sufficient references to earlier research articles and reviews to provide entry points into this vast literature. At the same time, I have also tried to indicate where future research is most needed and most likely to lead to significant new insights. The book is sufficiently self-contained for it to be understandable to individuals with little previous knowledge about hemispheric asymmetry, although some passing knowledge of cognitive psychology or other aspects of cognitive neuroscience should help. At the same time, I have attempted to make the present review timely enough to be of interest and value to active researchers who study hemispheric asymmetry and interhemispheric interaction.

I could not have written this book without the assistance of many people. Steve Kosslyn, who encouraged me to write a book on hemispheric asymmetry, also made generous comments on an earlier version. Angela von der Lippe, Editor for the Behavioral Sciences at Harvard University Press, guided me through the writing and publication process with skill and patience. Kate Schmit did an excellent job of copyediting the manuscript. Much of my own research discussed in this book was supported in part by research grants from the National Science Foundation, and for that I am very grateful. I owe thanks to the many colleagues, postdoctoral fellows, graduate students, and undergraduate students with whom I have collaborated on studies of hemispheric asymmetry and from whom I have learned a great deal: Gretchen Bauer, Darlene Bibawi, Mike Bloch, Barbara Cherry, Steve Christman, Bill Corwin, Elizabeth Cowin, Pam Cox, Tami Eng, Zohar Eviatar, Lisa Gohl, Jon Jonsson, Dan Kee, Kathy Kellum Hirschman, Fred Kitterle, Santosh Kumar, Lidia Litvac, Buck Longstreth, Chikashi Michimata, Bryan Moseley, Michael O'Boyle, Ron Saul, Justine Sergent, Vicki Sergent, Karen Stevens, Annette Taylor, Ron Webster, Cheryl Wolcott, Tony Wong, and Judy Zatkin. Several of these individuals read and commented on portions of the manuscript, and their comments were very helpful in preparing an improved version. I especially want to thank three of these colleagues for reading and commenting in detail on the penultimate version of the entire manuscript: Steve Christman, Michael O'Boyle, and Justine Sergent. Our discussions led me to important insights that are reflected in the version that you are reading. Thanks also go to Vicki Sergent for preparing several of the original figures published here.

Very special thanks go to my wife, Colleen, and to my three daughters, Erin, Katie, and Bridget. They have all been very patient with me as I worked on this book during the last two years or so. They have also helped me to keep things in perspective, to keep my sense of humor, and to refrain from taking myself too seriously. My older daughters, Erin and Katie, often looked after their baby sister when I needed to sit down at home and work, and for that I am very grateful. In her own way, Bridget also helped—by keeping her older sisters busy and reminding us all just how much fun simple things can be. The biggest thanks of all go to Colleen, without whose love and support I would never have undertaken this project. I will be indebted to her forever for this and for the many other things that she has given me.

Hemispheric Asymmetry

1

Introduction

The cortex of the brain is the capstone of human evolution. It is made up of more neurons than any other brain structure and is necessary for the higher-order mental processes that make us unique among species. Consequently, it should come as no surprise that attempts to understand the cortex have come to occupy a central place in cognitive neuroscience. The cortex is divided anatomically into two hemispheres—the right and the left. Although the two hemispheres are roughly symmetrical in appearance, they are not completely equivalent in their information-processing abilities and propensities. The resulting *hemispheric asymmetry* is the focus of attention in this book.

Particularly vivid demonstrations of hemispheric asymmetry were provided in the 1960s and early 1970s in studies with so-called split-brain patients, which captured the imagination of both scientists and the lay public (e.g., Gazzaniga, 1970, 1985; Sperry, Gazzaniga, and Bogen, 1969). During the last three decades we have seen a nearly geometric increase in the number of articles and books dealing with hemispheric asymmetry. In fact, ideas about hemispheric asymmetry have become so widespread that it seems safe to assume that virtually all readers of this book have encountered popularized accounts of varying degrees of detail and accuracy. One of the most common beliefs is that the left hemisphere controls symbolic processing and rational thinking whereas the right hemisphere is more artistic, intuitive, and creative. This idea has become so familiar that it is even used in product marketing. For example, a few years ago a number of magazines carried ads promoting Saab as a "car for the left side of your brain" by citing a variety of favorable statistical details as well as "a car for the right side of your brain" by picturing the sleek automobile speeding along a ribbon of undulating highway. You may also recall James Garner, the actor, promoting beef as a food for both sides of your brain—the left because it is so good for you and the right because it tastes so good. (Of course, these commercials disappeared shortly after Garner underwent heart bypass surgery!) Popularized accounts of hemispheric asymmetry were followed quickly by the idea that the thoughts and feelings of an individual may be dominated more or less completely by one hemisphere, the dominant side being correlated with a host of individual differences in cognitive ability. In fact, I have heard some students excuse their poor academic performance because they were doomed as "a right-brain person in a left-brain world."

With so many oversimplified accounts of hemispheric asymmetry available in the popular press, it is not surprising that it has become difficult to sort the facts from the fantasy of hemispheric asymmetry—to sort out what's right and what's left in terms of hemispheric dominance. One purpose of this book is to help you do just that. Of course, our understanding of hemispheric asymmetry is still being shaped by new observations of the behavior of neurological patients and by experiments conducted with neurologically intact individuals. This being the case, it is not possible to paint a portrait that gives the final answers to all important questions about hemispheric asymmetry. Instead, I will try to present a freeze-frame view of our current state of understanding, with sufficient historical perspective to show why the current frame looks the way it does and sufficient speculation about tomorrow's discoveries to anticipate what the next frame might look like.

The task of sorting fact from fantasy is not always easy because research on the topic of hemispheric asymmetry is often controversial and experimental results are sometimes contradictory. Nevertheless, I will try to tell a coherent story while remaining true to the corpus of empirical results. My approach is to emphasize those empirical findings that are most firmly established and for which there are converging data from a variety of research techniques. At the same time, promising new leads are reviewed in order to provide the most up-to-date picture of what we know about hemispheric asymmetry. I note where there has been controversy about a specific result and where I believe more data are needed either to bolster an empirical finding or to clarify the theoretical interpretation. In this way, I hope to help you see what's right (correct) about the concept of hemispheric asymmetry by focusing on what's left after critical analysis of the theories advanced so far and by noting what is left to be discovered.

Five general themes emerge from my view of the topic, and it is useful for you to know what they are at the outset. Accordingly, I have set them out here at the beginning. In order to understand and evaluate contemporary views about hemispheric asymmetry, it is necessary for you to know something about the various research techniques that are used to study asymmetry. With this in mind, I next turn to a brief survey of the different populations and methods used to learn about behavioral asymmetries in humans. This is followed by an overview of the remainder of the book. Finally, like all the others in this book, this chapter ends with a brief summary of the main points I have tried to make in the chapter, emphasizing the conclusions to which those points lead.

Five Recurring Themes

The themes that I will try to bring out throughout this book have been influenced both by the history of research on hemispheric asymmetry and by recent data that have led to a re-thinking of the subject. Knowing what these themes are from the beginning may help you to organize information as it is presented and to anticipate where we are going.

1. Hemispheric asymmetries exist and infuence behavior. There are a great many cognitive and behavioral asymmetries in human beings, many of which can be attributed to hemisphere or brain asymmetries. This theme may seem obvious, in the sense that the entire book deals with hemispheric asymmetry, but I point it out for a reason. So many popularized accounts of hemispheric asymmetry have taken so much liberty with existing data that they have become misleading at best and fictitious at worst. As a result, the entire concept of hemispheric asymmetry is sometimes treated with skepticism (if not derision) by some members of the scientific community. While it is true that so-called laterality studies have been of uneven quality and that many of the popularized accounts are misleading, it is important to understand the facts of hemispheric asymmetry and to consider what the existence of behavioral asymmetry implies for cognitive neuroscience.

2. *We have one brain, not two.* When thinking about hemispheric asymmetry, it is critical to be mindful of the unity of the brain. That is, the two hemispheres are parts of a much larger, anatomically extensive system that encompasses both hemispheres, the connections between them, numerous subcortical structures, and more. All too often this fact is forgotten in the enthusiasm about hemispheric asymmetry and what has come to be called "hemispheric specialization." In recent years, a new awareness of the fact that information processing is anatomically extensive has led to more interest in topics like the manner in which the two hemispheres interact in the normal brain to produce unity of thought and action.

3. *Other species have asymmetries too.* There are many reliable behavioral and biological asymmetries in nonhuman species. Furthermore, some of these asymmetries are analogous or homologous to asymmetries in humans. This was not widely known or believed as recently as ten years ago. The discovery of so many animal models has opened the door to new kinds of investigation and has implications for how we think about the evolution of hemispheric asymmetry.

4. *Individuals differ in asymmetry.* Individuals differ from each other in patterns of hemispheric asymmetry and in the ways that the two hemispheres interact. In the past, much of this individual variation was treated as random error and ignored. In fact, many of the individual differences are reliable, and we have a few enticing leads about how these individual differences relate to cognitive abilities and propensities. While it is true that popular accounts of hemispheric asymmetry have suggested that we know far more about individual differences than we do, these overzealous popularizations should not lead us to ignore the genuine differences that do occur.

5. *Asymmetry unfolds over (human and evolutionary) time.* We need to consider new ideas about why and how hemispheric asymmetry unfolds over time. This is an important question whether we are asking about the life span of a single individual or evolutionary

time. Across the life span of an individual, functional hemispheric asymmetries are shaped by the complex interaction of many biological and environmental factors, beginning with the fetus as it develops *in utero* and continuing into old age. In terms of evolution, the recent documentation of so many asymmetries in other species leads to questions about whether tool use and language were of central importance for the evolutionary emergence of hemispheric asymmetry in humans and to new scenarios about the distant seeds of present-day hemispheric asymmetries.

Learning about Behavioral Asymmetries in Humans

Claims about hemispheric asymmetry and about the foregoing themes are only as good as the empirical data on which they are based. As you read the rest of this book and as you read about tomorrow's discoveries, you need to understand the advantages and limitations of the techniques used to study the various facets of hemispheric asymmetry.

The notion that the left and right sides of the brain are functionally asymmetric came originally from observations regarding the behavior of humans. Behavioral asymmetries in humans have been studied in a variety of populations and with several research techniques. When the goal is to discover functional asymmetries between the cerebral hemispheres, each population and strategy has advantages and disadvantages. Taken together, the various techniques provide the opportunity for strong converging tests of hypotheses about hemispheric symmetry and asymmetry. Associated with each technique are a host of logical and methodological issues that must be considered in the design and interpretation of experiments. While a detailed discussion of methodological issues is beyond the scope of the present book, I will outline in this section the techniques that have been used and point out what can and what cannot be learned from well-done studies. The approaches discussed include examination of (1) the consequences of unilateral brain injury, (2) split-brain patients and the positive competence of each hemisphere, (3) perceptual asymmetries in neurologically normal individuals, (4) response asymmetries in neurologically normal individuals, and (5) measures of localized brain activity. As a way of illustrating the kind of data that are provided by each approach, I will give examples of results that indicate a special role for the left hemisphere in a variety of linguistic and verbal tasks.

The Consequences of Unilateral Brain Injury

If the human brain were completely symmetrical, so that the left and right cerebral hemispheres were functionally equivalent, then injury to homologous areas of the two hemispheres should have equivalent effects. Even a cursory review of the neuropsychological literature indicates dramatic instances where this is not the case. In fact, Broca (1861) and Wernicke (1874) were led to postulate the importance of specific areas of the left hemisphere for certain aspects of language as a result of observing the correlation between injury to particular regions of the left hemisphere and specific language disorders. The areas implicated by Broca and by Wernicke are illustrated in Figure 1.1, which presents a diagram of the surface of the human left hemisphere. Details of the specific language deficits associated with injury to each of these areas will be reviewed later. For the moment, it is sufficient to note that similar disorders of language are much less likely after injury to the homologous areas of the right hemisphere, indicating an asymmetry of function.

In view of these often dramatic asymmetries in behavioral deficits, it is not surprising that patients with unilateral brain injury are an important source of data arguing against functional symmetry. Accordingly, such data will constitute one source of converging information throughout the remainder of this book. However, a host of methodological issues arise in considering such studies (for discussion see Bradshaw, 1989; Bradshaw and Nettleton, 1983; Efron, 1990; Hellige, 1983a). In addition, even when the studies are done perfectly from a methodological point of view, there are important logical issues that must be considered in the interpretation of the asymmetries that are discovered.

In order to illustrate the logical issues, assume that we have unequivocal evidence that ablation of a specific area in the left hemisphere eliminates a person's ability to speak fluently, whereas ablation of the homologous area in the right hemisphere does not. What can and what cannot be concluded from this finding?

With respect to the issue of symmetry versus asymmetry, the interpretation of such findings is quite clear—the hemispheres must be functionally asymmetric. While that is not a trivial conclusion, it is silent about the specific nature of the asymmetry or the mechanisms responsible for it. One popular and inviting conclusion is that the left hemisphere contains an area that is somehow "specialized" for producing fluent speech, but this conclusion does not follow with necessity or even high probability from the data that have been described so far.

For example, it could be that the results have nothing whatsoever to do with any *specific* cognitive or motor ability. Instead, left-hemisphere injury may *always* lead to greater impairment than does injury to homologous areas of the right hemisphere—with

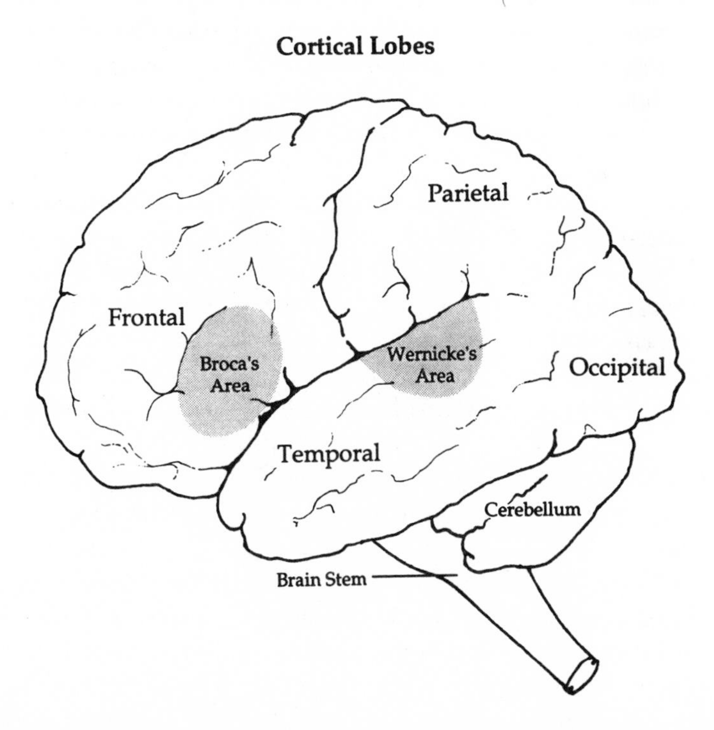

Figure 1.1. In this diagram of the left hemisphere of the human brain, the approximate locations of the major lobes are labeled. Two areas that are especially important for certain aspects of language, Broca's area and Wernicke's area, are shaded.

the specific ability involved being determined by exactly where within the hemisphere the injury occurs. One way to rule out this possibility is to show that there are other abilities that are spared when the left hemisphere is injured and that some of those abilities are impaired after right-hemisphere injury. This type of hemisphere-by-task interaction has been referred to as a *double dissociation* in the neuropsychological literature (e.g., Teuber, 1955).

Even with the demonstration of double dissociations, the interpretation of hemispheric asymmetry is not resolved completely. One set of questions deals with the nature of the specific deficits that are observed in an individual. For example, an activity like "fluent speech" could be impaired by a variety of different cognitive and motor deficits. For example, the patient's memory for the names of objects might be impaired, the ability to use grammar might be impaired, use of the musculature of the face might be impaired, and so forth. Therefore, it is always a challenge to test individuals in a way that identifies *specific* subprocesses that have been influenced. Another set of questions concerns the distinction between brain areas that are *necessary* for performing a specific process and brain areas that are *sufficient* for performing a specific process. At its best, the observation of a specific impairment after localized brain injury can tell us that brain tissue in that area is somehow necessary for normal function. It does not tell us that the tissue in the injured area is specialized for the now-impaired process, that the now-impaired process is in any sense "housed" or "localized" in that area of the brain, or that the tissue in the injured area is normally involved at all in the now-impaired process.

What we are left with is the conclusion that studies of unilateral brain injury can provide unequivocal evidence for the existence of hemispheric asymmetry. With proper care, they can even lead to generalizations about the functions that are more impaired after left-hemisphere as opposed to right-hemisphere injury. This is an important clue and should not be underestimated. It does not follow from these studies, however, that some function is *completely localized* in one hemisphere or the other—the localization or dominance or specialization hypothesis is one possibility, but we need to look for other types of evidence to test it.

As noted earlier, for well over a hundred years neuropsychologists have proposed that the left hemisphere plays a special role in both the production and perception of language. It is often said that the left hemisphere is "dominant" for linguistic or verbal processing or for "language." This conclusion was reached initially on the basis of observing the incidence of language disorders after unilateral brain damage (e.g., Broca, 1861; Wernicke, 1874). It is now well established that *aphasia* (the acquired loss of language) is far more likely after left-hemisphere than after right-hemisphere injury and that the specific nature of the symptoms depends on which regions of the left hemisphere are injured (see Berndt, Caramazza, and Zurif, 1983; Bradshaw, 1989; Ellis and Young, 1988). A more detailed review of these findings is contained in Chapter 2. For now, it is sufficient to note that one of the most reliable asymmetries concerns the production of speech. In fact, studies of patients with unilateral brain injury have led to estimates that the left hemisphere is dominant for speech in approximately 95 percent of right-handed adults, with the right hemisphere being dominant for speech in the other 5 percent of right-handed adults (e.g., Segalowitz and Bryden, 1983). In line with the logical issues noted earlier, such results demonstrate that the integrity of certain areas within the left hemisphere is *necessary* for the production of speech and certain other language-related activities.

Split-Brain Patients and the Positive Competence of Each Hemisphere

In the normal brain, the left and right hemispheres are richly interconnected, with the corpus callosum and anterior commissures being the largest tracts of interconnecting fibers. Dramatic evidence for hemispheric asymmetry in humans comes from studies with patients who have had the corpus callosum and other connecting fibers surgically severed in order to control the spread of epileptic seizures. In fact, it is the dramatic asymmetries exhibited by these so-called split-brain patients that created the present *Zeitgeist* in which the study of hemispheric asymmetry has flourished, and it was for his pioneering work with these patients that Roger Sperry received a Nobel Prize in 1981.

What makes these patients so interesting is that the left and right hemispheres are disconnected and no longer communicate with each other. Consequently, with a variety of clever techniques, it is possible to examine how each hemisphere functions in isolation—what Zaidel (1983a) has called the *positive competence* of each hemisphere. We saw earlier how difficult it can be to infer the functions of an area of the brain by observing what functions are impaired when that area is injured. This being the case, evidence of the positive competence of each hemisphere in isolation provides important converging data for understanding the functions for which each isolated hemisphere is *sufficient.*

In order to examine the competence of a single hemisphere, it is necessary to present stimulus information to only one hemisphere at a time and to give one hemisphere the opportunity to respond. This can be a tricky business (several methodological issues are discussed in Efron, 1990; Gazzaniga, 1970; and Hellige, 1983a). Some techniques take advantage of the fact that information from each half of the body or each half of space typically projects directly only to the contralateral cerebral hemisphere. For example, tactile information presented to the fingers of the left hand projects primarily or exclusively to the right hemisphere, whereas tactile information presented to the fingers of the right hand projects primarily or exclusively to the left hemisphere. Consider also the case of vision. In humans, neurons from the temporal half of each retina project to the ipsilateral visual cortex (e.g., optic neurons from the temporal half of the right eye project to the visual cortex of the right hemisphere), whereas neurons from the nasal half of each retina project to the contralateral visual cortex (e.g., optic neurons from the nasal half of the right eye project to the visual cortex of the left hemisphere). As a result of this anatomical arrangement, when the two eyes are fixated directly on a point in space, visual information from each side of the fixation point (each *visual field*) projects directly to the contralateral visual cortex (e.g., information from the right visual field projects to the visual cortex of the left hemisphere). This anatomical arrangement is illustrated in Figure 1.2.

It is possible to take advantage of this anatomical arrangement in the following way. The patient is instructed to fixate his gaze on a dot in the center of a screen. When the eyes are fixating appropriately, a visual stimulus (e.g., a word) is flashed very briefly to the left or right visual field. The duration is so brief (less than 200 milliseconds or so) that the stimulus disappears before the eyes can move to look directly at it. Consequently, in the split-brain patient only one hemisphere "sees" the stimulus on each trial—the hemisphere contralateral to the visual field that is stimulated. By varying the nature of the stimuli that are used, the nature of the task that is to be performed, and the nature of the response that is to be given, it is possible to use this technique to learn whether the two hemispheres have equivalent abilities.

With respect to hemispheric asymmetry for linguistic or verbal

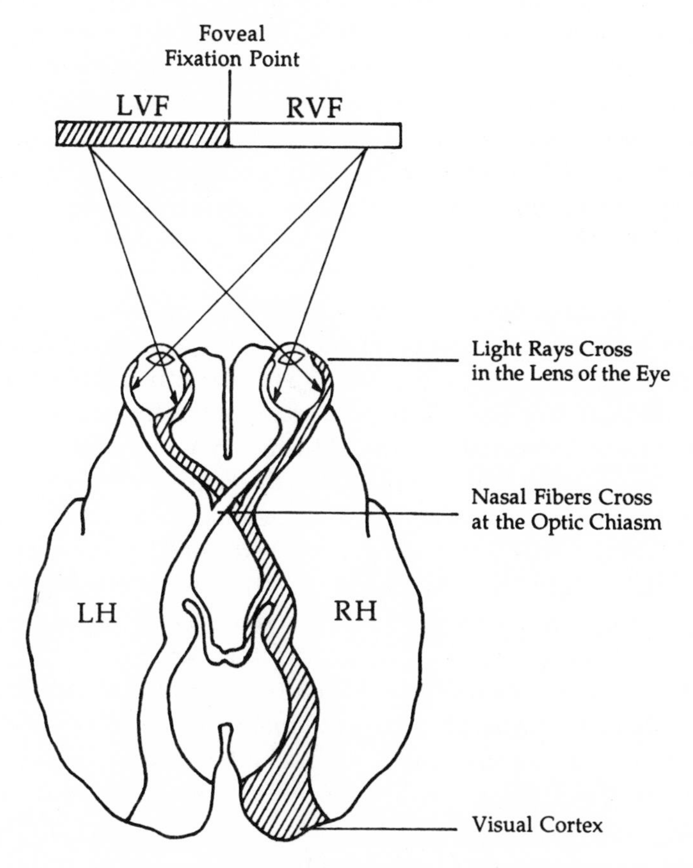

Figure 1.2. The anatomical arrangement of the visual projection system in humans. Note that information from each visual half-field is projected directly to the visual cortex of the contralateral hemisphere.

processing, results obtained from split-brain patients have supported the conclusion that the left hemisphere plays a special role (especially for speech production and for the identification of phonetic information). For example, if a common object (e.g., a comb) is placed into a patient's right hand (with tactile sensation projecting to the left hemisphere), the patient can say the name of the object (e.g., "comb"). However, if the same object is placed into the left hand (with tactile sensation projecting to the right hemisphere), the patient is unable to say the name of the object. (The objects, of course, are held behind a screen so the patient is unable to see them.) Likewise, if either a picture of an object (e.g., a picture of a comb) or the written name of the object (e.g., *comb*) is flashed briefly to the right visual field (RVF)/left hemisphere, the patient can name the object or say the word. However, when the same information is flashed to the left visual field (LVF)/right hemisphere, the patient is unable to do so (Gazzaniga, 1970, 1985; Gazzaniga and LeDoux, 1978). Such results indicate that the left hemisphere is not only *necessary* for certain linguistic or verbal processes, it is also *sufficient.*

Additional tests with these patients indicate that it is not that the right hemisphere does not "know" what object has been presented. For example, when only the right hemisphere has received information about an object (e.g., via the left hand or via the LVF), patients are typically able to pick out the correct object with the left hand (the hand controlled by the right hemisphere). However, the right hemisphere is unable to speak and, therefore, the patient is unable to say the name of the object. In fact, in over 40 split-brain patients that had been studied until 1983, only 5 were thought by Sidtis and Gazzaniga (1983) to have any right-hemisphere language ability, and most of those could not *speak* from the right hemisphere. Hemispheric asymmetry for the *perception* of phonetic information is indicated by the fact that only the left hemisphere can typically make judgments that require phonetic decoding—for example, indicating whether two words with different spellings rhyme with each other (e.g., Zaidel and Peters, 1981).

Although the study of split-brain patients is a particularly intriguing converging technique to study hemispheric asymmetry, it is important to acknowledge certain limitations that make it complicated to generalize from split-brain patients to neurologically normal individuals (e.g., Whitaker and Ojemann, 1977). For one thing, all of these patients suffer from severe epilepsy and are virtually certain to have neurological abnormalities. Furthermore, the abnormalities have probably been present for long enough so that they may have led to atypical patterns of hemispheric asymmetry in at least some individuals. In addition, each patient has a unique history of behavioral and neurological problems, so we cannot consider the collection of split-brain patients a homogeneous group. Of course, the large individual differences make it that much more impressive when similar findings are obtained from a number of such patients.

Despite the potential problems just noted, the careful study of individual split-brain patients can tell us a great deal about the positive competence of each hemisphere in isolation; that is, about the processes for which each isolated hemisphere is sufficient. However, the hemispheric asymmetries that are discovered in split-brain patients may not operate as they do in the intact brain, with the full potential of interhemispheric interaction. For example, it is possible that the efficiency of interhemispheric communication is so great that the functional significance of hemispheric asymmetry for a particular task is very small in neurologically intact individuals relative to split-brain patients. Alternatively, as I will discuss later, it is possible that in the intact brain activity in one hemisphere inhibits activity in homologous regions of the other hemisphere. Such reciprocal inhibition in the intact brain (which is eliminated in the split-brain) might serve to magnify the left/right performance asymmetries in neurologically intact individuals relative to split-brain individuals. In view of these considerations, it is important to examine hemispheric asymmetry in neurologically intact individuals as well as in various patient populations. It is to such techniques that we now turn attention.

Perceptual Asymmetries in Neurologically Intact Individuals

In the neurologically intact brain, it is not possible to test the competence of each cerebral hemisphere in isolation, but it is possible to examine the speed and accuracy with which people perform a variety of tasks as a function of which hemisphere receives a stimulus directly and must, at least, initiate processing. Such examinations of perceptual laterality are possible in various stimulus modalities and are patterned after the techniques used to lateralize stimuli to a single hemisphere in split-brain patients. Of course, in the neurologically intact brain, information presented directly to one hemisphere can be shared with the other hemisphere far more easily than in split-brain patients. In order to illustrate some of the issues involved in these perceptual laterality studies, I will discuss the use of visual-half-field presentation outlined earlier for split-brain patients. Following this, I will note how similar logic applies to other stimulus modalities.

Because of the anatomy of the human visual system, information flashed to one side of the visual field projects directly to the visual cortex of the contralateral cerebral hemisphere (see Figure 1.2). There are several ways in which hemispheric asymmetry has been hypothesized to lead to performance differences as a function of the visual field that is stimulated. Despite the potential for transfer of information from one hemisphere to the other, one possibility is that the hemisphere that receives the stimulus information directly (e.g., the left hemisphere when the right visual field is stimulated) carries out all of the processing and shares, at most, only the final result with the other hemisphere. This possibility, which has been referred to as the "direct access" model by Zaidel (1983a), seems to characterize what happens in some tasks (such as deciding whether or not a string of letters spells a word). According to this model, performance on each trial reflects the processing style and competence of only one hemisphere—much as in the split-brain patient.

An alternative way in which hemispheric asymmetry might lead to perceptual asymmetry has been termed the "callosal-relay" model (Zaidel, 1983a). On this view, processing for certain tasks always takes place in the same hemisphere (e.g., the left hemisphere) regardless of which visual field is stimulated. Thus, the stimulus information either reaches the "appropriate" hemisphere directly (by being presented in the visual field that projects directly to that hemisphere) or indirectly (by being transferred from the opposite hemisphere, which received the information directly). It is usually assumed that performance on a task will be better if the stimulus information projects directly to the appropriate hemisphere than if it travels a longer, indirect anatomical route. This might occur because the longer route is "noisy" and degrades the input to the appropriate hemisphere.

A third possibility, considered by Kinsbourne (1970, 1975), is based on ideas about the balance of activation between the two cerebral hemispheres. On this view, tasks that involve one hemisphere more than the other lead to greater biological activation of the more involved hemisphere. Kinsbourne hypothesizes that, when the two hemispheres are activated unequally, attention is directed more easily to the side of space contralateral to the more activated hemisphere. Consequently, performance will be better when a stimulus is presented to the visual field contralateral to the hemisphere that is more activated by the task being performed.

Detailed discussion of how we might distinguish among these alternatives for a particular task is beyond the scope of the present chapter (but see Bouma, 1990; Hellige, 1983a; Zaidel, 1983a). In the remainder of this book, I will note whenever it is important to make assumptions about which of these three mechanisms is dominating performance. For present purposes, it is sufficient to note that it is likely that hemispheric asymmetry influences perceptual laterality in a variety of ways. As a result, it is possible to test various hypotheses about hemispheric asymmetry in the normal brain by examining perceptual laterality.

The fact that hemispheric asymmetry is likely to produce perceptual asymmetries does not mean that perceptual asymmetries cannot be influenced by a host of other factors. For example, visual-field differences can be influenced by voluntary differences in the distribution of attention to the two sides of space, by post-exposure scanning biases, by the order of reporting multiple stimuli, by asymmetries in the peripheral visual pathway, and so forth. When the goal is to use visual-field asymmetries to study hemispheric asymmetry, therefore, it is necessary to rule out explanations in terms of other factors. Although this can be difficult, it is not impossible (the interested reader should consult methodological discussion in Beaumont, 1982; Bryden, 1982; Hellige, 1983a; and a special issue of *Brain and Cognition,* 1986, vol. 5, no. 2).

The study of perceptual laterality is not restricted to vision. One of the most popular techniques involves examining how performance on tasks using auditory stimuli depends on the ear to which the stimulus is presented. As illustrated in Figure 1.3, each ear sends information to both cerebral hemispheres. However, the pathway from each ear to the contralateral hemisphere contains more fibers than does the pathway from each ear to the ipsilateral hemisphere. In addition, when two items are presented simultaneously to both ears (a procedure called *dichotic* presentation), the pathway from each ear to the ipsilateral hemisphere is inhibited, especially if the two items are acoustically similar. Thus, with dichotic presentation, information presented to one ear projects primarily or exclusively to the contralateral cerebral hemisphere. Accordingly, ear differences in audition (like visual-field differences) are likely to be influenced by hemispheric asymmetries and

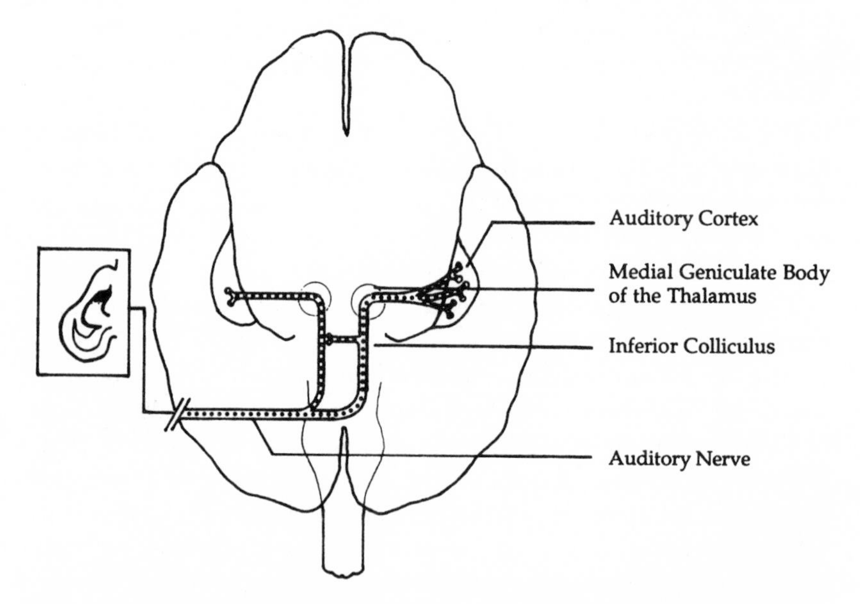

Figure 1.3. The anatomical arrangement of the auditory projection system in humans. Note that each ear sends projections to both cerebral hemispheres, but that projections to the contralateral hemisphere are more numerous than projections to the ipsilateral hemisphere. As discussed in the text, the ipsilateral projections are inhibited when two different (but acoustically similar) items are presented simultaneously, one to each ear.

so have become an important converging technique with both split-brain patients and neurologically intact individuals. For additional methodological discussion about the use of dichotic presentation, see Bryden (1982), Hellige (1983a), and Hugdahl (1988).

Although most studies of perceptual asymmetries have used visual or auditory stimuli, there have been some investigations of tactile asymmetry that follow the same general logic. Such studies take advantage of the fact that tactile information presented to one hand is presented primarily or exclusively to the contralateral cerebral hemisphere (e.g., O'Boyle, van Wyhe-Lawler, and Miller, 1987).

Investigations of perceptual asymmetry provide additional support for the hypothesis that the left hemisphere is in some sense dominant for certain linguistic or verbal functions. For example, when words or pronounceable nonwords are flashed briefly to one visual field or the other, there is a robust RVF/left-hemisphere advantage for right-handed observers, even when aspects of the procedure make it unlikely that any of the artifacts noted earlier could influence performance. To illustrate this sort of result, consider the following experimental paradigm.

In a series of experiments, my colleagues and I required right-handed observers to identify a consonant-vowel-consonant (CVC) nonsense syllable (e.g., *DAG*) projected briefly to the LVF/right hemisphere or RVF/left hemisphere on each trial. On each trial, the CVC was followed by a masking stimulus to help control the time available for processing, and the duration of the CVC was varied throughout the experiment so that performance averaged across the two visual fields was approximately 50 percent. Figure 1.4 shows the visual-field differences for each of 100 observers. For this group as a whole, there was a robust RVF/left-hemisphere advantage for CVC recognition, with the mean number correct on RVF/left-hemisphere and LVF/right-hemisphere trials being 21.7 (of 36 possible) and 14.5 (of 36 possible), respectively. For each observer, the number correct on LVF/right-hemisphere trials was subtracted from the number correct on RVF/left-hemisphere trials in order to produce the difference scores plotted on the horizontal axis of Figure 1.4. What the figure shows is the percentage of subjects with difference scores of various magnitudes. Notice that not only is there an RVF/left-hemisphere advantage in the group mean, but 85 out of 100 observers recognized more items on RVF/left-hemisphere trials than on LVF/right-hemisphere trials, 12 observers showed the opposite visual-field advantage, and 3 observers showed no difference between the two visual fields.

In dichotic-listening experiments, there is typically a right-ear (RE)/left-hemisphere advantage for the identification of auditorially presented words and syllables. For example, Figure 1.5 shows the results for 100 right-handers who attempted to identify consonant-vowel syllables presented on earphones. The syllables used were /*ba*/, /*da*/, /*ga*/, /*pa*/, /*ta*/ and /*ka*/. On each of 120 trials, two of these syllables were presented simultaneously and partici-

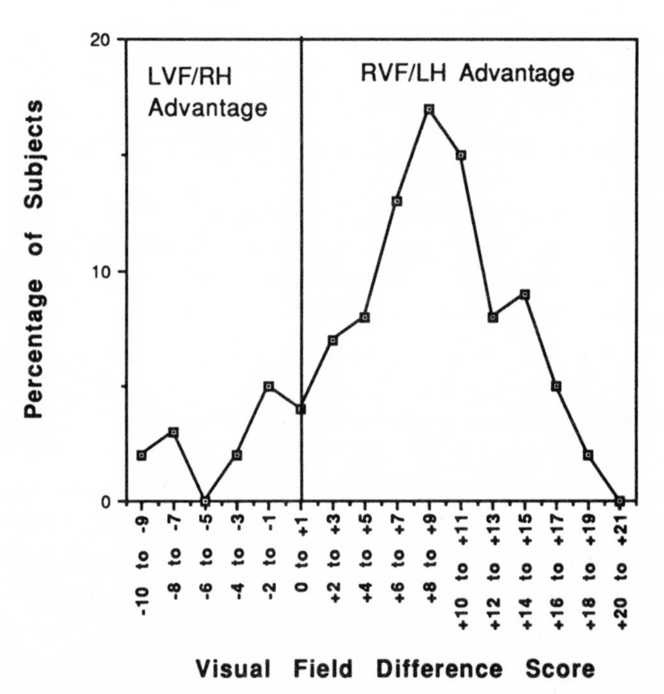

Figure 1.4. Visual-field difference scores on a syllable-identification test for 100 right-handed observers. The syllables were shown to the left or right visual fields (LVF or RVF) on each trial, and the difference score was calculated by subtracting the number of correct responses to LVF stimuli from the number of correct responses to RVF stimuli (with a maximum possible score of 36). Difference scores are plotted along the horizontal axis and the percentages of observers showing difference scores of each magnitude are shown on the vertical axis.

pants attempted to identify which two from the set of six had been presented. The results were typical for this paradigm. On the average, more syllables were identified from the right than from the left ear (RE mean = 84.7 correct identifications out of 120 possible; LE mean = 67.4 correct identifications out of 120 possible). For each participant, I have computed a difference score by subtracting the number of correct identifications of LE stimuli from the number of correct identifications of RE stimuli. Figure 1.5 shows the percentage of subjects whose difference scores fell into categories of various magnitudes. Note that the RE score was greater than the LE score for 80 participants, the ear difference was reversed for 19 participants, and 1 participant showed exactly equal performance for the two ears. The RE advantage for iden-

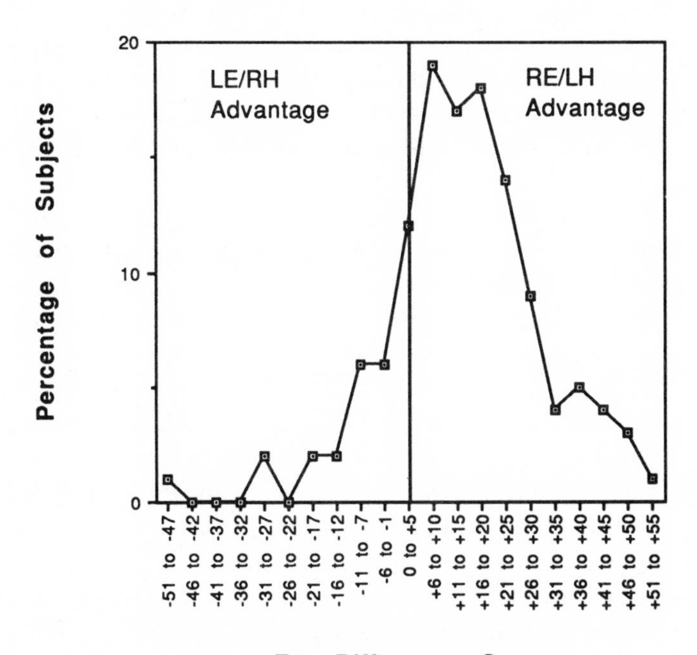

Figure 1.5. Ear difference scores on a dichotic-listening test for 100 right-handed listeners. Syllables were presented simultaneously to both the left and right ears (LE and RE), and the difference score was calculated by subtracting the number of correctly identified syllables to the LE from the number of correctly identified syllables to the RE (with a maximum possible score of 120). Difference scores are plotted along the horizontal axis and the percentages of listeners showing difference scores of each magnitude are shown on the vertical axis.

tifying words and syllables is even more dramatic in split-brain patients (e.g., Zaidel, 1983a).

Response Asymmetries in Neurologically Intact Individuals

Some overt behavioral responses are produced primarily or exclusively by only one cerebral hemisphere. Therefore, certain types of response asymmetry have also been used to provide converging information about hemispheric asymmetry. The approach that is usually taken is to examine how responses controlled by each hemisphere are influenced by the performance of a variety of concurrent tasks. The idea is to determine whether some tasks involve more processing from one hemisphere than from the other.

In typical experiments of this type, individuals are asked to perform a manual task with either the left or right hand, sometimes performing the manual task alone and sometimes performing the manual task along with an additional *concurrent* task. An effort is made to use a manual task, such as tapping of the index finger, for which there is good reason to believe that the movements of one hand are controlled by the contralateral hemisphere. In such studies, the interest is in the extent to which concurrent tasks interfere differentially with the manual activity of the two hands. The assumption is that when the concurrent task requires more processing resources from one hemisphere, the hand controlled by that hemisphere will show more interference (e.g., slowing of tapping rate) when the concurrent task is imposed (see Chapter 3). For now, it is sufficient to note that concurrent verbal activities (e.g., reciting a nursery rhyme, repeating nonsense syllables, reading aloud and silently, solving anagram problems, and so forth) have been shown to interfere more with manual activity performed by the right hand than with manual activity performed by the left hand (for reviews see Hellige and Kee, 1990; Kinsbourne and Hiscock, 1983).

Measures of Localized Brain Activity

Another strategy for learning about behavior and the brain in general and hemispheric asymmetry in particular is to examine the extent to which different regions of the brain are "activated" while an individual is performing some cognitive or motor task. The assumption that underlies this approach is that those areas of the brain that are more involved in the ongoing processing will be more active while that processing is in progress. With respect to hemispheric asymmetry, the goal is to determine whether some tasks involve more activation from an area within one hemisphere than from the homologous area within the other hemisphere.

Before we review briefly some of the specific measures of activation that are taken, it is instructive to consider what a perfect measure of localized brain activation can and cannot reveal. For the moment, assume that by "activation" we mean that the neurons within an area are firing at high rates, thereby using a great deal of metabolic resources. While it is undoubtedly useful to know what specific areas are more or less activated at a given moment, knowing that a particular area is more active does not reveal anything about exactly *what* that area is doing or *how well* it is doing it. In fact, it is entirely possible that an area that can perform a specific cognitive function efficiently will be less active and use fewer metabolic resources than an area that performs that function less efficiently. In addition, many of the neural connections in the brain are inhibitory, which means that a decrease in activation within a particular area is likely to be as important a clue for localization of function as an increase in activation. Another problem is that even simple tasks involve a number of specific operations or subprocesses and it is not trivial to determine which specific subprocesses change the activation of a particular area of the brain. Furthermore, the whole idea that brain activation changes when an individual is performing a task necessitates the measure of a baseline level of activation against which to measure change. Choosing the most appropriate baseline is not always straightforward.

The oldest and still most widely used techniques to study brain activity involve a variety of electrical measures, with the two most popular being electroencephalograms (EEG) and measures of event-related potentials (ERPs). Briefly, the EEG is a continuous recording of electrical activity from which the ongoing activity is broken down into several frequency ranges. The frequency of greatest interest is usually 8 to 12 Hz (the *alpha* frequency);

suppression of activity in the alpha range is thought to indicate ongoing processing. The ERP is a recording that is time-locked to a particular stimulus event. The electrical responses to several repetitions of the stimulus event are averaged and the resulting average ERP is examined for the magnitude and latency of various positive and negative shifts in electrical potential. With both of these techniques, electrodes are placed at various (usually standardized) sites on the scalp and the results from different recording sites are compared. When the goal is to study hemispheric asymmetry, particular interest is placed on differences between the electrical activity measured at homologous sites on the two sides of the scalp. One advantage of these electrical measures is that they have excellent temporal resolution (e.g., they can register changes during a fraction of a second), but they have poor spatial resolution—at best, they might localize activity with a resolution of a few inches (but see Gevins, 1983).

A variety of other techniques offer better spatial resolution (1–2 mm) but much lower temporal resolution (ranging from 30 seconds or so to over 30 minutes). In fact, some of the techniques, like computerized axial tomography (CAT) scans, are useful for studying morphological asymmetries in the brain but are not able to provide a dynamic picture of how activation changes during task performance. Other methods, however, such as positron emission tomography (PET) scans, nuclear magnetic resonance imaging (NMRI) scans, and various other measures of regional cerebral bloodflow (rCBF), are able to provide dynamic pictures. The logic behind rCBF measures is as follows. When a particular area of the brain has been metabolically active, localized increases in carbon dioxide lead to a dilation of blood vessels serving that area. As a consequence, blood flow is increased to that area. Now that we have techniques to measure the relative amounts of blood flow to localized brain areas, it is possible to determine how the regions of greater activation change with differences in task demand. Of course, as noted earlier, depending on the specific technology employed, an individual must engage in the same task for anywhere from 30 seconds to several minutes in order for the measures to be taken, and this requirement places limits on the types of tasks that can be used and the inferences that can be made (for a review and discussion of important issues see Bradshaw, 1989; Wood, 1983; Wood, Flowers, and Naylor, 1991).

Various asymmetries in brain activation will be discussed in more detail later. For now, it is sufficient to note that studies using a number of these measures of localized activation are consistent with the conclusion that areas of the left hemisphere are more activated by certain linguistic and verbal tasks than are homologous areas of the right hemisphere. For example, EEG studies have shown suppression in alpha activity (thought to be indicative of greater information processing) over the left than over the right hemisphere during a variety of verbal tasks, and the opposite has been found for certain nonverbal tasks (e.g., Galin and Ornstein, 1972). In addition, during verbal tasks rCBF has been found to be greater for specific areas of the left hemisphere than for homologous areas of the right hemisphere (e.g., Risberg, Halsey, Wills, and Wilson, 1975; see also Posner, Petersen, Fox, and Raichle, 1988). It is important to note, however, that these measures indicate that *many* areas of *both* hemispheres are activated to some extent by all the tasks that have been used in tests to this point. While this does not argue in favor of complete hemispheric symmetry, it reinforces one of the themes listed earlier: processing in the unified brain is anatomically extensive and hemispheric asymmetry must be understood in this context.

Additional techniques are sometimes used to study hemispheric asymmetry, but their use is infrequent and they will not be outlined in this chapter. Instead, details of those methods will be reviewed when results are described from particular studies. One example is the use of the so-called Wada amytal test, which involves injection of barbituate anesthetic into the left or right carotid artery. The carotid artery on each side provides the blood supply to the ipsilateral hemisphere; the effect of a unilateral injection is to anesthetize one hemisphere. This technique is used prior to certain types of neurosurgery to help identify which hemisphere in a particular individual is dominant for language. It is interesting that studies with this technique suggest that the left hemisphere is dominant for speech in 90–95 percent of right-handed adults (e.g., Rasmussen and Milner, 1977)—an estimate very close to that obtained from surveys of studies of patients with unilateral brain injury.

Even from this brief overview of research strategies and techniques it is apparent that no single approach can provide a complete picture of hemispheric asymmetry. This being the case, the strongest conclusions will emerge when a number of different techniques are brought to bear on the same issues. To the extent possible, the remainder of this book will focus on questions that have been addressed in a number of different ways and on answers that receive converging support from studies of different types.

The Plan of the Book

Hemispheric asymmetry has been studied most extensively in humans, and many of the most important new findings continue to come from investigations of neurologically normal humans and of patients with brain injuries of various sorts. Accordingly, the first several chapters of the book focus on hemispheric asymmetries in humans. You have already learned something about the techniques used to study hemispheric asymmetry and some of what they have taught us about hemispheric asymmetry for language. Chapter 2 begins with a review of several of the more widely discussed behavioral asymmetries that have been discovered using those methods. Of particular importance is a discussion of the search for the fundamental nature of behavioral asymmetries. This discussion will include criticism of the once-popular idea that all behavioral asymmetries can be reduced to a single fundamental dichotomy. To replace this idea, I present in Chapter 3 a componential approach to the study and understanding of hemispheric asymmetry. Briefly, the componential approach acknowledges the fact that even relatively simple tasks require the coordination of a number of information-processing subsystems or modules, and recent research shows that hemispheric asymmetry differs for some of the modules postulated by componential models of perception, cognition, and action. A very important aspect of this componential approach is that the discovery that different processing subsystems have different neurological correlates lends support to theories in which such subsystems are treated as separable and independent. Chapter 4 supplements the discussion of *behavioral* asymmetries with a review of recently discovered *biological* asymmetries in the human brain and a consideration of their relationship to the behavioral asymmetries.

Chapter 5 deals with behavioral and biological asymmetries in nonhuman species. Until recently, systematic left/right asymmetries in the behavior and brains of nonhumans were considered rare, unimportant, and unrelated to hemispheric asymmetry in humans. Lately, however, the increased interest in functional hemispheric asymmetry in humans and the discovery that such asymmetries are not restricted to language have led to a renewed search for animal models of brain asymmetry. The resulting (and very recent) literature documents the ubiquity of both behavioral and biological asymmetries in nonhumans. In many cases, these asymmetries are reminiscent of those found in humans and provide important clues about the biological mechanisms of asymmetry.

Chapter 6 addresses the important question of how the two hemispheres, with their different processing abilities and propensities, interact with each other to form an integrated information-processing system. It contains a review of the techniques used to study interhemispheric interaction and illustrates the types of interactions that have been discovered. In particular, I illustrate the cooperation of the two hemispheres on tasks that involve subsystems for which the hemispheres are asymmetric, consider when it is and is not beneficial to spread processing across both hemispheres, and discuss the concept of metacontrol for tasks that the hemispheres are predisposed to handle in qualitatively different ways.

Chapter 7 focuses on individual differences in hemispheric asymmetry. It begins by discussing the various ways in which individuals might differ in aspects of hemispheric asymmetry; that is, the dimensions of individual variation. These dimensions will then be used to consider the relationship of hemispheric asymmetry to a variety of subject variables: right- versus left-handedness, sex, intellectual abilities, and psychopathology. The chapter ends with a discussion of the merits and demerits of the notion of "hemisphericity" (e.g., in what sense, if any, are there "right-brained" versus "left-brained" people?) and what might be the consequences of hemispheric dominance for cognitive abilities and propensities.

Chapter 8 deals with the important topic of how hemispheric asymmetry unfolds across the life span, from conception through old age. An important conclusion that emerges is that hemispheric asymmetry may not "develop" in the traditional sense of that term. Instead, the seeds of hemispheric asymmetry seem to be sown long before birth, with the asymmetry beginning to unfold during gestation and continuing after birth, perhaps even into old age.

Chapter 9 considers the evolutionary history of hemispheric asymmetry. It begins with a discussion of evolutionary pressures that favor bilateral symmetry versus bilateral asymmetry and with a discussion of the extent to which certain asymmetries might be continuous across species. Following this is a discussion of several milestones in hominid evolution—walking upright, tool manufacture and use, language, and prolonged immaturity—that may have led to the asymmetries that are characteristic of present-day humans. Important clues about the manner in which various hemispheric asymmetries are related seem to be buried within our evolutionary past.

The final chapter, Chapter 10, returns to the five themes described earlier and uses them to discuss what would be required of a comprehensive model of hemispheric asymmetry. This discussion serves as a framework for suggesting fruitful directions for future research.

Summary and Conclusions

The two hemispheres of the human cerebral cortex have different information-processing abilities and propensities. The purpose of this book is to review our current understanding of this hemispheric asymmetry by discussing what's right and what's left in terms of hemispheric dominance and by emphasizing what's right (correct) about the concept of hemispheric asymmetry by focusing on what's left after critical analysis of existing theory and data.

Five themes emerge from the present review and recur throughout the book. (1) There are many cognitive and behavioral asymmetries in humans that can be attributed to hemispheric asymmetries. (2) The two hemispheres are parts of a much larger, anatomically extensive system, and it is the whole system that must be taken into account when considering hemispheric asymmetry. (3) There are many behavioral and biological asymmetries in nonhuman species; these differences open the door to new kinds of investigation and have important implications for how we think about the evolution of hemispheric asymmetry. (4) Individuals differ from each other in patterns of hemispheric asymmetry and in the ways that the two hemispheres interact. (5) Questions about the unfolding of hemispheric asymmetry over the life span of an individual and about the evolution of hemispheric asymmetry in the human species have important implications for understanding the nature of hemispheric asymmetry and the mechanisms that create it.

The notion of functional hemispheric asymmetry came originally from observations of behavioral asymmetries in humans. These asymmetries have been studied in a variety of populations and with many research techniques. The most commonly used approaches include investigation of (1) the behavioral consequences of unilateral brain injury, (2) split-brain patients and the positive competence of each hemisphere in isolation, (3) perceptual asymmetries in neurologically normal individuals, (4) response asymmetries in neurologically normal individuals, and (5) measures of localized brain activity. When the goal is to make inferences about functional asymmetries between the hemispheres, each population and each experimental strategy has advantages and disadvantages. Although none of the approaches is perfect by itself, taken together they provide the opportunity for strong converging tests of hypotheses about hemispheric symmetry and asymmetry.

Results obtained with each of these approaches indicate a special role for the left hemisphere in a variety of linguistic and verbal tasks, especially in speech production and the perception of phonetic information. Debate continues about the specific processing subsystems responsible for these asymmetries, but the consistency of findings across such a wide range of approaches is very impressive.

2

Behavioral Asymmetries in Humans

Literally hundreds of behavioral asymmetries have been reported in humans during the last thirty years and many of them have been attributed to hemispheric asymmetry. From this large database have emerged several widespread generalizations about the tasks for which each hemisphere is dominant, generalizations that are often regarded as well established and relatively noncontroversial in the neuropsychological literature. For example, virtually all reviews of hemispheric asymmetry contain the ideas that the left hemisphere is dominant for several aspects of motor control and language whereas the right hemisphere is dominant for nonverbal, visuospatial processing, distributing attention across space, and processing emotion. An attempt to catalog all of the behavioral asymmetries that support (and sometimes contradict) these views is far beyond the scope of the present book. Consequently, in the first section of this chapter I have chosen to concentrate on reviewing those behavioral asymmetries that are used most often to support claims about the functions for which each hemisphere is dominant, with the goal of being illustrative rather than exhaustive. In this review, I emphasize asymmetries that have been replicated sufficiently often to have been established beyond a reasonable doubt. At the same time, I will note certain inconsistencies that still remain and, wherever possible, consider how those inconsistencies might be resolved.

For a time, one primary theoretical goal of research on hemispheric asymmetry was to discover *the* fundamental information-processing dimension along which the hemispheres differ and from which all of the specific behavioral asymmetries could be derived. In the second section of this chapter, I will review the search for this dimension and will argue that this quest has been largely unsuccessful and that it should be abandoned.

Before beginning a review of behavioral asymmetries, I want to highlight two important conclusions to which the review will lead. One is that, with the possible exception of overt speech and one or two other tasks, it is rarely the case in the intact brain that one hemisphere can perform a task normally whereas the other hemisphere is completely unable to perform the task at all. Instead, even when hemispheric asymmetry for a task exists, it is typically the case that both hemispheres have some ability to perform the task, even though one may do a better job than the other. For some tasks, one hemisphere performs better because the task is approached in different ways as a function of which hemisphere is injured or which hemisphere receives the stimulus information—one *approach* leads to better performance than the other. A second important conclusion is that even simple tasks consist of a number of components or subprocesses, and there is no guarantee that hemispheric asymmetry for one component is the same as that for another component. Thus, it is not usually easy to state which hemisphere is "superior" for a multicomponent task.

A Review of Behavioral Asymmetries

The asymmetries reviewed here are divided into categories according to the nature of the cognitive or motor processes that are likely to be involved. Although the boundaries between categories cannot always be precise, the division serves a useful historical and heuristic function. The categories include (1) handedness and the control of motor activities, (2) language, (3) visuospatial processing (including the processing of human faces), and (4) emotion.

Handedness and the Control of Motor Activities

Voluntary motor control in humans involves several areas of the cerebral cortex, including the primary motor cortex, secondary motor cortex, and premotor cortex, as well as various subcortical structures (e.g., Rosenbaum, 1990). The primary motor cortex is located along a band shaped like an arch that extends from ear to ear across the top of the brain. Most of the body can be matched to the part of the primary motor cortex that controls it—making a kind of "map." The area of the cortex devoted to a particular part of the body corresponds to the precision with which that part of the body can be controlled. One way of representing this mapping is illustrated in the right half of Figure 2.1. Note, for example, that a much larger area of the primary motor cortex is devoted to the hands and fingers than to the trunk. Of particular importance for the study of hemispheric asymmetry is the fact that the primary motor cortex of each hemisphere controls voluntary movement primarily in the contralateral side of the body, with the extent of contralateral motor control increasing for movement of body parts further away from the center of the trunk. Thus, whereas movements of one side of the trunk may involve equal control by both the contralateral and ipsilateral hemi-

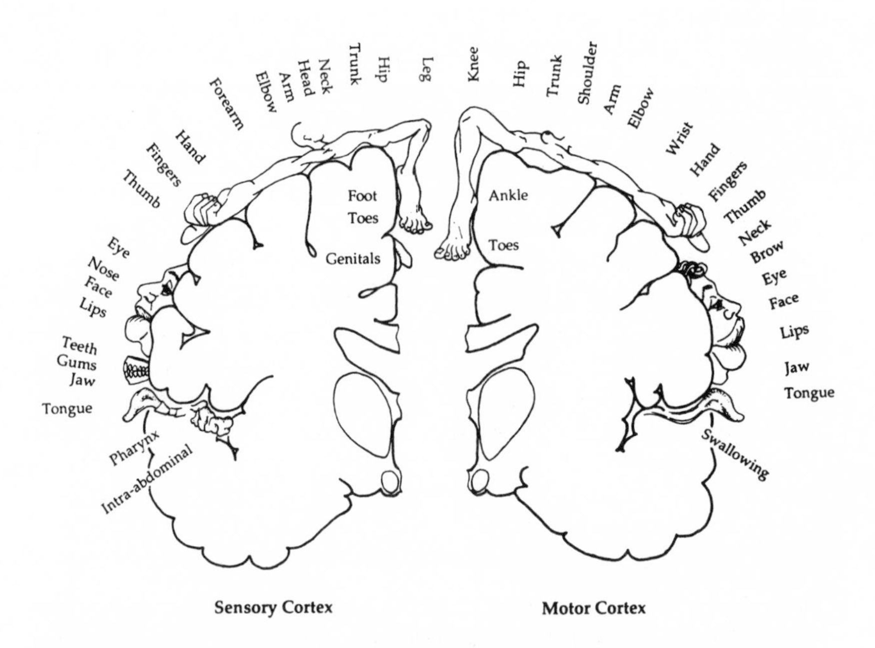

Figure 2.1. Diagram of the area of sensory and motor cortex devoted to different portions of the body. Note that some areas (such as the face and hands) have a large cortical representation relative to their size, whereas other areas (such as the trunk) have a much smaller cortical representation.

spheres, movements of the fingers of one hand involve either exclusive or superior control by the contralateral hemisphere.

Immediately behind the primary motor cortex is a second arch (the somatic sensory cortex) that receives input from the skin, bones, joints, and muscles. Once again, most of the body can be mapped onto the somatic sensory cortex, and again the area of cortex devoted to a particular part of the body corresponds to the sensitivity of that part of the body to incoming stimuli (see the left half of Figure 2.1). It is important to note that the somatic sensory cortex of each hemisphere receives sensations primarily from the contralateral side of the body, although ipsilateral connections are not always completely absent (e.g., Trope, Fishman, Gur, Sussman, and Gur, 1987).

The most obvious behavioral asymmetry in humans is handedness. Most of us write proficiently with only one hand and classify ourselves as right- or left-handed on the basis of which hand is used for writing. In fact, within an individual, proficiency of the two hands differs for many other tasks as well, and often the hand difference is dramatic. In view of the predominantly contralateral control of voluntary movements, handedness indicates that one hemisphere's hand control is either better than or preferred to the other. It will come as no surprise to you that approximately 90 percent of humans prefer to use the right hand rather than the left for a variety of tasks. Thus, the left hemisphere is dominant for hand control for a large majority of humans. In Chapter 7, I will consider the relationship of hand dominance to other motor and cognitive asymmetries. For the moment, it is sufficient to know that there are some important relationships of this sort, although they are far from perfect. Consequently, unless I state otherwise, the remaining asymmetries to be discussed in the present chapter are based on research with right-handed individuals.

Right-handed individuals display a number of other motor asymmetries. For example, the left hemisphere is superior to the right in the learning and execution of a sequence of movements. Kimura and Archibald (1974) tested patients with unilateral brain injury on a test that involved copying hand postures that were demonstrated by the experimenter. Both left- and right-hemisphere-injured groups were able to copy a single posture without difficulty. However, the left-injured patients had difficulty reproducing a *sequence* of hand postures whereas the right-injured patients did not. The fact that the left-injured patients could tell whether the sequence of hand postures produced by the experimenter was correctly reproduced by someone else indicates that they had no trouble perceiving and remembering the sequence. The problem was in the ability to *produce* a sequence of movements (see also Kimura, 1977; Kolb and Milner, 1981). It is interesting that this production problem was present for both hands in these patients, suggesting that the integrity of the left hemisphere is important for sequential movements on both sides of the body (see also Corballis, 1991; Wyke, 1971). Left-hemisphere injury is also more disruptive than right-hemisphere injury when the task is to produce a sequence of articulatory movements of the mouth and tongue (e.g., Mateer and Kimura, 1977; see also Sussman, 1979). These findings have led to the conclusion that an important function of the left hemisphere is the control of changes in limb or articulatory posture. In fact, Kimura (1977) has suggested that more complex left-hemisphere superiorities (such as those for speech) derive from this asymmetry of motor control.

Although the preceding asymmetries are generally considered to be well established, the complete picture of hemispheric asymmetry for limb movements is more complex. For example, the results depend on exactly what task is being performed and on exactly which aspects of movement are measured; there is sufficient inconsistency in the empirical findings to prevent a strong conclusion about specific mechanisms at the present time. On the basis of a programmatic series of studies using patients with injury to one hemisphere or the other, Haaland and Harrington (1989a; Haaland, Harrington, and Yeo, 1987; Harrington and Haaland, 1991) propose that the left hemisphere is dominant for controlling open-loop movements, which are rapid, programmed movements performed with little or no modification by sensory input. This seems consistent with many of the findings related to left-hemisphere dominance for sequencing movements of the limbs and articulatory muscles. By way of contrast, Haaland and Harrington (1989a) report no evidence of hemispheric asymmetry for controlling closed-loop movements, which are slower and modified from moment to moment by sensory feedback. However, using different tasks, Haaland and Harrington (1989b) found some indication of right-hemisphere dominance in programming closed-loop movements (see also Carson, Goodman, and Elliott, 1992). In addition, in their own study of brain-injured patients, Fisk and Goodale (1988) found that right-hemisphere injury produced impairments in initiating a closed-loop movement toward a visual target whereas left-hemisphere injury produced impairments in the final stages of guiding movements to the target (the corrective components characteristic of closed-loop movements). In future attempts to resolve inconsistencies in motor-performance studies, it is important to consider to what extent the specific tasks used are able to separate hemispheric differences in processing visual location from hemispheric differences in movement per se. In addition, some of the inconsistency found in existing studies may also be attributable to other differences in procedure and to differences in the specific etiology of patients (for discussion, see Haaland and Harrington, 1989a,b).

When neurologically normal individuals speak, there is an asymmetry in the hand movements that they make. Right-handed individuals are equally likely to make self-touching movements with both hands but they make more free-movement gestures (gestures of the hands and arms away from the body) with their right hand (e.g., Kimura, 1973; Lavergne and Kimura, 1987). Because the production of free movements is related to the occurrence of speech, Kimura and her colleagues suggest that the movements result from activation of areas of the left hemisphere that program both the sequential articulatory movements necessary for speech and movements involving changes in limb posture.

Additional support for these hypotheses comes from studies using methods of brain stimulation first introduced by Penfield (1958) and refined by Ojemann and his colleagues (1983; Ojemann and Mateer, 1979). In this technique, a small electrode is used to explore the exposed cortex of a conscious patient prior to neurosurgery. Applying a small current to the probing electrode sometimes results in the inability of the patient to perform some task. The primary clinical value of this technique is that it helps the surgeon to identify those areas most critical to speech, so that they can be avoided as much as possible during the subsequent surgery. Of relevance to the present discussion is the finding that both *movements* of the facial musculature surrounding

Figure 2.2. Photographs of well-known personalities showing a variety of asymmetrical facial expressions. Note that in this sample the expressions are more pronounced on the left side of the face in approximately 80 percent of the cases. [Reprinted from M. Moscovitch and J. Olds, "Asymmetries in Sponteneous Facial Expression and Their Possible Relation to Hemispheric Specialization," *Neuropsychologia,* 20 (1982):71–82. Copyright 1982 by Pergamon Press plc. Reprinted by permission.]

the mouth and the ability to *identify* acoustically presented phonemes are altered by stimulation of the same locations. Such findings are consistent with the hypothesis that there is a common system for production and perception of phonemes.

Various motor asymmetries have also been observed in the faces of neurologically intact individuals. Wolf and Goodale (1987) studied asymmetries in the amplitude and velocity of oral movements in right-handed individuals as they either pronounced syllables or produced nonverbal movements of the mouth (e.g., blowing). By analyzing videotapes frame by frame, they found that the right side of the mouth opened wider and faster than the left side for both verbal and nonverbal movements, and that the magnitude of this asymmetry increased with the complexity of the movements. Because the muscles on each side of the mouth are stimulated more directly by the contralateral hemisphere, Wolf and Goodale relate these results to the hypothesis that the left hemisphere plays a dominant role in the control of complex motor behavior.

There is reason to believe that facial asymmetries depend on the task performed and on the measures that are taken. For example, when an individual smiles or expresses other emotions, the two sides of the face are rarely mirror images of each other (see Figure 2.2). Furthermore, the two sides of the face are not usually rated as equal in the intensity of emotional expression. Typically, the left side of the poser's face (controlled more by the right hemisphere) is seen as more emotionally expressive. For example, Moscovitch and Olds (1982) videotaped people relating an emotional experience from their lives and later counted the number of expressive facial movements. For right-handed (but not left-handed) individuals there were more expressive movements made by the left side of the face than by the right side (see also Borod, Koff, and White, 1983; Borod, St. Clair, Koff, and Alpert, 1990). It seems, therefore, that asymmetries of facial movement are related to hemispheric asymmetry for the cognitive aspects of the task being performed.

Language

Just as handedness is the most obvious external manifestation of behavioral asymmetry, left-hemisphere dominance for many aspects of language is the most obvious and widely cited cognitive asymmetry. As we have already seen in Chapter 1, the evidence of left-hemisphere dominance for the production of *overt speech* and for *phonetic processing* of incoming auditory and visual information comes from a variety of research techniques and a variety of neurological patients and intact individuals. This section reviews additional language-related functions for which one hemisphere or the other is thought to be dominant.

Early characterizations of different language disorders were based on grossly defined language tasks. For example, injury to an area of the left hemisphere anterior to the fissure of Rolando and adjacent to an area that is important for controlling facial musculature (Broca's area in Figure 1.1) was thought to impair the ability to produce fluent speech. Consequently, the resulting "Broca's aphasia" was referred to as "expressive" or "nonfluent" aphasia. In contrast, injury to an area of the left hemisphere posterior to the fissure of Rolando near an area that is important for auditory perception (Wernicke's area in Figure 1.1) was thought to impair the ability to understand speech, with the resulting "Wernicke's aphasia" referred to as "receptive" or "fluent" aphasia. From the point of view of hemispheric asymmetry, it is important to note that injury to homologous areas of the right hemisphere has a much less disruptive effect on language.

Although problems with production versus reception of language continue to be used as markers of Broca's aphasia versus Wernicke's aphasia, recent neurolinguistic research shows that the problems are not related to output versus input per se. As noted by Berndt, Caramazza, and Zurif (1983; see also Caramazza and Martin, 1983), patients with injury to Broca's area are deficient in using *syntactic* information (e.g., syntactically important grammatical morphemes such as auxiliary verbs, modals, prepositions, verb tenses, number markers, etc.) in both the production and understanding of language. One reason that the problem of production is more noticed in these individuals is that speech devoid of grammatical structure sounds unusual and telegraphic. Furthermore, in normal conversation much of the meaning can be understood on the basis of semantic and pragmatic information—even if little syntactic information is processed. Anyone who has learned a second language well enough to develop a reasonable vocabulary but has not mastered the syntactic structure knows that this is the case. By way of contrast, patients with injury to Wernicke's area have difficulty using *semantic* information in both the production and understanding of language, but their ability to use syntactic information correctly is relatively intact. As a result, their speech sounds "normal" but it doesn't make any sense and there are clear problems of understanding.

Language functions for which the left hemisphere is dominant are not restricted to speech output and understanding of the spoken word. One indication of this comes from interesting studies with individuals who have been deaf since birth and who are proficient at producing and understanding signs made with the hands. Injury to areas of the left hemisphere that are known to be important for language in hearing individuals disrupts the ability of deaf patients to produce and understand sign language. That the "language" aspect is critical is illustrated by the fact that the ability to produce and perceive nonlinguistic visuospatial relationships is spared (see Bellugi, Poizner and Klima, 1983). Converging evidence about language asymmetry in the congenitally deaf comes from visual-half-field investigations in neurologically intact individuals. Whereas there is an LVF (right-hemisphere) advantage for the recognition of meaningless hand positions, there is an RVF (left-hemisphere) advantage for meaningful signs (see Poizner, Klima, and Bellugi, 1987, for a review).

Another indication that left-hemisphere dominance for aspects of language is not restricted to speech output and understanding of the spoken word comes from studies of reading problems in patients with brain injury and in visual-half-field studies of split-brain patients and neurologically intact individuals. Some of the evidence was reviewed in Chapter 1. Additional evidence comes from the fact that various acquired dyslexias (reading problems that result from brain injury) are far more likely after left- than after right-hemisphere injury.

A particularly interesting form of acquired dyslexia, *deep dyslexia,* is almost always accompanied by extensive injury to the left hemisphere. Perhaps the most interesting and characteristic symptom of deep dyslexia is the occurrence of semantic paralexias—reading aloud a printed word as another word that is similar in meaning (e.g., "stone" for *rock* or "money" for *cost*). Deep-dyslexic patients also tend to: (1) have difficulty reading nonwords (e.g., *gife*), (2) have more difficulty reading aloud abstract words like *virtue* than concrete words like *elephant,* (3) have more difficulty reading function words like *quite* than content words like *queen,* (4) make frequent visual errors (e.g., misread *decree* as "degree"), and (5) make morphological errors (e.g., misread *edition* as "editor"). (See Ellis and Young, 1988.) One reason for the intense interest in deep-dyslexic patients is the possibility that, in view of the extensive left-hemisphere injury, the reading performance of these patients is mediated by the right hemisphere (e.g., Coltheart, 1983, 1985; Zaidel and Peters, 1981). From this point of view, it is interesting that the performance of deep dyslexics is qualitatively similar to the performance of the isolated right hemisphere in split-brain patients (e.g., Zaidel and Peters, 1981). However, the reading performance of deep dyslexics is typically better than that of split-brain patients relying on the right hemisphere, a difference that is consistent with the alternative hypothesis that reading in deep dyslexics is mediated in part by the injured left hemisphere (e.g., Hinton and Shallice, 1991; Patterson and Besner, 1984; Seidenberg and McClelland, 1989). For additional discussion about the existence and possible nature of right-hemisphere language and reading, see Baynes, Tramo, and Gazzaniga (1992), Bradshaw (1989), Gazzaniga (1983), Levy (1983), and Zaidel (1983b,1985).

At this point, there is no doubt that the left hemisphere is typically dominant for a number of important aspects of language: overt speech, phonetic decoding, syntactic and semantic processing—although with the exception of overt speech there is little evidence that the right hemisphere is completely without ability. It is reasonable to ask whether there are additional aspects of language for which the right hemisphere is typically dominant and the answer seems to be "yes." For example, right-hemisphere injury is more likely than left-hemisphere injury to produce difficulties in understanding and utilizing the context in which an utterance occurs, in appreciating necessary presuppositions and in comprehending metaphors—things that involve the *pragmatic* aspects of language (for reviews, see Gardner, Brownell, Wapner, and Michelow, 1983; Brownell, Simpson, Bihrle, Potter, and Gardner, 1990; Hough, 1990; Kaplan, Brownell, Jacobs, and Gardner, 1990).

By way of example, consider which of the endings would be the best punch line for the following joke:

> The neighborhood borrower approached Mr. Smith on Sunday afternoon and inquired: "Say Smith, are you using your lawnmower this afternoon?"
> "Yes, I am," Smith replied warily.
> The neighborhood borrower then replied:
> (a) "Fine, then you won't be needing your golf clubs. I'll just borrow them."
> (b) "You know, the grass is greener on the other side."
> (c) "Do you think I could use it when you're done?"
> (d) "Gee, if only I had enough money, I could buy my own."

You will probably agree that ending (a) is correct, although you may question whether the joke is all that funny in the best case! Brownell, Michel, Powelson, and Gardner (1983) presented this kind of multiple-choice task to neurologically intact individuals and to patients with right-hemisphere injury. They found that patients with right-hemisphere injury were impaired in the ability to select the correct endings, especially because they were also attracted to incorrect non-sequitur endings, such as choice (b). They suggest that the patients appreciated that a good joke should have a surprise ending, but they were unable to establish an interpretation of the first few sentences that tied the correct ending coherently to the body of the joke. Other studies show similar effects of right-hemisphere injury on tasks that examine narrative-level linguistic performance—extracting humor from nonverbal cartoons, understanding stories, interpreting utterances in context, and so forth—and demonstrate that left-hemisphere injury produces much smaller effects (e.g., Gardner et al., 1983; Hough, 1990; Kaplan et al., 1990). In addition, recent studies show that right-hemisphere injury also causes more disruption than left-hemisphere injury in appreciating the metaphoric meanings of individual words (e.g., *warm* could refer to temperature, but also to a human emotion) (e.g., Brownell et al., 1990).

In addition to performing better on narrative-level linguistic tasks, the right hemisphere appears to be dominant for the processing of intonation and prosody. For example, injury to the right more than the left hemisphere disrupts the use of inflections in speech, both in output and in perception (e.g., Ross, 1985), and in dichotic-listening studies with neurologically intact individuals there is a left-ear (right-hemisphere) advantage for identifying the emotional tone of short sentences (e.g., Ley and Bryden, 1982). This is likely related to a more general involvement of the right hemisphere in the production and perception of emotion, which will be discussed in more detail later.

Even this brief review of research on hemispheric asymmetry for "language" indicates that activities that are so complex consist of many parts and subparts. Furthermore, since hemispheric asymmetry is not equivalent for the various parts, a more *componential* approach is needed to understand hemispheric asymmetry and its implications for information processing in the intact brain (see Chapter 3). Consequently, we must be very cautious in identifying one hemisphere as dominant for something as complex and multifaceted as language. The reality is likely to be that the hemispheres play complementary roles in the intact brain.

Visuospatial Processing

Behavioral asymmetries are not restricted to tasks involving language. In fact, statements that the left hemisphere is dominant for many aspects of language are typically followed by the statement that, in something of a complementary fashion, the right hemisphere is dominant for many nonverbal "visuospatial" tasks. In this section, I review some of the behavioral asymmetries that have led to this generalization. In the preceding section, we saw that "language" can be decomposed into a variety of specific subprocesses and it is incorrect to suppose that the same hemisphere is dominant for all of them. So, too, it is with so-called visuospatial processing. (The most recent thinking about how the hemispheres may be asymmetric for various visuospatial subprocesses will be discussed in Chapter 3).

When patients with unilateral brain injury are asked to draw an object, such as a cube, the nature of the drawings depends on which hemisphere has been injured (e.g., Bouma, 1990; Bradshaw, 1989; Gainotti, 1985). In particular, when the right hemisphere is injured, the drawings lack spatial coherence. Although most of the individual parts may be present, they are not arranged in the appropriate spatial configuration and they lack a three-dimensional quality. Drawings produced by the right hand of split-brain patients, the hand controlled by the left hemisphere, show some of these same characteristics (e.g., Gazzaniga, 1970, 1985). In contrast, when the left hemisphere is injured, the drawings are often lacking in detail, but an appropriate spatial organization is clearly present. Indeed, the same thing is true of drawings produced by the left hand of split-brain patients, the hand controlled by the right hemisphere (see Figure 2.3). Such findings suggest that the right hemisphere is both necessary and sufficient for appreciating fully the spatial or configurational properties of the visual world.

Effects such as those just reviewed are not restricted to drawing. For example, consider the block-design task illustrated in Figure 2.4. To perform this task, an individual is given a set of cubes, each of which is white on two sides, red on two sides, and half-red and half-white (divided diagonally) on two sides. A design is shown to the individual, who must arrange the cubes to reproduce the design. Performance on this task is much worse after unilateral right-hemisphere injury than after unilateral left-hemisphere injury (e.g., Nebes, 1978). In fact, when the right hemisphere is injured, the patient often fails to arrange the cubes in a square at all, much less reproduce the correct arrangement of the cubes within a square. Likewise, in split-brain patients the left hand (right hemisphere) is typically better than the right hand (left hemisphere) for this task (e.g., Bogen and Gazzaniga, 1965; Gazzaniga, 1970; Gazzaniga and LeDoux, 1978). Furthermore, solving a block-design problem produces greater disruption of alpha activity over the right than over the left hemisphere of neurologically intact individuals (Galin and Ornstein, 1972) and interferes more with tapping of the left than of the right index finger (Hellige and Longstreth, 1981).

Given that the tasks described so far demand both visual analysis and manipulation, it could be argued that whatever right-hemisphere dominance is illustrated is for the ability to produce correct spatial arrangements manually. In fact, Gazzaniga and LeDoux (1978) make exactly this argument and refer to right-hemisphere dominance for *manipulospatial* processing. Although asymmetries may be magnified when active manipulation is involved, there is ample evidence that hemispheric asymmetry is often present even when no manipulation is required.

A variety of research with both neurological patients and neurologically intact individuals suggests that the right hemisphere is also dominant for such things as indicating the precise location of a dot in coordinate space (at least when the opportunity for verbal coding is minimized, when there is some delay between seeing the stimulus and making a response, and when precise information

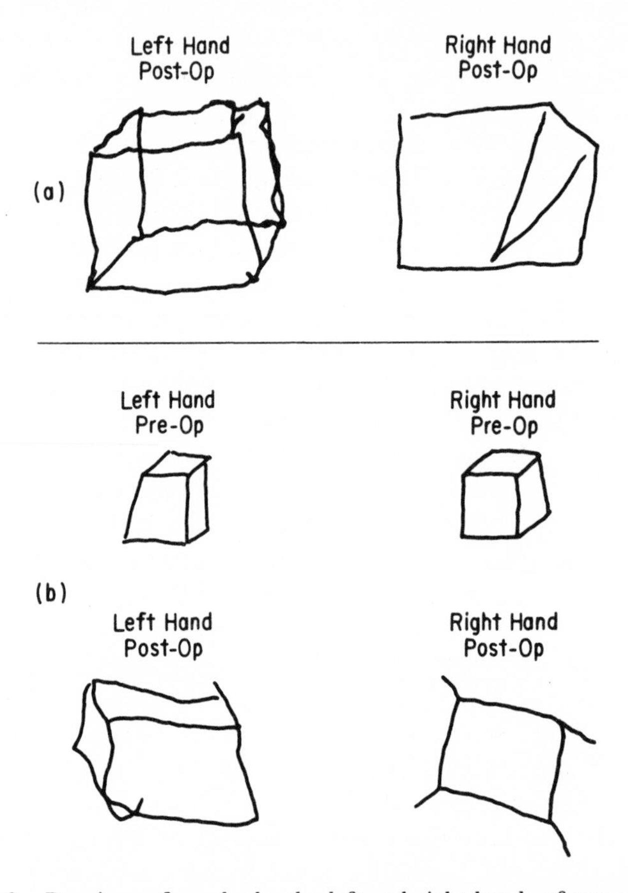

Figure 2.3. Drawings of a cube by the left and right hands of two split-brain patients. Section (a) shows postoperative drawings made by patient W. J. and section (b) shows preoperative and postoperative drawings made by patient P. S. Note that, after surgery, only the drawings made by the left hand (controlled by the right hemisphere) convey accurate configural, 3-dimensional information. [Reprinted from M. S. Gazzaniga, *The Social Brain: Discovering the Networks of the Mind,* (New York: Basic Books, 1985). Copyright 1985 by Basic Books, Inc. Reprinted by permission of Basic Books, a division of HarperCollins Publishers, Inc.]

about distance must be processed) (e.g., Levy and Reid, 1978; see also Chapter 3), determining the orientation of lines and other stimuli (e.g., Benton, Hannay, and Varney, 1975), recognizing three-dimensional objects presented in unusual orientations (e.g., Warrington and Taylor, 1973; Warrington, 1985), and recognizing stimuli that have been perceptually degraded (e.g., Sergent and Hellige, 1986).

For example, Levy and Reid (1978) had neurologically intact individuals identify the location in a rectangle of a single dot flashed briefly to the LVF/right hemisphere or RVF/left hemisphere on each trial. The dot could appear in any one of 20 locations within the rectangle, defined by a 5-by-4 matrix of positions. After the single dot was flashed on each trial, the subjects chose the position they thought was correct from a response card containing a rectangle with the 20 positions labeled. For right-handed individuals, this task was performed better on LVF/right-hemisphere trials than on RVF/left-hemisphere trials; each of the 24 participants showed a difference in this direction. That this LVF advantage is task-specific is indicated by the fact that all of

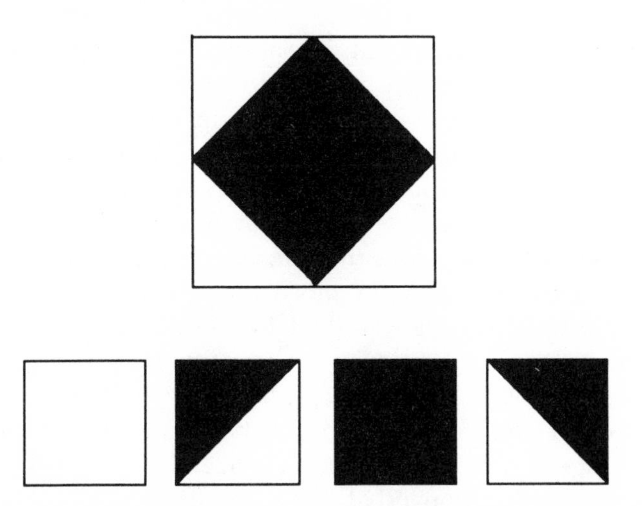

Figure 2.4. An illustration of the block-design task. The goal is to arrange the four blocks in the lower portion in the target pattern shown in the upper portion. Note that, in this example, the two blocks with a single color on the side showing must be turned so that one of the sides with two colors faces up. Then the four blocks can be properly rearranged.

these individuals also showed an RVF/left-hemisphere advantage for recognizing CVC nonsense syllables. Despite the strength of the results obtained by Levy and Reid, other dot-localization tasks have not always supported the conclusion of hemispheric asymmetry (see Bryden, 1982). Possible reasons for this will be considered in Chapter 3 in a discussion of what have been termed "categorical" versus "coordinate" aspects of spatial information.

If the right hemisphere is superior to the left for extracting visuospatial information, then one would expect the right-hemisphere advantage to be demonstrated most clearly when the viewing conditions make visuospatial analysis difficult. With this in mind, a number of researchers have investigated the effects of perceptual degradation on hemispheric asymmetry. The general finding to emerge from these investigations is that the right hemisphere is superior to the left for the identification of perceptually degraded visual material. Put another way, several forms of perceptual degradation are more detrimental to left-hemisphere processing than to right-hemisphere processing. By way of illustrating the nature of these findings, consider the following experiments.

In a study of letter identification in neurologically intact individuals, Bryden and Allard (1976) reported an RVF/left-hemisphere advantage for identifying normal printed letters but an LVF/right-hemisphere advantage for unusual and elaborate scripts. Similar effects of print versus script on letter identification have been reported in patients with unilateral brain injury (e.g., Faglioni, Scotti, and Spinnler, 1969). Bryden and Allard suggest that the right hemisphere is superior to the left for the visual preprocessing that removes the effects of perceptual degradation and unusual orientation. In more direct studies of perceptual degradation, my colleagues and I have examined the effects of different types of degradation on identification of single letters (e.g., Hellige, 1980; Hellige and Webster, 1979) and on tasks that require the comparison of two simultaneously presented letters (e.g. Hellige, 1976; Jonsson and Hellige, 1986) or two simultaneously presented nonverbal letter-like figures (e.g., Michimata and Hellige, 1987). In particular, we have degraded stimuli with an overlay mask or by moderate blurring (see Figure 2.5 for examples). The general finding to emerge is that both types of degradation disrupt performance more on RVF/left-hemisphere trials than on LVF/right-hemisphere trials. That this effect is attributable to hemispheric asymmetry is confirmed by the fact that exactly these same forms of perceptual degradation have larger effects on the letter-comparison performance of patients with right-hemisphere strokes than of patients with left-hemisphere strokes (Wolcott, Saul, Hellige, and Kumar, 1990). In Chapter 3, I will discuss the extent to which these effects of perceptual degradation can be accounted for by a difference in the extent to which the two hemispheres utilize information carried by high versus low visual-spatial frequencies (e.g., Sergent, 1983; Sergent and Hellige, 1986).

The right hemisphere also seems to be dominant for processing the visual stimuli presented by human faces, although recent studies suggest that such dominance is restricted to certain aspects of faces. For example, early studies of patients with unilateral brain injury suggested that *prosopagnosia,* the inability to identify familiar faces, was associated with unilateral injury to the posterior area of the right hemisphere (e.g., Milner, 1968). However, later studies and reviews suggested that prosopagnosia actually requires bilateral lesions, rather than a lesion only of the right hemisphere

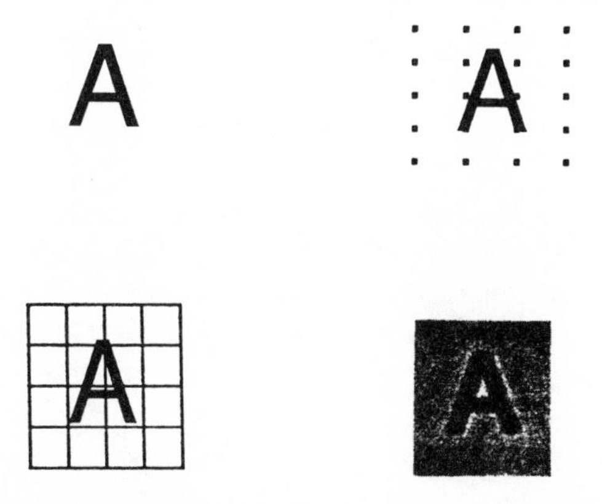

Figure 2.5. Illustrations of a clear letter and types of perceptual degradation similar to those that have been shown to interfere more with the processing of letters presented to the RVF/left hemisphere than with the processing of letters presented to the LVF/right hemisphere.

(e.g., Benton, 1980; Damasio and Damasio, 1986). Such results suggest that both hemispheres make contributions to face recognition, although the contributions may differ. For example, it has been suggested that patients with right-hemisphere injury often recognize faces by focusing on distinctive features, such as a moustache, whereas patients with left-hemisphere injury often use more holistic or configural properties (for reviews see Bradshaw, 1989; Levine, 1989), although the empirical evidence in favor of this particular distinction is, at best, rather weak.

An important issue concerns whether the right hemisphere is more involved than the left in processing the configural aspects of faces in the neurologically intact brain. Evidence that it is is provided in a positron emission tomography (PET) study of regional cerebral blood flow reported by Sergent, Ohta and MacDonald (1992). Neurologically normal adults participated in several tasks while PET scans were being taken and a subtraction procedure was used to find which cortical regions received additional activation associated with one task compared with another. Of course, all tasks produced activation of many areas of both hemispheres. With respect to asymmetries, classifying unfamiliar faces as male or female (a gender-classification task believed to be carried out on the basis of configural properties) produced activation of extra-striate areas of the right hemisphere. When compared with this gender-classification task, a face-recognition task (classifying each face of a famous person as that of an actor or not) produced additional activation in several areas of both hemispheres and asymmetrical activation in the ventro-medial areas of the right hemisphere. Thus, the study produced converging evidence of right-hemisphere dominance for gender classification and for at least certain aspects of face recognition over and above asymmetries that were present for gender identification. Of additional importance is the fact that cerebral activation produced by an object-recognition task (classifying an object as "natural" versus "man-made") did not involve the areas of the right hemisphere activated by the face-recognition task. Instead, this particular object-recognition task tended to activate temporal areas of the left hemisphere. Thus, there is a dissociation of hemispheric asymmetry for face and object processing (for additional discussion of this dissociation, see also Farah, 1990).

In visual-half-field studies of face recognition in neurologically normal individuals, there tends to be an LVF/right-hemisphere advantage under many conditions but not all (e.g, Levine, 1989; Sergent, 1987). To a large extent, the conditions that favor such an advantage are those that either require processing of configural aspects of faces or are performed best by focusing on such aspects. For example, an LVF/right-hemisphere advantage has been reported by Sergent et al. (1992) for a gender-classification task using the same stimuli and subjects from her PET study. She reported a similar LVF/right-hemisphere advantage for the actor-classification task at a very early stage of practice—namely, the first time the subject saw each specific stimulus. This result is consistent with the PET result because each subject saw each specific stimulus only once during the PET portion of the study. However, an RVF/left-hemisphere advantage was obtained for the same task when the same stimuli were repeated—perhaps because the now familiar stimuli could be recognized on the basis of distinctive features. More specific generalizations reached in a review by Sergent and Bindra (1981) indicate additional methodological factors that seem to be relevant for determining visual-field differences in face-processing tasks. They argue that an LVF/right-hemisphere advantage tends to be found when the following conditions hold: (1) faces are perceptually degraded, (2) faces to be compared are highly discriminable, (3) a small set of unfamiliar faces is used, and (4) the task allows a lax criterion of recognition. To the extent that opposite conditions hold, there tends to be an RVF/left-hemisphere advantage. On this view, the early suggestions about right-hemisphere dominance for face recognition came from studies that happened to use just the right set of conditions to favor right-hemisphere processes.

To the extent that each hemisphere is dominant for different subprocesses that are relevant for face recognition, it may be misleading to postulate that a region of the right hemisphere is specialized specifically for face recognition. Instead, it is necessary to look for hemispheric asymmetries in those subprocesses that seem relevant and to consider how those asymmetries might produce the complex results that have been observed. Further discussion of this componential approach is contained in Chapter 3.

An additional indication of the importance of the right hemisphere for spatial processing comes from the fact that injury to the posterior parietal area of the right hemisphere often leads patients to exhibit *hemineglect;* that is, to neglect the left side of objects, drawings, and other stimuli, despite the absence of primary sensory or motor deficits. The neglect of a stimulus on the left side is much stronger when another stimulus is presented simultaneously on the right side (a phenomenon known as *extinction to simultaneous stimulation*). Although the opposite direction of hemineglect is sometimes found after left-hemisphere injury, it is argued to be less likely and generally less severe. However, some caution must be used in reaching conclusions about asymmetries of this kind because the incidence of hemineglect after left-hemisphere injury may be underestimated to the extent that other, more severe problems overshadow the symptoms of neglect.

Right-hemisphere patients with left hemineglect may bump into objects on the left side, read only the right side of a page, copy only the right side of a drawing, bisect a line too far to the right (because they ignore the left-most portion), and so forth. One illustration that the problem is not sensory is that patients ignore the "left half" of the stimulus even when the entire stimulus is contained within the "intact" RVF. In fact, Farah, Brunn, Wong, Wallace, and Carpenter (1990) have shown that the neglected hemifield is defined both with respect to a viewer-centered frame of reference (left or right of body midline) and also with respect to an environment-centered frame of reference (left or right along the horizontal axis of the environment). There is also some indication that the neglected side can be defined with respect to an object-centered frame of reference (left or right side of the center of an object), at least for objects with a canonical left-right axis (e.g., Behrmann and Moscovitch, 1992). Another indication that the problem is not sensory is that patients also neglect the left half of their own mental images. For example, Bisiach and his colleagues (Bisiach and Luzzatti, 1978; Bisiach, Capitani, Luzzatti, and Perani, 1981) asked their Italian hemineglect patients to describe what they would see if they were in the central square (Piazza del Duomo) of Milan, facing the cathedral. Patients with left hemineglect mostly described buildings that would be on the right side of the square from the patient's imagined viewpoint;

that is, buildings on the left side of the square were ignored. The same patients were then asked to imagine themselves taking the opposite viewpoint; that is, standing on the steps of the cathedral, looking away from it. They now described quite readily the buildings they had previously neglected (which would now be on the right side of the square from the patient's new viewpoint) and neglected the buildings they had previously described (which would now be on the left side of the square from the patient's new viewpoint)!

Hemineglect is not a unitary disorder and is more accurately described as a clinical syndrome consisting of several component deficits. For example, studies of the effects of parietal injury on covert orienting of attention (Posner, Walker, Friedrich, and Rafal, 1984, 1987) have shown specific deficits in the ability to disengage attention from its current location so that it can be moved in a direction contralateral to the lesion (see Chapter 3 for additional discussion). Direction-specific deficits have also been described for limb movements (e.g., Heilman, Watson, and Valenstein, 1985), oculomotor control (e.g., Girotti, Casazza, Musicco, and Avanzini, 1983), and visual orienting (e.g., Reuter-Lorenz and Posner, 1990).

To the extent that hemineglect is more likely after right- than after left-hemisphere injury, it has been suggested that the right hemisphere represents extracorporeal space both contralaterally and ipsilaterally (that is, for both the left and right sides of the body midline) whereas the left hemisphere represents only the contralateral (i.e., right) half of space. On this view, if the left hemisphere is injured, the intact right hemisphere will continue to represent both sides of space and hemineglect should be absent or minimal. In contrast, if the right hemisphere is injured, the intact left hemisphere can maintain a high level of performance only for the right half of space, and thus left-sided hemineglect results. Although this hypothesis can account for asymmetry in the occurrence of hemineglect (including hemineglect in visual images), it is not clear how it deals with the problems that right-hemisphere patients have in shifting attention toward the left, even though the stimuli are contained entirely in the presumably intact right half of space.

Emotion

The emotional behavior of patients with unilateral brain injury depends on which hemisphere is injured. Left-hemisphere injury leads to more catastrophic reactions (e.g., anxiety, sudden bursts of tears) and right-hemisphere injury leads to more indifference reactions (e.g., a cheerful acceptance of disability and indifference toward failure) (e.g., Gainotti, 1987; Tucker, 1987). In addition, studies of several sorts indicate impaired right-hemisphere performance when individuals are depressed and impaired left-hemisphere performance when individuals are anxious (see Tucker, 1987; Liotti and Tucker, 1992). Along with a variety of other findings, such results suggest hemispheric asymmetry for the production of emotion. Two specific hypotheses have received a great deal of attention. The first hypothesis is that the right hemisphere is dominant over the left for emotional expression (and perception) of all types. The second hypothesis is that hemispheric asymmetry for production and perception of emotions depends on emotional valence: right-hemisphere dominance is most clearly established for "negative" emotions and is weaker or even replaced with left-hemisphere dominance for "positive" emotions. Despite certain noteworthy inconsistencies in the literature, both hypotheses have received at least some empirical support.

A variety of behavioral evidence is more consistent with the hypothesis that the right hemisphere is dominant for producing and perceiving emotion, regardless of valence, rather than with the alternative. For example, earlier in the present chapter, I noted that right-hemisphere injury is more likely to lead to disruptions of prosody and intonation in speech, and this is true whether the emotion being expressed is positive or negative. In addition, the left side of the face is more emotionally expressive than the right side and there is little evidence of opposite asymmetries for positive versus negative emotions. The right hemisphere also seems to be superior to the left for the *perception* of both positive and negative emotions. For example, right-hemisphere injury causes more difficulty than left-hemisphere injury in identifying positive and negative emotions shown in faces, and this is true over and above any right-hemisphere dominance for visuospatial processing more generally (Bowers, Bauer, Coslett, and Heilman, 1985). So, too, in neurologically normal individuals, there is an LVF/right-hemisphere advantage for identifying emotions in faces, but little evidence that the visual-field advantage reverses for positive versus negative emotions (e.g., Bryden, 1982). Furthermore, various control tasks suggest that this LVF/right-hemisphere advantage is present over and above any right hemisphere advantage for face recognition or visuospatial processing in general, although this is always a very difficult alternative to rule out completely. In addition, when looking at faces in free vision, both neurologically normal individuals and neurological patients are more influenced by the emotional expression shown on the half of the face toward the viewer's left than on the half of the face toward the viewer's right (e.g., Levy, Heller, Banich, and Burton, 1983b; Borod et al., 1990).

The right hemisphere is also superior to the left for identifying the emotional tone of spoken material. One of the clearest indications comes from an experiment reported by Ley and Bryden (1982). Their subjects listened to a number of short sentences that were spoken with happy, sad, angry, or neutral intonation (the verbal content of the sentence did not indicate emotion). The sentences to be identified were dichotically paired with neutral sentences of similar semantic content. Subjects were instructed to attend to one ear for an entire block of trials and to report both the emotional tone of the target sentence and the verbal content. There was a left-ear/right-hemisphere advantage for identifying which emotion was conveyed by the tone of voice and, at the same time, a right-ear/left-hemisphere advantage for identifying the verbal content. Furthermore, there was no indication that hemispheric asymmetry for identifying emotion depended on whether the emotion was positive or negative. In a similar experiment using two-syllable words instead of sentences, Bryden and MacRae (1989) obtained a similar result. However, in their study, the left-ear/right-hemisphere advantage for identifying the emotional tone of the words was stronger for negative than for positive emotions—offering some support for the hypothesis that hemispheric asymmetry for perceiving emotions depends on their positive/negative valence.

The strongest and most consistent evidence that emotional valence is important has been obtained in studies that examined the effects of emotional experience on electrophysiological responses recorded from the left and right sides of the scalp (for review, see Davidson, 1992). For example, Davidson, Schwartz, Saron, Bennett, and Goleman (1979) recorded EEG responses from electrodes placed over the frontal and parietal areas of the two hemispheres while right-handed subjects watched videotaped segments of television programs designed to elicit positive or negative emotion. As the subjects watched, they continuously rated their emotional reaction. For frontal locations, there was relatively more left-hemisphere activation during the 30-second period rated by the subject as leading to the most positive emotion and relatively more right-hemisphere activation during the 30-second period rated by the subject as leading to the most negative emotion. No such asymmetry was present in the activation patterns at parietal locations. These results have been replicated and extended in a variety of ways, including the observation of similar results in children and infants (for review, see Davidson, 1988, 1992; Fox, 1991; see also Chapter 8). On the basis of these electrophysiological results, Davidson and Fox argue that, beginning in infancy, the arousal of positive, approach-related emotions is associated with selective activation of the left frontal region of the cortex whereas the arousal of negative, withdrawal-related emotions is associated with activation of the right frontal region of the cortex.

It should be obvious from the foregoing review that neither of the two hypotheses about hemispheric asymmetry for emotion receives consistent support. On one hand, there is more evidence of right-hemisphere dominance for the *perception* of both positive and negative emotion than for the alternative hypothesis. On the other hand, there is evidence that the balance of activation between the frontal lobes of the two hemispheres depends on the positive/negative valence of *experienced* emotion. It is instructive to consider how these different findings might be reconciled.

One possibility is that there is a genuine dissociation between emotional experience (leading to selective activation of the frontal regions) and the perception of emotions (and, perhaps, the production of emotional behaviors). On this view, cerebral activation correlated with emotional experience is asymmetric in the manner proposed by Davidson (1992), with emotional valence related to the direction of asymmetric activation. However, the perception of emotion from faces and voices as well as the production of certain emotional behaviors involves other regions of the cortex in addition to (or instead of) the frontal regions. This being the case, hemispheric asymmetry for such things as the perception of emotion need not depend on valence. For example, van Lancker and Sidtis (in press) have found that patients with right-hemisphere injury make less use of a particular type of acoustic information known as fundamental frequency in classifying the emotional content of speech than do patients with left-hemisphere injury. This may have nothing to do with the experience of emotion per se. However, to the extent that a specific auditory emotion-perception task demands the analysis of fundamental frequency, regardless of positive/negative valence, there might well be a right-hemisphere advantage that is independent of valence and that, indeed, has nothing whatsoever to do with emotional experience. And, in fact, this right-hemisphere advantage could co-exist with a valence-related asymmetry for the experience of emotion. Similar arguments might well be developed for those aspects of vision that are most useful for identifying emotion as opposed to other aspects of faces.

A second possibility is that the right hemisphere is in some sense dominant for emotional experiences but that whether the emotional experience is positive or negative depends on the level of right-hemisphere activation. On this view, negative emotional experiences are associated with overactivation of the right hemisphere whereas positive emotional experiences are associated with underactivation of the right hemisphere. Given that the two hemispheres are in something of a mutually inhibitory relationship (see Chapter 6), changes in activation of one hemisphere could be accompanied by opposite changes in the other hemisphere (see Levy, Heller, Banich, and Burton, 1983a,b). From this perspective, either the activation level of the left hemisphere is a byproduct of the activation level of the right hemisphere, or it serves to moderate the emotional reactions for which the right hemisphere is dominant. Although this possibility is highly speculative, it is interesting that a similar view has been taken of hemispheric asymmetry for emotional reaction in rats and chicks (see Chapter 5).

The Quest for a Fundamental Dichotomy

The review of behavioral asymmetries in the preceding section of this chapter illustrates the range of hemispheric differences that have been documented. As noted earlier, there are now so many additional tasks that show asymmetries that there is little hope of ever compiling a complete list. Given such a wide range of specific asymmetries, it makes sense to try and reduce the findings to a small set of dichotomies that characterize hemispheric asymmetry, a set from which all of the specific findings can be derived. In fact, the most parsimonious set would contain only a *single fundamental dichotomy* that would underlie all the rest. Consequently, it should come as no surprise that a great deal has been written about the nature of such a dichotomy.

The idea that there could be a fundamental dichotomy is closely related to the idea of *hemispheric specialization.* To illustrate this concept, recall that the left hemisphere has been shown to be dominant for several aspects of language. Early findings of this sort led to the idea that the left hemisphere of humans has evolved to be *specialized* for language. Now, the fact is that referring to a left-hemisphere specialization for language (or anything else) is little more than a restatement of the fact that there is a correlation between the side of unilateral brain injury and impaired language function or between visual hemifield and performance on a language task (see Efron, 1990). The problem is that the word *specialization* implies much more. For example, it implies that the left hemisphere (or subparts of the left hemisphere) are both necessary and sufficient for the language functions in question and that it is for those functions in particular that the left hemisphere is especially well adapted. We have seen that even relatively simple tasks require the coordination of a number of specific subprocesses. Consequently, when one hemisphere is superior for a task it is natural to ask which subprocess is responsible for that superiority. This is equivalent to asking about the fundamental, underlying nature of hemispheric specialization.

A Sampling of Suggested Dichotomies

Verbal/nonverbal. A number of dimensions have appeared as candidates for the title of "fundamental dichotomy." One of the earliest maintained that whereas the left hemisphere is specialized for processing verbal stimuli, the right hemisphere is specialized for processing nonverbal stimuli. One problem with this dichotomy is that it is defined in terms of stimulus properties. Consider a printed word. Viewed as a word, it is clearly a "verbal" stimulus; but viewed differently it is simply a collection of luminance changes across space. Thus, it was rather quickly realized that "verbal" and "nonverbal" were more accurate as descriptions of the *processing* needed for a task than as descriptions of a stimulus per se. As a result, the dichotomy was reformulated in terms of verbal versus nonverbal processing.

When research began to show left-hemisphere dominance for certain "nonverbal" tasks (e.g., production and perception of temporal sequences; Kimura, 1977; Efron, 1990) and right-hemisphere involvement for certain "verbal" tasks (e.g., Bradshaw and Nettleton, 1981), the adequacy of a verbal/nonverbal dichotomy was questioned. As a result, a host of other dichotomies emerged. While several of them seemed promising for a time, none has been easy to sustain for very long (for reviews see Allen, 1983; Bouma, 1990; Bradshaw and Nettleton, 1981, 1983). Rather than go through a list of the various dichotomies, I will review two that are particularly interesting because they generally encompass most of the others (and are even similar to each other in some ways) and because, like the Phoenix of mythology, time after time they rise from the ashes.

Focal/diffuse organization. In a provocative article, Semmes (1968) suggested that elementary functions are represented *focally* in the left hemisphere but *diffusely* in the right hemisphere. This claim comes from her studies of the sensory and motor capacities of the two hands in brain-injured patients. In patients with left-hemisphere injury, sensorimotor deficits for both hands were found only after injury to the primary sensorimotor projection areas. In patients with right-hemisphere injury, however, sensorimotor deficits were found even when injury was outside of the primary projection areas. In addition, she reported that injury to a specific area of the left hemisphere typically leads to disruption of a specific function whereas injury to the right hemisphere that is great enough to disrupt one function typically disrupts others as well. The idea that the two hemispheres are organized differently in this way is extremely interesting and it is for this reason that Semmes' arguments continue to be invoked. Given the attention the theory has received, it is surprising that attempts to replicate the specific results reported by Semmes have not been reported.

Semmes (1968) went on to argue that a focal representation of elementary functions in the left hemisphere would favor the integration of similar units and lead to specialization for behaviors that demand fine sensorimotor skill, including manual skills and speech. In contrast, she argues, a diffuse representation of elementary functions in the right hemisphere would favor the integration of dissimilar units and lead to specialization for behaviors that demand multimodal coordination and integration of information across wide areas of space. The logic that takes us from the idea of hemispheric differences in focal versus diffuse organization to these particular functional specializations seems arbitrary. For example, it is never made clear why a "focal" organization would favor the integration of similar units. Still, her results and the theoretical ideas they led to are very interesting and, in other domains, recent computer simulations of connectionist models are making it clear that performance depends on the way in which a network is organized (e.g., Kosslyn, Chabris, Marsolek, and Koenig, 1992; Rueckl, Cave, and Kosslyn, 1989).

Analytic/holistic. During the 1970s, investigators proposed semi-independently a number of dichotomies that eventually led to the hypothesis that the left hemisphere is well adapted for *analytic* processing whereas the right hemisphere is well adapted for *holistic* processing. The most comprehensive and convincing presentation of this particular dichotomy was given by Bradshaw and Nettleton (1981, 1983). They did a very thorough and thoughtful job of reviewing the many asymmetries that had been reported and the dichotomies that had been suggested. By looking for similarities among the tasks that reliably produced a left-hemisphere advantage and contrasting them with similarities among the tasks that reliably produced a right-hemisphere advantage, they suggested that a critical difference was the extent to which a task demanded analytic versus holistic processing. Although it is difficult to define *analytic* and *holistic,* the following statements indicate how these terms have generally been used:

> The left hemisphere is characterized by its mediation of discriminations involving duration, temporal order, sequencing, and rhythm, at the *sensory* (tactual, visual, and, above all, auditory) level, and especially at the *motor* level (for fingers, limbs, and, above all, the speech apparatus). Spatial aspects characterize the right, the mapping of exteroceptive body space, and the positions of fingers, limbs, and perhaps articulators, with respect to actual and target positions. Thus there is a continuum of function between the hemispheres, rather than a rigid dichotomy, the differences being quantitative rather than qualitative, of degree rather than of kind. (Bradshaw and Nettleton, 1981, p. 51)

Despite a certain intuitive appeal, the analytic/holistic dichotomy was criticized from the moment it was suggested (e.g., the commentaries that follow Bradshaw and Nettleton, 1981). Although many points can be raised, the primary problem is that the analytic/holistic distinction has never been operationalized with sufficient precision to make empirical tests possible. Consequently, although the dichotomy has served a useful purpose in helping to organize a vast literature, its theoretical usefulness is limited. What is interesting is that, in spite of all the criticism, the analytic/holistic dichotomy is still frequently presented as capturing the fundamental difference between the two hemispheres (e.g., Anderson, 1990; Stillings, Feinstein, Garfield, Risslandy, Rosenbaum, and Weisler, 1987). In fact, even critics of the distinction have sometimes used it to "interpret" some of their own empirical findings. Thus, the distinction has taken on a life of its own, being sufficiently complex to permit a variety of *post hoc* conclusions and sufficiently vague to prevent clear disconfirmation.

Why should there be a single dichotomy? Some of the commentaries that follow the Bradshaw and Nettleton (1981) article question whether we should expect a single information-processing dichotomy or principle to underlie hemispheric asymmetry. As pointed out by Bertelson (1981), in order to understand the importance of hemispheric differences we must consider the evolutionary advantages that resulted in hemispheric asymmetry for certain functions. In addition to this consideration of phylogenetic development, I would add the consideration of the ontogenetic development of an individual. These considerations lead to underlying principles of a sort different from those discussed here and they will be discussed in Chapters 8 and 9. The final section of the present chapter provides additional reasons to question whether any single processing dichotomy of the sort considered so far is likely to account for all hemispheric asymmetries.

Multitask Studies and the Quest for a Fundamental Dichotomy

To a large extent, the quest for a fundamental dimension of hemispheric asymmetry has been unsuccessful. It is possible that such a dimension exists but has not yet been discovered. It is also possible that no single information-processing dimension will ever be able to account for all hemispheric asymmetries. In fact, empirical evidence favors the latter alternative.

One research strategy is to determine whether an individual's asymmetry for one task is related to that individual's asymmetry for other tasks. Suppose that each of two tasks, *A* and *B,* produce valid and reliable measures of hemispheric asymmetry. Furthermore, suppose that each of these tasks produces left-hemisphere superiority in 80 percent of the right-handed individuals tested and right-hemisphere superiority in the remaining 20 percent. Is it likely that the overall left-hemisphere advantage for these two tasks occurs because both tasks require processes at one end of a single, fundamental dichotomy? At first glance, this conclusion seems reasonable because the two tasks produce exactly the same laterality effect. However, much can be learned by having the same individuals perform both tasks. If both tasks produce the same overall left-hemisphere advantage for the same reason, then those individuals who show a left-hemisphere advantage for task *A* should also show a left-hemisphere advantage for task *B* and those individuals who show a right-hemisphere advantage for task *A* should also show a right-hemisphere advantage for task *B.* This may or may not be the case.

To illustrate this, suppose that the same 100 individuals perform both tasks and, as expected, for each task there is a left-hemisphere advantage for 80 individuals and a right-hemisphere advantage for the remaining 20. Because we have tested the same individuals on both tasks, we can look at the number of individuals falling into each cell of a 2-by-2 table defined by whether an individual showed a left- or right-hemisphere advantage for each task. Table 2.1 illustrates two possible outcomes that are both consistent with all that has been said about these two tasks.

In one outcome (the *correlated* outcome), asymmetry for the two tasks is perfectly correlated. That is, for every subject the direction of hemispheric advantage is the same for both tasks—so that knowing the direction of asymmetry for one task is perfectly predictive of asymmetry for the other task. This is the kind of outcome that should be found if both tasks produce equivalent hemispheric asymmetries for the same reason; that is, if both tasks are different ways of tapping the same fundamental processing dimension. A strong version of the single-dichotomy view predicts that *all* tasks that show the same direction of asymmetry should

Table 2.1. Hypothetical outcomes from 100 patients tested on two tasks: number of subjects showing each possible outcome

Correlated Outcome

	Hemispheric advantage for task *A*		
	Left	Right	Left plus right
Hemispheric advantage for task *B*			
Left	80	0	80
Right	0	20	20
Left plus right	80	20	

Independence Outcome

	Hemispheric advantage for task *A*		
	Left	Right	Left plus right
Hemispheric advantage for task *B*			
Left	64	16	80
Right	16	4	20
Left plus right	80	20	

be perfectly correlated with each other in this way. In the other outcome (the *independent* outcome), asymmetry for the two tasks is completely uncorrelated, even though they both produce the same overall pattern of asymmetry. That is, the number of subjects showing the same direction of asymmetry for both tasks (64 show a left-hemisphere advantage for both and 4 show a right-hemisphere advantage for both) is exactly the number predicted by assuming that hemispheric asymmetry for one task is completely independent of hemispheric asymmetry for the other. This pattern of independence makes it unlikely that hemispheric asymmetries for the two tasks are produced by a single underlying information-processing dichotomy. Thus, it is theoretically important to determine from multi-task experiments whether asymmetries are correlated or independent. Of course, it is possible to test for independent versus correlated outcomes by a variety of other measures, such as correlation coefficients between laterality indexes (e.g., left-hemisphere score minus right-hemisphere score), that preserve both the direction and magnitude of asymmetry. Techniques such as factor analysis and principal-components analysis are also useful (e.g, Boles, 1991; Kim, Levine, and Kertesz, 1990).

The idea that the hemispheres are asymmetric for a single, fundamental information-processing dimension also predicts a relationship between tasks that lead to overall asymmetries in opposite directions. For example, consider an individual who shows a relatively large left-hemisphere advantage for a task that produces an overall left-hemisphere advantage in the population. According to a single-dichotomy viewpoint, such an individual is strongly asymmetric for the fundamental dimension and should, therefore, show a relatively large right-hemisphere advantage for a task that produces an overall right-hemisphere advantage in the population. This negative correlation has been discussed by Bryden (1982) in terms of "complementary specialization" that goes beyond what would be expected on the basis of statistical independence.

In view of the potential value of studies that have the same individuals perform more than one laterality task, it is surprising that only a few have reached the literature. Many of the studies using neurologically intact individuals show only weak relationships between asymmetries for two tasks or no relationships at all (e.g., Bryden, 1982; Dagenbach, 1986; Hellige, Bloch, and Taylor, 1988; Nestor and Safer, 1990). Furthermore, when relationships have been found, they are often opposite in direction from what would be expected if all hemispheric asymmetries are the result of a single information-processing dimension (e.g., Kim et al., 1990; Boles, 1991; see also Chapter 7). On the basis of similar findings from neurological patients, Bryden (1982) suggests that the search for a unifying description of hemispheric asymmetry may be futile. Instead, he suggests that specific functions may become lateralized to one hemisphere or the other independently and somewhat randomly, with some functions being strongly biased in one direction or the other at the population level. Clearly, this is an idea that deserves further investigation.

Summary and Conclusions

Hundreds of behavioral asymmetries have been identified in humans, many of which can be attributed to hemispheric asymmetry. (Remember theme 1 identified in Chapter 1.) From this large database have emerged several generalizations about the types of tasks for which each hemisphere is dominant.

The most obvious behavioral asymmetry in humans is handedness. With approximately 90 percent of humans being right-handed, it is clear that the left hemisphere is dominant for hand control in most individuals. In addition, there are several other motor asymmetries. For example, the left hemisphere is superior to the right for learning and executing a sequence of movements, including articulatory movements and certain nonverbal oral movements (e.g., blowing). In contrast, the left side of the face (controlled more by the right hemisphere) is more emotionally expressive, consistent with the idea that the right hemisphere plays a greater role than the left in the production of emotion and overt emotional expression.

Left-hemisphere dominance for many aspects of language is the most obvious and most often cited cognitive asymmetry. In particular, the left hemisphere seems dominant for the production of overt speech, for the perception of phonetic information, for using syntactic information, and for certain aspects of semantic analysis. However, the right hemisphere seems dominant for certain other aspects of language, including the use of pragmatic aspects of language (e.g., narrative-level linguistic information) and the use of intonation and prosody to communicate emotional tone.

The right hemisphere seems superior to the left for a variety of nonverbal tasks that demand visuospatial processing. In particular, the right hemisphere is better able to perceive and produce the configural properties of visual stimuli, locate stimuli in coordinate space, recognize three-dimensional objects in unusual orientations, and recognize visual stimuli that have been perceptually degraded. Although it was originally believed that the right hemisphere was dominant for the recognition of faces, it now appears that both hemispheres can contribute to face recognition but may emphasize different types of visual information. The clinical disorders of hemineglect and extinction to simultaneous stimulation are also more likely and more severe after right-hemisphere injury than after left-hemisphere injury, suggesting that the right hemisphere may be more involved in representing extracorporeal space and orienting attention to specific locations within that space.

There is also evidence of hemispheric asymmetry for the perception, production, and experience of emotion. The right hemisphere seems to play a greater role than the left in the perception of emotion from faces and voices and, perhaps, in the production of appropriate emotional behaviors. Although this seems generally true for all emotions, there is some indication that right-hemisphere dominance is stronger for negative emotions than for positive emotions. At the same time, the balance of activation between the frontal lobes of the two hemispheres depends on the valence of experienced emotion. Specifically, there is relatively more left-hemisphere activation while positive emotions are experienced and relatively more right-hemisphere activation while negative emotions are experienced.

Given this wide range of hemispheric asymmetries, it makes sense to look for common threads that would identify a small set of information-processing dichotomies from which all of the specific behavioral asymmetries could be derived. In fact, the most parsimonious set would contain only a single fundamental dichotomy. Several dichotomies have been suggested and rejected, either because of contradictory evidence or because the dichotomies were not defined with sufficient precision to lead to empirical tests. Among them have been the ideas that the left and right hemispheres are (1) specialized for verbal versus nonverbal processing, (2) specialized for processes that are better carried out with a focal versus diffuse neural organization, and (3) specialized for analytic versus holistic processes.

In view of the fact that the quest for a fundamental dimension of hemispheric asymmetry has been unsuccessful, it is important to consider whether it is likely that any single processing dimension will ever be able to account for all hemispheric asymmetries. In fact, the idea that all hemispheric asymmetries are a reflection of a single processing dichotomy predicts that individuals who show a large asymmetry for one task should also show large asymmetries for other tasks (relative to other individuals). That is, hemispheric advantages should be correlated strongly across tasks in a certain way. In fact, there is little evidence for the existence of such correlations, either in studies with neurologically intact individuals or in studies with brain-injured patients. This suggests that the search for a single fundamental dichotomy may be futile. Instead, specific processing components may become lateralized to one hemisphere or the other independently or in a manner that is governed by an entirely different sort of principle.

This review of behavioral asymmetries and the quest for a fundamental dichotomy is consistent with two important conclusions. The first is that it is rarely the case in the intact brain that one hemisphere can perform a task normally whereas the other hemisphere is completely unable to perform the task at all. Instead, both hemispheres often have some ability to perform a task, but they may go about it in different ways and one is sometimes better than the other. Consequently, for many complex activities (e.g. language or face recognition), both hemispheres play a role, and these roles are often complementary. This point is related to the second conclusion. Even simple tasks consist of number of specific subprocesses, and there is no guarantee that hemispheric asymmetry for one subprocess involved in a task is the same as that for another subprocess that is also involved in that task. Consequently, is it usually impossible to state in simple terms that one hemisphere or the other is superior or specialized for a multiprocess task. It may make more sense to look for hemispheric asymmetries for specific subprocesses or components, and to look for them while being guided by existing componential models of perception, cognition, and action. Such an approach is illustrated and discussed in Chapter 3.

3

Hemispheric Asymmetry and Components of Perception, Cognition, and Action

It is abundantly clear that even relatively simple tasks require the coordination of a number of information-processing subsystems, components, or modules. In fact, the identification of separable components is a hallmark of the information-processing or computational approach that has come to characterize cognitive psychology and cognitive neuroscience (e.g., Anderson, 1990; Kosslyn, 1987; Stillings et al., 1987). It is also clear that the direction and magnitude of hemispheric asymmetry is often different from subsystem to subsystem, precluding simple statements about hemispheric superiority for an entire task (e.g., Allen, 1983; Hellige, 1980, 1990). The fact that hemispheric asymmetry can vary from subsystem to subsystem within a task suggests that it would be worthwhile to study asymmetry in terms of relevant subsystems. In order to do so, we need to have some principled conceptualization about the subsystems involved in a task and an empirical means of distinguishing among them. In recent years, it has proven particularly worthwhile to consider the subsystems, modules, or processing stages postulated by various models of perception, cognition, and action.

By way of example, consider two of the processing steps or subsystems that have been proposed for determining whether a visually presented probe letter is contained currently in short-term memory (e.g., Sternberg, 1969, 1975). When the probe letter is presented the stimulus must undergo sufficient visual analysis to be *encoded.* The result of encoding makes available the name or some other identity code for the letter that has been presented. According to Sternberg, the stimulus-encoding process can be slowed by perceptually degrading the visual probe stimulus. Consequently, effects of perceptual degradation provide a useful means of examining the efficiency of the encoding processes. A *memory-comparison* operation compares the encoded representation with the information currently held in short-term memory. This operation takes longer when there are more items to scan in short-term memory (that is, as the size of the memory set increases). Consequently, effects of memory set size provide a useful means of examining the efficiency of the memory-comparison operation.

In a memory-scanning task with letters, I required observers to indicate as quickly as possible whether a probe letter flashed to the LVF/right hemisphere or RVF/left hemisphere was contained in a set of letters held in short-term memory (Hellige, 1980). On different trials, the probe letter was either presented clearly or degraded by being embedded in a matrix of dots. In addition, the size of the memory set was varied from 2 to 5 items across trials. Averaged across all of these conditions, the results showed no advantage for one visual field or the other. The equality of performance on LVF and RVF trials indicates that neither hemisphere was superior for the memory-scanning task considered as a whole. However, the detrimental effect of perceptually degrading the probe letter was smaller on LVF/right-hemisphere trials than on RVF/left-hemisphere trials, suggesting that the right hemisphere was superior to the left for at least some aspect of stimulus encoding. In contrast, the increase in reaction time (RT) associated with increase in memory set size was smaller on RVF/left-hemisphere trials than on LVF/right-hemisphere trials, suggesting that the left hemisphere was superior to the right for at least some aspect of memory comparison. Additional aspects of the results suggested that this advantage may hold because memory comparison was based on different types of representation for LVF and RVF trials. Specifically, the pattern of results for LVF/right-hemisphere trials suggested a visual representation (perhaps a visuospatial representation of what the letters looked like) whereas the pattern of results for RVF/left-hemisphere trials suggested a more abstract representation (perhaps the names of the letters).

In the remainder of this chapter, I will review several additional examples of what could be called a *componential approach* to the study of hemispheric asymmetry. In each case, a processing distinction that was proposed initially in studies of human cognition, perception, and action is used to generate predictions about asymmetry for specific processing subsystems. Finding asymmetry for specific subsystems sheds light on the nature of hemispheric asymmetry and also provides converging evidence that models that decompose tasks into those subsystems are on the right track. For example, the fact that hemispheric superiority was different for stimulus encoding than for memory comparison (Hellige, 1980) supports the assertion that these are separable components of the memory-comparison task.

The componential approach illustrated here has several positive features. One is that it makes no attempt to reduce all instances of hemispheric asymmetry to a single fundamental processing dimension, an enterprise that has heretofore proven unsuccessful in practice and of questionable merit in principle. A second is that it incorporates the study of hemispheric asymmetry into the larger enterprise of trying to understand how the brain produces cognition. The fact is that our understanding of hemispheric asymmetry has been hindered rather than helped by being carried out in the past as a separate enterprise. A third advantage is the possibility that hemispheric asymmetry for a specific processing component or subsystem will contribute to asymmetry for all tasks that involve that subsystem, even though the tasks may be quite different in other respects. Finally, the emphasis on subsystems or components reinforces the idea that both hemispheres are normally involved in performing nearly all tasks, even though their contributions may differ. Thus, to fully understand hemispheric asymmetry it is necessary to understand interhemispheric interaction.

The research reviewed in this chapter is divided into sections dealing with four general topics: (1) language, (2) vision, (3) imagery, and (4) attention. This is not meant to imply that these are the only topics for which a componential approach to the study of hemispheric asymmetry is relevant. The componential approach is illustrated most clearly by research in these areas, however, and this range of topics is sufficiently broad to permit an evaluation of the approach.

Language

The view of hemispheric asymmetry for language has changed considerably in the last decade. As reviewed in Chapter 2, the idea that the left hemisphere is dominant for all aspects of language has been replaced by the view that hemispheric dominance differs for different components of language. For example, injury to different areas of the left hemisphere selectively disrupts the ability to use syntactic or semantic information. In addition, injury to the right hemisphere disrupts the use of context and the ability to integrate information across sentences and larger units of language (the pragmatic aspects of language). These particular components of language were taken from contemporary psycholinguistic theory and the empirical techniques for studying them were adapted from experimental paradigms in psycholinguistics. Being rooted in contemporary psycholinguistics, the studies of hemispheric asymmetry can provide converging information about the theories of language that motivated them. For example, finding that syntactic processing, semantic processing, and the processing of pragmatic cues depend on different neurological substrata offers support for theories that treat them as distinct subsystems.

More detailed componential models of certain aspects of language receive support from additional studies of patients with localized brain injury. For the most part, these models are based on data from patients with interesting dissociations of cognitive function, regardless of the locus of injury, and they have not been designed with hemispheric asymmetry in mind. Consequently, they will not be reviewed in detail here. However, it is instructive to review briefly the kind of componential model that has been proposed and to note that most of the supporting data come from patients with injury to areas of the left hemisphere.

In their introduction to human cognitive neuropsychology, Ellis and Young (1988) provide an extensive review of the data related to recognizing spoken and printed words. They propose a model that illustrates some of the component processes and their interconnections (see Figure 3.1). As an illustration, consider how the printed word *DOG* might lead to the correct spoken output. One route to correct pronunciation is via grapheme-to-phoneme conversion (path 5, 15, 9 in Figure 3.1). That is, the word is divided up into letters or letter clusters and those units are translated into phoneme strings. Note that this route would also lead to a correct pronunciation of a nonword such as *DEG.* An alternative route for naming printed words is via the visual input lexicon, the semantic system, and the speech output lexicon (path 5, 6, 7, 4, 8, 10, 9 in Figure 3.1). That is, the visual stimulus leads directly to an activation of the meaning of the word *dog* and the spoken

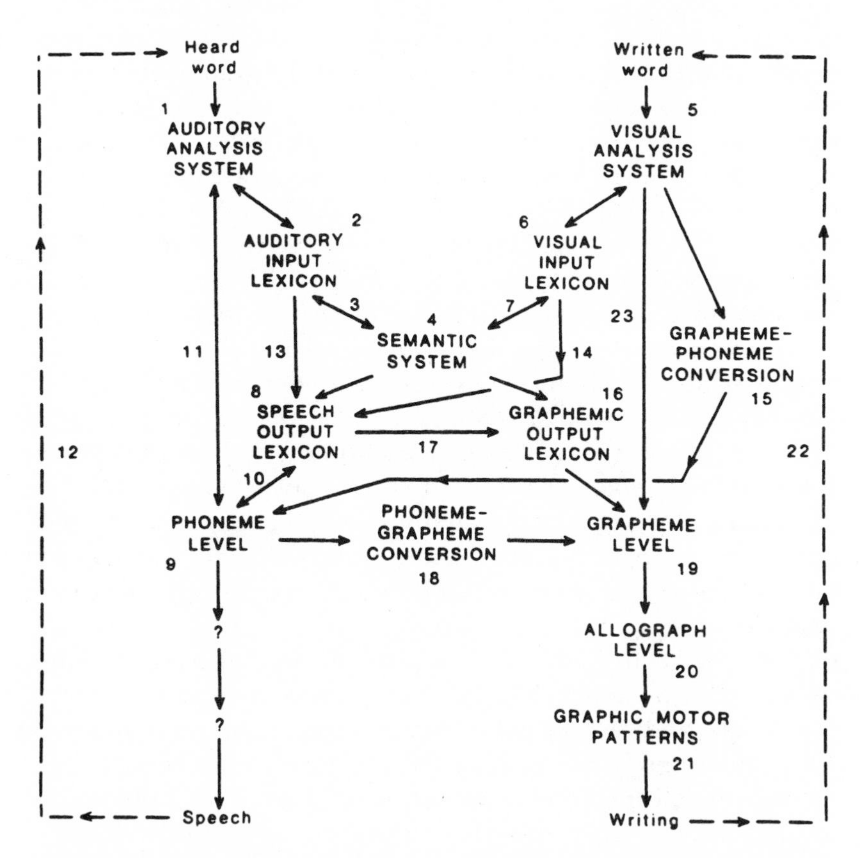

Figure 3.1. A composite model for the recognition and production of spoken and written words. Note that there are several pathways that lead from auditory and visual input to speech and writing output. [Reprinted from A. W. Ellis and A. W. Young, *Human Cognitive Neuropsychology* (Hove, U.K.: Lawrence Erlbaum Associates Ltd., 1988). Copyright 1988 by Lawrence Erlbaum Associates, Ltd. Reprinted by permission.]

output corresponds to the phonemes associated with that semantic concept. Note that this route would not permit the correct pronunciation of nonwords. Ellis and Young are led to postulate the specific components that are shown because data from specific patients can be accounted for by assuming that they have a deficit in a particular set of these components (perhaps a single component) or in one or more connections between components.

As I have noted, the development of this model has little to do with hemispheric asymmetry per se. However, it is interesting to note that the phenomenon of deep dyslexia (see Chapter 2), which is accompanied by extensive injury to the left hemisphere, has been hypothesized to involve disruption of the grapheme-to-phoneme conversion component. On this view, the deep-dyslexic patient must rely on the orthographic-semantic route for pronouncing printed material. Consequently, the patient cannot pronounce nonwords and makes paralexic errors (e.g., saying "stone" for *rock*). Note also that one hypothesis is that the remaining reading performance of deep dyslexics is mediated by the right hemisphere, which would mean that the right hemisphere has some ability to traverse the semantic route shown in Figure 3.1 (e.g., Coltheart, 1983; Ellis and Young, 1988; Zaidel and Peters, 1981). However, as noted in Chapter 2, the idea that reading in deep dyslexics is mediated by the right hemisphere is controversial.

It should be noted that there is also controversy surrounding the criteria that are appropriate for distinguishing between such things as dual-route and single-route models of reading and about the manner in which various routes might operate. Of particular importance are neural-network or connectionist computer models that are meant to flesh out or replace some of the components contained in models like the one illustrated in Figure 3.1 (e.g., Seidenberg and McClelland, 1989; see also Hinton and Shallice, 1991). Such models hold additional promise for learning about the way in which certain components of language are processed within the left hemisphere. However, at the present time they do not deal directly with the issue of which hemisphere is dominant for specific components of language.

Evidence that both hemispheres can access semantic information on the basis of printed words comes from studies of lexical priming (see Chiarello, 1988, 1991, for a review). In a lexical decision task, an observer is required to indicate as quickly as possible whether or not a string of printed letters spells a word. Not surprisingly, there is a robust RVF/left-hemisphere advantage for responding correctly to words, even when observers are encouraged to pay more attention to the LVF than to the RVF (e.g., Hardyck, 1991; Hardyck, Chiarello, Dronkers, and Simpson, 1985). A target word (e.g., *DOG*) is recognized more quickly (and sometimes more accurately) when the target is preceded by a prime word that is semantically related to the target (e.g., *CAT*) than when the target word is preceded by an unrelated word (e.g., *BLANK*). This facilitation is thought to indicate spread of activation through a semantic network. That is, presentation of the prime word tends to activate semantically related words and when one of those words is presented as a target, it is recognized more quickly. Therefore, it is useful to study the magnitude of this facilitation in patients with unilateral brain injury and when the stimuli are presented to the LVF or RVF of split-brain patients or of neurologically intact individuals.

Hemispheric asymmetry for semantic priming depends on whether the spread of activation is preattentive and automatic or more purposeful and controlled. Specific tests of this hypothesis were motivated, in part, by the report of Milberg and Blumstein (1981) that patients with Wernicke's aphasia showed normal semantic priming, even though they were impaired in making explicit semantic judgments about the same stimuli. For example, *DOG* successfully primed *CAT* but the patients were unable to tell that *DOG* and *CAT* were related words. Thus, there appears to be a dissociation between tasks that measure the nature and integrity of structural relations in the lexicon and tasks that require active decisions about semantic relations between words.

In order to examine this possibility explicitly, Chiarello (1985, 1988) took advantage of the distinction made in the cognitive literature between *automatic priming* and *controlled priming* (e.g., Neely, 1977; Posner and Snyder, 1975). Automatic priming requires little or no attention and is likely to occur even when conditions prevent or discourage the observer from actively using the prime word to aid in making a lexical judgment about the target. Thus, automatic priming is thought to reflect the passive spread of activation through the mental lexicon. Automatic priming is likely when the prime word is visually masked, the prime-target interval is too brief to allow controlled processing (less than 500 msec or so), and there is a low probability that the prime will be related to the target. In contrast, controlled priming requires attention to the prime so that it can be actively used to aid in making a lexical decision about the target. Controlled priming is most likely when the observer is aware that primes and targets are very likely to be related, there is an attempt by the observer to identify the prime and use it as a clue, the prime is presented clearly, and the prime-target interval is long enough to permit controlled processing. It is thought to provide a measure of active or directed lexical access (Chiarello, 1985).

Chiarello (1985) examined the extent of semantic priming when stimuli were presented to the RVF/left-hemisphere or LVF/right-hemisphere of neurologically normal observers. One experiment favored controlled priming by making the probability high (.75) that the prime and target would be related and by encouraging the observers to use the prime to predict the target word. In this case, the priming was larger on RVF/left-hemisphere trials than on LVF/right-hemisphere trials. Another experiment favored automatic priming by making the probability low (.25) that the prime and target would be related and by telling the observers that the prime was a nonspecific warning cue. In this case, semantic priming was larger on LVF/right-hemisphere trials than on RVF/left-hemisphere trials. Similar results have been obtained by Michimata (1988), who manipulated the likelihood of automatic versus controlled priming by varying the prime-target interval.

Chiarello (1985) also examined automatic and controlled priming in experiments where the prime-target relationship was orthographic (e.g., *BEAK* and *BEAR*) or phonological (e.g., *JUICE* and *MOOSE*) rather than semantic. Automatic priming with orthographically similar words was restricted to LVF/right-hemisphere trials whereas automatic priming with phonologically similar words was restricted to RVF/left-hemisphere trials. Controlled priming with orthographically similar words was also larger on LVF/right-hemisphere trials than on RVF/left-hemisphere trials whereas controlled priming with phonologically similar words was equal for the two visual fields. On the basis of her entire pattern of results, Chiarello argues that there are hemispheric asymmetries in the availability of lexical information, with orthographic relationships more available to the right hemisphere, phonological relationships more available to the left hemisphere, and semantic information available to both. With respect to semantic information, however, hemispheric differences in availability depend upon whether the spread of activation is automatic or controlled. Moreover, the results of additional experiments that examine the spread of inhibition as well as the spread of activation have led Chiarello (1991) to suggest that word meanings are accessed differently by the two hemispheres. She argues that when a word is presented in context, its various meanings are activated rapidly in the left hemisphere and only those meanings that best fit the context remain activated. That is, other meanings are suppressed very quickly. In something of a complementary fashion, a larger set of word meanings is activated in the right hemisphere and the entire set remains activated for a much longer time than is the case in the left hemisphere.

Although there is still much to be learned about hemispheric asymmetry and lexical access, a good deal of order has been brought to the literature by the distinction between automatic and controlled spread of activation. In fact, without that distinction the literature on lexical priming in the two hemispheres is confusing and filled with equivocal findings. With it, order is introduced by noting whether earlier studies were likely to have involved automatic or controlled priming (see Chiarello, 1985, 1988). Thus, it is clear that this particular distinction, which was imported from the cognitive literature on lexical processing, has proven very useful. At the same time, the fact that hemispheric asymmetry differs for automatic versus controlled priming reinforces the view that the distinction between the two types of spreading activation is meaningful. The further use of paradigms to study the spread of inhibition has produced additional insights into the complementary ways in which the two hemispheres might activate word meanings.

Vision

As noted in Chapter 2, at various times it has been suggested that the left hemisphere is dominant for verbal processing whereas the right hemisphere is dominant for nonverbal processing, including the processing of nonverbal, visuospatial information. In fact, the pattern of hemispheric asymmetry for visuospatial processing is far more complex: each hemisphere is now hypothesized to be dominant for processing different *types* of visual information. This section begins with a review of three distinctions that have been imported from the visual-processing literature in an effort to bring order to models of hemispheric asymmetry for visual processing: (1) global versus local processing, (2) low versus high visual-spatial frequencies, and (3) coordinate versus categorical spatial relations. Following this review is a discussion of the relationships among these three distinctions.

Global versus Local Processing

The concept of *hierarchical organization* has come to be very important for theories of visual information processing. The visual system is said to have a hierarchical organization because visual stimuli and the internal representations that they give rise to contain many levels of embedded structure, with smaller *(local)* patterns or "parts" contained within larger *(global)* patterns or "wholes." Whether a particular pattern is treated as the "whole" or as one of the "parts" depends on the context in which it is considered. For example, a human hand might be considered a whole object, with the thumb being a part of that whole. At the same time, the hand is also a part in the context of an entire human body. Therefore, parts versus wholes (or local versus global patterns) are best defined by their relative place in a hierarchy of levels (e.g., Lamb, Robertson, and Knight, 1990; Palmer, 1977; Robertson, 1986).

There is now strong evidence of hemispheric asymmetry for the processing of global versus local levels of information in hierarchical visual stimuli. Specifically, the right hemisphere is well suited or predisposed for the processing of global aspects of the visual world, whereas the left hemisphere is well suited or predisposed for the processing of local aspects of the visual world. The clearest evidence comes from a series of elegant studies of patients with unilateral brain injury, with some converging support provided by visual-half-field studies using neurologically intact individuals.

In an informative pilot study, Delis, Robertson, and Efron (1986) presented brain-injured patients with the sort of hierarchical stimuli shown in Figure 3.2 (*a* and *b*) and asked the patients to draw the stimuli from memory after performing a distractor task for 15 seconds. As illustrated in Figure 3.2, patients with right-hemisphere injury often produced the correct local information (e.g., the *z*'s of Figure 3.2*a*) but missed completely the global pattern of which the local details are parts (e.g., the *M* of Figure 3.2*a*). In marked contrast, patients with left-hemisphere

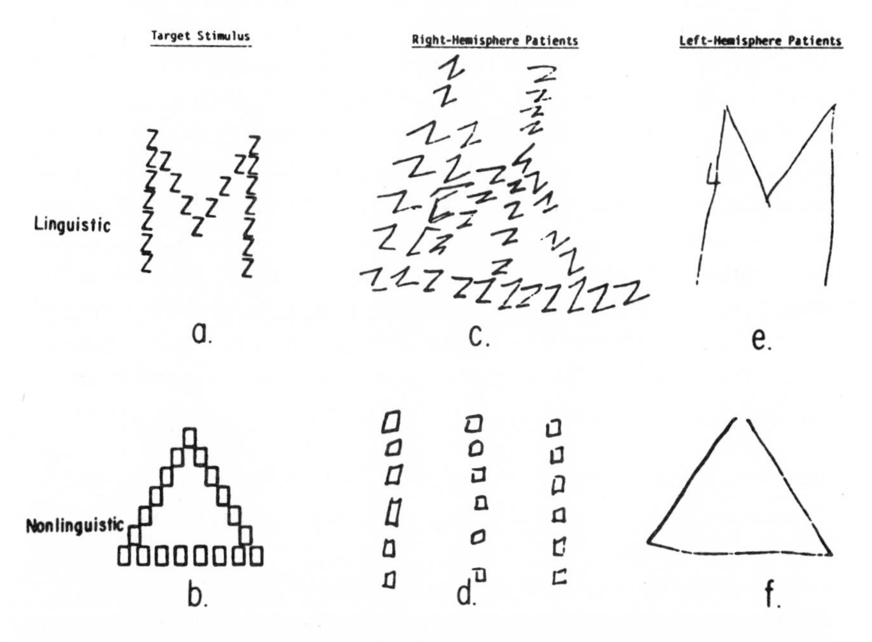

Figure 3.2. Examples of hierarchical stimuli and patients' recalled drawings in the pilot study reported by Delis et al. (1986). (a) Linguistic symbols; (b) nonlinguistic symbols; (c) and (d) drawings of patients with right-hemisphere injury, which illustrate correct reproduction of the smaller elements but not the larger configuration; (e) and (f) drawings of patients with left-hemisphere injury, which illustrate correct reproduction of the larger configuration but not the smaller elements. [Reprinted from D. Delis, L. Robertson, and R. Efron, "Hemispheric Specialization of Memory for Visual Hierarchical Stimuli," *Neuropsychologia,* 24 (1986):205–214. Copyright 1986 by Pergamon Press plc. Reprinted by permission.]

injury did just the opposite. That is, they often produced the global pattern correctly but missed the local detail. Such results suggest that the right hemisphere is necessary for normal processing of global aspects of visual stimuli whereas the left hemisphere is necessary for normal processing of the local details.

This pattern of hemispheric asymmetry is not restricted to tasks that require patients to draw the stimuli. In a recognition-memory experiment, Delis et al. (1986) presented hierarchical stimuli similar to those shown in Figure 3.2 and required patients to choose the correct stimulus from a set of four alternatives. Suppose the to-be-remembered stimulus was the pattern shown in Figure 3.2*a*, a large *M* made up of small *z*'s. One alternative would be the correct stimulus. A second alternative would contain the correct global information but incorrect local detail (e.g., a large *M* made up of small *T*'s). Note that mistaking this alternative for the correct answer would indicate the greater influence of global than of local information. A third alternative would contain incorrect global information but correct local detail (e.g., a large *O* made up of small *z*'s). Note that mistaking this alternative for the correct answer would indicate the greater influence of local than of global information. A fourth alternative would contain both incorrect global and incorrect local information (e.g., a large *O* made up of small *T*'s). The most theoretically important finding was that, when they made errors, patients with right-hemisphere injury tended to choose the third alternative, indicating a disruption of the ability to use global information. In contrast, when patients with left-hemisphere injury made errors, they tended to choose the second alternative, indicating a disruption of the ability to use local information.

In extensions of these studies, Lamb, Robertson, and Knight (1989) required brain-injured patients and neurologically intact control subjects to identify as quickly as possible the letter at either the global level (e.g., the *M* in Figure 3.2*a*) or the local level (e.g., the *z* in Figure 3.2*a*), with global versus local instructions used on separate blocks of trials. In a replication of earlier studies with neurologically intact individuals, the control subjects responded faster to the global level than to the local level. Relative to these control subjects, the global advantage in reaction time (RT) was increased in patients with left-hemisphere injury and decreased in patients with right-hemisphere injury (see Robertson, Lamb, and Knight, 1988, for a similar pattern of results on a divided-attention task).

In an effort to investigate further the component mechanisms underlying the processing of hierarchically organized patterns, Lamb et al. (1990) required subjects to indicate which of two target letters occurred in a hierarchical display, but in this experiment the target occurred randomly and unpredictably at either the global or the local level. When observers must divide their attention across both global and local levels, they typically respond faster to local than to global targets—an outcome that is opposite the global advantage observed when attention on a block of trials can be restricted to one level or the other. Lamb and colleagues varied the size of the stimulus displays in an effort to separate genuine effects of hierarchical structure from the effects of retinal size and stimulus discriminability. In addition to neurologically intact control subjects, the study included patients with focal injury to the posterior superior temporal gyrus (STG) of either the right or left hemisphere and control patients with posterior lesions that did not involve the superior temporal gyrus.

In a replication of earlier experiments that have used this task, the neurologically intact control subjects showed an RT advantage for local rather than global targets, and this local advantage increased with increases in visual angle. Patients with injury to the right STG showed an even larger local advantage than neurologically intact control subjects, whereas patients with injury to the left STG showed a global advantage. This unusual effect for the left-STG group occurred even when the visual angle increased, indicating that the results are attributable to the target's relative position in a hierarchy rather than to retinal size per se. The neurologically normal control subjects also showed a typical pattern of interference of global distractors on responses to local targets. Patients with injury to either STG showed little evidence of this type of interference. Control patients (whose injury did not involve the STG) performed normally, indicating the importance of the STG in particular. Given these results and those of earlier studies, the investigators argue that an analysis of hierarchical stimuli involves several component mechanisms linked to specific anatomical regions:

> One mechanism, associated with the left posterior superior temporal lobe, favors the processing of more local levels of structure. Another mechanism, associated with the right posterior superior temporal lobe, favors the processing of more global levels of structure. Interference between levels occurs as the result of the integration of information from different levels into a coherent whole, and this integration process is disrupted by lesions of the posterior superior temporal lobe or by lesions of afferent pathways to this area. The relative speed with which local and global levels of structure are processed can be modulated by an attentional mechanism associated with the rostral inferior parietal lobe. These component mechanisms act together so that different levels of structure can be perceived as distinct yet organized. (Lamb et al., 1990, p. 482)

Additional evidence for the anatomic specificity of these effects is provided by Robertson, Lamb, and Knight (1991), who demonstrate that patients with injury confined to prefrontal areas of one hemisphere or the other show the same global versus local effects as neurologically intact control subjects.

Given such clear-cut hemispheric asymmetries for the processing of global versus local information, one would expect to find that visual half-field asymmetries in neurologically intact individuals depend on the level of information to be processed. In fact, some studies have found this to be the case. For example, Sergent (1982) used stimuli similar to those shown in Figure 3.2 and found an LVF/right-hemisphere advantage for the identification of the large letter and an RVF/left-hemisphere advantage for the identification of the small letters. Martin (1979) and Alwit (1982) also found some evidence of interaction of global versus local factors by visual field, although only the RVF/left-hemisphere advantage for the identification of small letters was significant. On the other hand, Alivisatos and Wilding (1982) and Boles (1984) failed to find any significant visual field differences for identifying either large or small letters. Nevertheless, on the basis of a meta-analysis of the results from these studies and from his own experiment, Van Kleeck (1989) concludes that there is a such an interaction in neurologically intact individuals. Some of the inconsistency in visual-half-field studies may reflect an inherent noisiness in the paradigm (e.g., Delis et al., 1986). Certainly, the results obtained with this paradigm are subject to a host of input and task factors that often vary unsystematically from study to study (e.g., Hellige and Sergent, 1986; Sergent and Hellige, 1986). It is also possible that, under some viewing conditions, interhemispheric transfer of information in the intact brain is so efficient that hemispheric asymmetries are masked. Despite the discrepancies in the studies, the results from Van Kleeck's meta-analysis and from studies using brain-injured patients leave little doubt about hemispheric asymmetry for processing global versus local information.

Several instances of right-hemisphere dominance for "visuospatial" tasks were reviewed in Chapter 2. It seems likely that at least some of them are related to hemispheric asymmetry for processing global versus local information. For example, consider the drawings of a cube by the two hands of split-brain patients (see Figure 2.3) and by brain-injured patients. When the right hemisphere is in control (when the left hand is used by split-brain patients and patients with left-hemisphere injury), the drawings maintain sufficient configural (global) information to convey a sense of three dimensions. When the left hemisphere is in control (when the right hand is used by split-brain patients and patients with right-hemisphere injury), many of the (local) details are represented but the global organization is missing. Likewise, patients with right-hemisphere injury frequently fail to arrange blocks in a square when trying to reproduce a design (see Figure 2.4), whereas patients with left-hemisphere injury arrange the blocks in a square but have difficulty with the inner arrangement of the blocks.

Low versus High Visual-Spatial Frequency

The nature of stimulus input is an extremely important variable in determining how the brain will process information. There is considerable evidence that each point in the visual field is multiply encoded by size-tuned filters corresponding to overlapping receptive fields. The different scales of resolution may be determined by outputs from neurons with receptive fields of different sizes (e.g., Marr, 1982) or by outputs from neurons that are tuned to

Figure 3.3. Three vertical sine-wave gratings. Note that the visual-spatial frequency increases from top to bottom. The spatial frequency of the middle grating is twice that of the top grating, and the spatial frequency of the bottom grating is twice that of the middle grating.

intensity variations over spatial intervals of different sizes (e.g., De Valois and De Valois, 1980). Although psychophysical evidence is generally consistent with both of these alternatives, the concept of spatial frequency has received more attention in studies of hemispheric asymmetry (e.g., Sergent, 1983; Sergent and Hellige, 1986). A stimulus composed of a single spatial frequency in a particular orientation consists of a regular sinusoidal variation of luminance across space and looks like alternating dark and light areas with fuzzy borders. Figure 3.3 shows three vertical sine-wave gratings that differ in spatial frequency; that is, they differ in how often the light and dark areas alternate per unit of space. The higher the spatial frequency, the more alternations per unit of space. In principle, it is possible to decompose (and to re-create) any complex image by analyzing the spatial frequencies that are present and their orientations, phase relationships, and so forth (e.g., Thomas, 1986). Consequently, spatial-frequency channels could serve as important components of visual perception and cognition.

It has been hypothesized that, at some level of processing beyond the sensory cortex, the left and right hemispheres are biased toward efficient use of higher and lower visual-spatial frequencies, respectively (e.g., Sergent, 1982, 1983, 1987a,b). This visual-spatial-frequency hypothesis came about as an attempt to explain the effects of input variables on hemispheric asymmetry in visual information processing. As reviewed in Chapter 2, for many tasks, reducing the perceptual quality of visual stimuli interferes with performance more when stimuli are projected to the RVF/left hemisphere than to the LVF/right hemisphere. The resistance of the right hemisphere to the effects of perceptual degradation was discussed initially in terms of the superiority of the right hemisphere for processing a variety of visuospatial relationships (see Chapter 2). The same effect could be explained by the spatial-frequency hypothesis if it is assumed that the manipulations used to reduce stimulus perceptibility resulted in the selective removal of higher spatial-frequency information. One advantage of the spatial-frequency hypothesis over other dichotomies is that it is, in principle, more amenable to operational definition and empirical test.

Since the initial formulation of the hypothesis, a number of experiments and literature reviews have demonstrated the importance of spatial frequency for determining visual hemispheric asymmetry. For example, in an extensive review of relevant studies from 1965 through 1987, Christman (1989) examines the effects on visual-half-field asymmetry of a number of perceptual variables that provide at least indirect tests of the spatial-frequency hypothesis. The review is based on the fact that availability to the brain of higher relative to lower spatial frequencies is reduced by moving stimuli further into the retinal periphery (i.e., increasing retinal eccentricity), decreasing luminance, decreasing exposure duration, increasing stimulus size and blurring, or using a computer to remove high spatial frequencies from the stimuli. Consequently, the spatial-frequency hypothesis predicts that all of these variables will produce greater impairment of RVF/left-hemisphere performance than of LVF/right-hemisphere performance.

Christman (1989) concludes that the accumulated results in the literature up until 1987 offer moderate support for the spatial-frequency hypothesis, with the support being stronger from those experiments that involved more direct manipulations of spatial frequency. For example, blurring (which fairly directly removes higher relative to lower spatial frequencies) and low-pass filtering of stimuli (the removal of higher spatial frequencies) have been show to produce greater impairment of RVF/left-hemisphere performance in studies involving a variety of tasks: letter processing (e.g., Jonsson and Hellige, 1986; Sergent, 1989; but see Peterzell, 1991; Peterzell, Harvey, and Hardyck, 1989), comparison of nonverbal line figures (Michimata and Hellige, 1987), face processing (e.g., Sergent, 1985), and temporal integration (Christman, 1990). The weakest support comes from studies using highly linguistic tasks like word identification and lexical decision (e.g., Chiarello, Senehi, and Soulier, 1986; Hardyck, 1991). This may indicate that hemispheric asymmetries for perceptual processes can be overridden by hemispheric asymmetry for the higher-level cognitive components of a task.

Various additional task factors are undoubtedly important in moderating the effects of input variables and probably account for some of the discrepant findings in the literature (e.g., Christman, 1989; Hellige and Sergent, 1986). A particularly important factor concerns the range of visual-spatial frequencies that are required for optimal performance of a particular task. Sergent

(1983) has suggested that a task that demands processing of the relatively high-spatial-frequency components of a stimulus will produce an RVF/left-hemisphere advantage whereas a task that demands processing of the relatively low-spatial-frequency components of a stimulus will produce an LVF/right-hemisphere advantage. With this in mind, Sergent (1985) required observers to perform three different tasks with the same set of 16 photographs of familiar faces as stimuli. She argued that two of the tasks (identifying the person pictured and classifying the person as a professor or not) are best accomplished on the basis of relatively high spatial frequencies, and for these tasks there was an RVF/left-hemisphere advantage. She argued that the other task (classifying a face as male or female) could be performed adequately on the basis of relatively low spatial frequencies, and for this task there was no visual-field or hemispheric asymmetry.

In a second experiment, Sergent (1985) had a new group of observers perform these same three tasks with the faces digitized and either broad-pass filtered (producing stimuli and results very much like those of her earlier experiment) or low-pass filtered (producing stimuli with spatial frequencies above 2 cycles per degree of visual angle filtered out). Samples of these stimuli are shown in in Figure 3.4. One important result was that filtering out high spatial frequencies increased RT more for the identification and "professor"-classification tasks than for the sex-identification task, providing converging evidence for Sergent's assertion that the first two tasks depended on higher spatial-frequency information than the third task. In addition, for all three tasks, removing the relatively high spatial frequencies increased RT more on RVF/left-hemisphere trials than on LVF/right-hemisphere trials, and the actual asymmetry for each condition was determined jointly by the range of spatial frequencies contained in the input and the range required for optimal task performance (see also Christman, 1989, 1990; Hellige, Jonsson, and Corwin, 1984).

The studies considered so far have used complex visual stimuli (e.g., alphanumeric characters, faces, and so forth) and varied characteristics of the stimulus display in an attempt to manipulate the proportion of higher and lower spatial frequencies in the input. The fact is that the specific manipulations used in most of these experiments have only crude or indirect effects on the spatial-frequency content of the input; consequently, the studies do not provide a straightforward manipulation of spatial frequency as opposed to other input characteristics, such as stimulus perceptibility or the total amount of visible information or energy contained in the stimulus (e.g., Jonsson and Hellige, 1986; Michimata and Hellige, 1987; Peterzell, 1991; Sergent and Hellige, 1986). In addition, even when spatial frequency is manipulated more directly (so there is explicit control of the spatial frequencies contained in the input), there is little or no control over which spatial frequencies convey the information needed to perform a task optimally.

One way to circumvent these problems is to use stimuli whose spatial-frequency spectra are both known and very simple (e.g., sine-wave gratings such as those shown in Figure 3.3). The use of

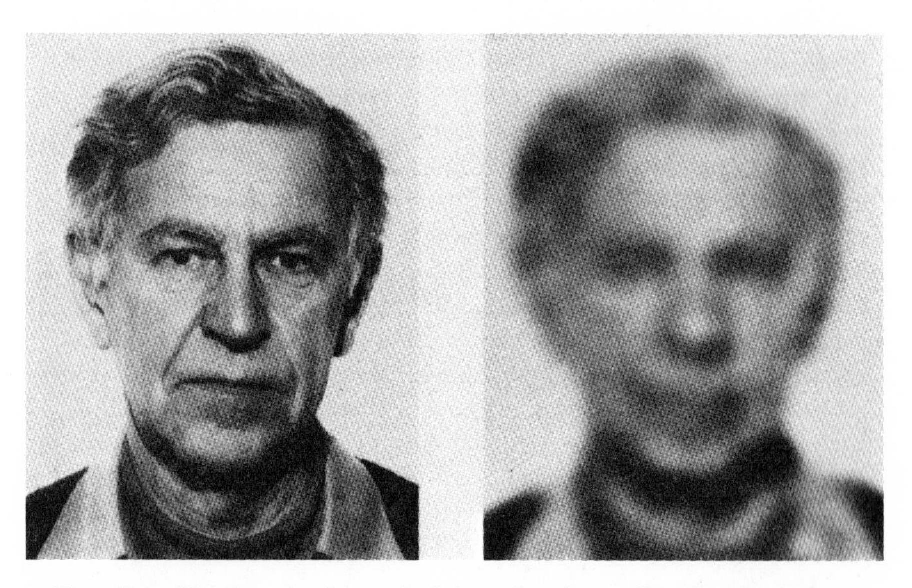

Figure 3.4. The face stimulus on the left was broad-pass filtered and contains a wide range of visual-spatial frequencies. The stimulus on the right was low-pass filtered so that relatively high visual-spatial frequencies have been removed. [Reprinted from J. Sergent, "Influence of Task and Input Factors on Hemispheric Involvement in Face Processing," *Journal of Experimental Psychology: Human Perception and Performance,* 11 (1985):846–861. Copyright 1985 by the American Psychological Association. Reprinted by permission.]

such stimuli allows both explicit control over the spatial frequencies *available* in the input and explicit knowledge of the spatial frequencies *required* for performing the experimental task. That is, in a sine-wave grating, the single frequency present in the input is also the only frequency that can be required for successful performance of the task. Recent experiments using such stimuli have provided strong evidence for certain aspects of the visual-spatial-frequency hypothesis and also point to important qualifications.

Kitterle, Christman, and Hellige (1990) presented sine-wave gratings of various spatial frequencies to the LVF/right hemisphere or RVF/left hemisphere of neurologically intact university students. In some experiments, the task of the observer was to respond as quickly and as accurately as possible to indicate whether or not a stimulus was presented on each trial. That is, the task involved only stimulus *detection.* It did not require the observer to indicate anything about the identity of the stimulus. In these detection experiments, there was no interaction between visual field/hemisphere and spatial frequency. That is, there was no indication that the left and right hemispheres are maximally sensitive to different spatial frequencies when the computational demands are minimized. This result replicates those of earlier studies based on stimulus-detection paradigms and is consistent with Sergent's (1983) position that any hemispheric differences in the processing of spatial frequency must result from processing taking place beyond the sensory level.

In contrast to the results for stimulus detection, Kitterle et al. (1990) found a marked interaction between visual field/hemisphere and spatial frequency when subjects were required to identify which of two sine-wave gratings was presented. Specifically, in this *identification* task, there was an LVF/right-hemisphere advantage for responding to a relatively low-spatial-frequency grating (1 cycle per degree of visual angle) and an RVF/left-hemisphere advantage for responding to a relatively high-spatial-frequency grating (9 cycles per degree of visual angle). Christman, Kitterle, and Hellige (1991) replicated these findings and extended them to show that the relationship also holds for complex gratings that are either made up of two low-frequency gratings (0.5 cycles per degree of visual angle and 1 cycle per degree of visual angle) or made up of two higher-frequency gratings (4 cycles per degree of visual angle and 8 cycles per degree of visual angle). Kitterle and Selig (1991) also obtained the same pattern of interaction in a *discrimination* task that required observers to indicate whether the second of two gratings was of higher or lower frequency than the first (see also Kitterle and Christman, 1991).

Note that the results for both the identification and discrimination tasks are consistent with the hypothesis that the left and right hemispheres are predisposed toward the processing of relatively high and relatively low ranges of spatial frequency when the task requires the use of information about spatial frequency. When there is no need to respond differently on the basis of spatial frequency (e.g., in stimulus-detection tasks), there are no hemispheric asymmetries that differ with spatial frequency. This may explain why no consistent interaction of visual field and spatial frequency was obtained for neurologically intact subjects in a study that required individuals to indicate whether two gratings of the same spatial frequency were oriented in the same direction (e.g., both vertical) or whether one was horizontal and the other vertical (Fendrich and Gazzaniga, 1990). That is, in judging the identity of orientation, there is no need to identify the spatial frequency of either grating or to discriminate between two stimuli on the basis of frequency. In the same study, split-brain patient V. P. showed a consistent LVF/right-hemisphere advantage for all spatial frequencies (but no trend toward an interaction of visual field and spatial frequency) and split-brain patient J. W. showed a trend toward a relative advantage for high frequencies in the LVF and low frequencies in the RVF.

The experiments reported by Kitterle et al. (1990), Christman et al. (1991), and Kitterle and Christman (1991) indicate that the spatial frequency contained in the stimulus input can be an important determinant of hemispheric asymmetry for visual processing. In addition, Kitterle, Hellige, and Christman (1992) have demonstrated that visual hemispheric asymmetries for a slightly more complex stimulus depend on which spatial frequencies in the stimulus are relevant for performing the experimental task. This was done by having observers respond to the same four stimuli during two different tasks. Two of the stimuli were sine-wave gratings like those shown in Figure 3.3—a "wide" stimulus (1 cycle per degree of visual angle) and a "narrow" stimulus (3 cycles per degree of visual angle). The other two stimuli were square-wave gratings whose fundamental frequencies corresponded to the two frequencies of the sine-wave gratings. Whereas the alternating light and dark bars appear to have fuzzy edges in a sine-wave stimulus, a square-wave stimulus consists of alternating light and dark stripes with sharp edges. For present purposes, it is important to know that a square-wave stimulus consists of several well-defined spatial frequencies. The lowest frequency is the fundamental frequency, which corresponds to the width of the bars. Information about the sharp edge is conveyed by much higher spatial frequencies, the odd higher harmonics of the fundamental frequency.

One task required the observers to indicate whether the single stimulus on a trial was one of the wide stimuli or one of the narrow stimuli, regardless of whether the bars appeared to be fuzzy or sharp. Note that this task requires observers to attend to the relatively low fundamental frequencies and ignore the higher harmonic frequencies. For this task, there was an LVF/right-hemisphere advantage. The other task required the observers to indicate whether the single stimulus on each trial contained fuzzy bars or sharp bars, a task that requires observers to attend to the higher harmonic frequencies and ignore the low fundamental frequencies. For this task, there was an RVF/left-hemisphere advantage. That is, the hemispheric advantage *for the very same stimuli* depended on whether the range of spatial frequencies that was relevant for the task was relatively low or relatively high.

The hemispheric advantage for processing a particular spatial frequency also depends on the context in which that frequency occurs. For example, Christman et al. (1991) have shown that a 2 cycle per degree of visual angle sine-wave grating is processed more efficiently on LVF/right-hemisphere trials when it is the *lowest* of three spatial frequencies in a complex pattern and is processed more efficiently on RVF/left-hemisphere trials when it is the *highest* of three spatial frequencies in a complex pattern. Such effects of *relative* frequency are reminiscent of the facts that a specific stimulus may be global in one context and local in another and that hemispheric asymmetry for processing global versus local information depends on the *relative* location of information in a hierarchy rather than on retinal size per se (Lamb et al., 1990).

In view of the fact that there is reasonable support for the spatial-frequency hypothesis, it is instructive to consider in less technical terms the kinds of visual information that must be conveyed by lower versus higher ranges of spatial frequency. If relatively high spatial frequencies are removed from a complex stimulus (e.g., a face), the stimulus looks blurred (see Figure 3.4). You can see this for yourself by looking at things through a set of reading glasses (assuming that you do not ordinarily use them). If your vision is ordinarily normal, wearing these glasses produces what is known as a dioptric blur—an effective means of selectively removing relatively high spatial frequencies. The more powerful the lenses, the greater the range of high frequencies removed. In such a blurred world, you can still make out the larger characteristics of objects (e.g., the outer contour of faces) because this information is carried by the moderate- to low-spatial-frequency channels, which are not influenced much by dioptric blur. Unfortunately, there is no easy way to illustrate for yourself what the visual world looks like when relatively low frequencies are selectively removed and relatively high frequencies remain. In general, however, the high spatial frequencies are most useful for processing small details (e.g., the inner features of a face) and are not particularly useful for extracting the larger configural properties.

Because different characteristics of a complex stimulus are conveyed by different ranges of visual-spatial frequency, it is understandable that hemispheric asymmetry for the processing of complex stimuli would be influenced by manipulations of spatial frequency. However, the extrapolation to complex stimuli is complicated by the fact that at least three aspects of spatial frequency are relevant: (1) the *absolute* range of spatial frequencies contained in the input, (2) the range of spatial frequencies that is most *relevant* for the task being performed, and (3) whether the frequencies that are most relevant are high or low *relative* to other frequencies contained in the stimulus. In addition, pattern recognition with complex stimuli depends not only on the analysis of which spatial frequencies are present but also upon the orientations of those frequencies, the phase relationships among them, temporal relationships, and so forth, and it cannot be assumed that the hemispheres are equivalent in their ability to deal with these additional pieces of information. For example, there is at least preliminary evidence that the right hemisphere is more efficient than the left for the processing of phase relationships (e.g., Fiorentini and Berardi, 1984). In addition, Rebai, Mecacci, Bagot and Bonnet (1986, see also 1989) have reported that evoked potentials recorded over temporal (but not occipital) leads show opposite hemispheric asymmetries depending on whether a single sine-wave grating reverses its phase at low (e.g., 4 Hz) or high (e.g., 12 Hz) temporal frequencies. Specifically, they report greater activity over the right hemisphere at low temporal frequencies and greater activity over the left hemisphere at high temporal frequencies. Thus, asymmetry for the processing of high versus low spatial frequencies must eventually be considered in view of asymmetries for other components known to be necessary for pattern recognition.

Coordinate versus Categorical Spatial Relations

The distinction between global and local processing and the distinction between low and high spatial frequencies both have to do with processing information about the *identity* of a visual stimulus, at least in the studies that relate those distinctions to hemispheric asymmetry. As indicated by the review in Chapter 2, there is a longstanding belief that the hemispheres are also asymmetric in their ability to *localize* a visual stimulus in space. The predominant view has been that the right hemisphere is superior for virtually all aspects of identifying spatial relations among objects. More recently, Kosslyn and his colleagues have developed a computational model of high-level vision that proposes two distinct subsystems for identifying spatial relations (1987; Kosslyn, Chabris, Marsolek, and Koenig, 1992; Kosslyn, Flynn, Amsterdam, and Wang, 1989; Kosslyn, Koenig, Barrett, Cave, Tang, and Gabrieli, 1989). Furthermore, they have proposed that each hemisphere is dominant for one of these subsystems.

Figure 3.5 shows the subsystems of high-level vision posited by Kosslyn, Flynn, et al. (1989). At a general level, their theory maintains that object properties (e.g., shape) and spatial properties (e.g., location) are processed by neurally distinct subsystems. In particular, they propose that object properties are processed by the so-called ventral system, which runs from the occipital lobe down to the inferior temporal lobe, whereas spatial properties are processed by the so-called dorsal system, which runs from the occipital lobe up to the parietal lobe (e.g., Ungerleider and Mishkin, 1982). The processing that takes place in each of these pathways is further subdivided: the encoding of object properties consists of preprocessing, feature detection, and pattern activation, and the encoding of spatial properties consists of spatiotopic mapping, which leads to *categorical*-relations encoding and to *coordinate*-relations encoding. It is these last two components that are hypothesized to be related to hemispheric asymmetry.

Kosslyn and his colleagues have proposed that the brain computes two different kinds of spatial-relation representations. One type of representation is used to assign a spatial relation to a category such as "on," "off," "outside of," or "above." The other

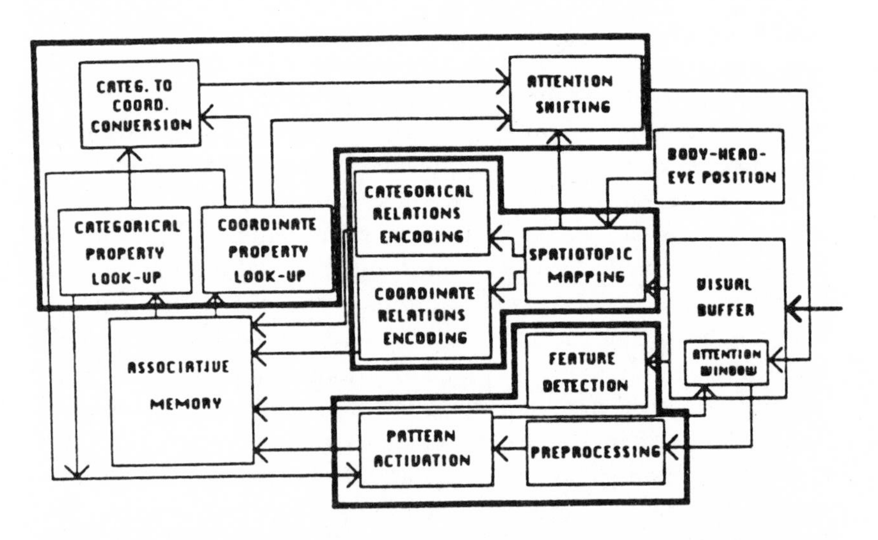

Figure 3.5. Subsystems of high-level vision according the the model presented by Kosslyn, Flynn, et al. (1989). Note especially the separate processing of categorical and coordinate spatial information. [Reprinted from S. M. Kosslyn, R. A. Flynn, J. B. Amsterdam, and G. Wang, "Components of High-Level Vision: A Cognitive Neuroscience Analysis and Accounts of Neurological Syndromes," *Cognition,* 34 (1990):203–277. Copyright 1990 by Elsevier Science Publishers BV. Reprinted by permission.]

type of representation preserves location information using a metric coordinate system in which distances are specified effectively. The distinction between these two types of spatial-relation representations was initially based on a variety of computational considerations that had nothing to do with hemispheric asymmetry. However, one way to obtain converging information about the plausibility of these two processing subsystems is to show that they have different neurological substrata. Consequently, it is important that Kosslyn (1987) hypothesized that the left hemisphere makes more effective use of the categorical processing subsystem whereas the right hemisphere makes more effective use of the coordinate processing subsystem.

Kosslyn's (1987) original hypothesis about hemispheric asymmetry for processing spatial relations was based on the assumption that each hemisphere is specialized for bilateral control over a specific function: the left hemisphere for the control of speech output and the right hemisphere for the control of rapid shifts of attention across space. He argued that these initial specializations provide a "seed" function for each hemisphere. The idea is that as new cognitive skills are added during the course of phylogenetic or ontogenetic development, they become more lateralized to one hemisphere or the other to the extent that they can be performed better by the neurological substrata laid down in one hemisphere as a result of previous development and lateralization. For a number of reasons, Kosslyn argued that the neurological substrata for speech control and for rapid shifts of attention across space are well adapted for categorization of many sorts and for processing coordinate information, respectively.

More recently, Kosslyn et al. (1992) have suggested that hemispheric asymmetry for processing spatial relations has to do with the nature of the visual information that is most useful for computing categorical versus coordinate information. In a set of neural-network simulations, Kosslyn and his colleagues show that networks with relatively large, overlapping "receptive fields" compute coordinate information (e.g., Is a dot within 3 mm of a line?) better than networks with relatively small, nonoverlapping "receptive fields," whereas exactly the reverse was found for the computation of categorical information (e.g., Is a dot above or below a line?). They suggest that the left hemisphere is predisposed toward efficient use of information from visual channels with small, nonoverlapping receptive fields whereas the right hemisphere is predisposed toward efficient use of information from visual channels with large, overlapping receptive fields. In support of this possibility, they cite Livingstone as having suggested that magnocellular ganglia (which have relatively large receptive fields) project preferentially to the right hemisphere. Note that both the original and revised versions of the hypothesis make similar predictions about hemispheric asymmetry for categorical versus coordinate tasks.

To support the predictions about hemispheric asymmetry for processing spatial relations, Kosslyn (1987) reported that categorical relations are computed faster when stimuli are projected to the RVF/left hemisphere whereas coordinate relations are computed faster when stimuli are projected to the LVF/right hemisphere. A more complete description of these results is contained in Kosslyn, Koenig, et al. (1989). For example, in one experiment observers were shown stimuli consisting of a line drawing of a "blob" and a dot that was either touching ("on") the blob or separated from ("off") the blob by various distances. There was an RVF/left-hemisphere advantage for indicating whether the dot was "on" or "off" the blob (a categorical task) and an LVF/right-hemisphere advantage for indicating whether the dot was within 2 mm of the blob (a coordinate or distance task). Interactions between task and visual fields were also reported for a task using a plus sign and a minus sign ("Is the plus to the right of the minus?" versus "Is the plus within 1 inch of the minus?").

Hellige and Michimata (1989) tested Kosslyn's (1987) predictions by having observers indicate whether a small dot was above or below a horizontal line (a categorical task referred to as the "above/below task") or whether the dot was within 2 cm of the line (a coordinate or distance task referred to as the "near/far task"). As illustrated in Figure 3.6, the dot could be in any one of 12 positions with respect to the line. There was an interaction of task and visual field of the sort predicted by Kosslyn's hypothesis, even though exactly the same stimuli were used for both tasks. Specifically, for the above/below task a RVF/left-hemisphere advantage approached statistical significance whereas for the near/far task there was a statistically significant LVF/right-hemisphere advantage. A similar interaction has been reported by Sergent (1991), but only when the stimuli were presented at the low level of luminance used by Hellige and Michimata.

Kosslyn, Koenig, et al. (1989) used the above/below and near/far tasks and found a similar interaction on an initial block of trials. However, the interaction disappeared quickly. The same pattern of results has also been reported by Rybash and Hoyer (1992) and by Koenig, Reiss, and Kosslyn (1990) in a study with children aged 5 and 7 years. Kosslyn, Koenig, et al. (1989) argue

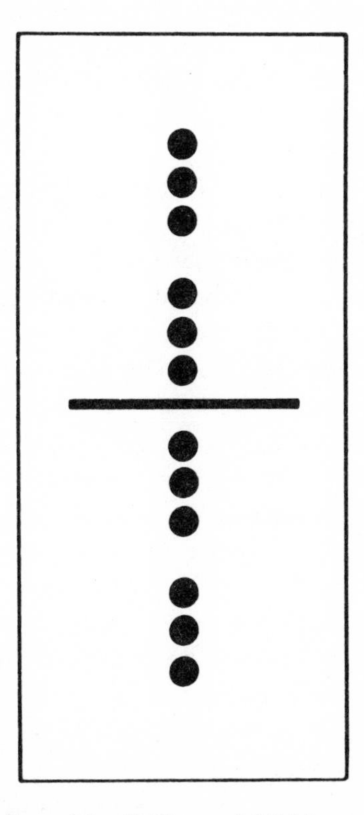

Figure 3.6. The stimuli used by Hellige and Michimata (1989). On each trial, observers saw the line and a single one of the 12 dots. For the categorical task, observers indicated whether the dot was above or below the line. For the coordinate task, observers indicated whether the dot was within 2 cm of the line (the six positions nearest the line on either side). [Reprinted from J. B. Hellige and C. Michimata, "Categorization versus Distance: Hemispheric Differences for Processing Spatial Information," *Memory & Cognition,* 17 (1989):770–776. Copyright 1989 by the Psychonomic Society Inc. Reprinted by permission.]

that when observers are presented with a coordinate, distance-judgment task they may begin to form stimulus categories after sufficient practice, thereby making coordinate processing unnecessary and eliminating the initial LVF/right-hemisphere advantage.

In their comparisons of above/below and near/far tasks, Michimata and Hellige (1989) included a condition in which two identical stimuli were presented simultaneously, one to the LVF/right hemisphere and one to the RVF/left hemisphere. The inclusion of this third visual-field condition allowed them to discover a "reversed association" between performance on the two tasks, which provides additional support for the assertion that the two tasks involve different processing subsystems (see Dunn and Kirsner, 1988). A similar pattern of reversed association was reported by Kosslyn, Koenig, et al. (1989).

Sergent (1991) has raised important criticisms of Kosslyn's (1987) original hypothesis and presented empirical results to demonstrate its limitations. At one level, she argues that the logic that leads to his hypothesis is not entirely compelling. For example, noting that what Kosslyn calls "categorical" spatial relations refers to judgments of relative position, she questions why it should be that the hemisphere specialized for speech output would provide a more suitable neurological base for processing such relationships. She also argues that a "coordinate" representation must convey information about relative as well as absolute locations of objects in space, eliminating the need to postulate two separate processing subsystems. In addition, she argues that the two kinds of spatial information that Kosslyn has proposed are so fundamental that it is very unlikely that each hemisphere would be bereft of one type of representation.

Of course, hypotheses must eventually stand or fall on the basis of empirical evidence, and Sergent (1991) argues that the evidence in favor of Kosslyn's (1987) hypothesis is no more compelling than the logic behind it. She argues that the experiments summarized by Kosslyn (1987) and reported in detail by Kosslyn, Koenig, et al. (1989) contain confounding factors that may preclude unequivocal conclusions about hemispheric asymmetry for two different types of spatial representation. Although Sergent replicated the results reported by Hellige and Michimata (1989) when the luminance of the stimuli was sufficiently low, she reports that the critical task-by-visual-field interaction disappears when the luminance is increased and the tasks become easier. As a result, she argues, the interaction may have more to do with the fact that hemispheric asymmetry depends on viewing conditions than with differences in categorical versus coordinate processing per se. In addition, she reports that both hemispheres of split-brain patients are able to perform both the above/below and near/far tasks at an above-chance level (with the luminance levels that produced no interactions in neurologically intact observers), with little indication of any hemispheric difference for either accuracy or RT.

Sergent also fails to find a task-by-visual-field interaction in neurologically intact individuals for a set of new tasks that are hypothesized to require judgments of relative or absolute locations. What she does find is that absolute distance between stimuli sometimes influences performance in tasks that require individuals to make decisions about relative location (i.e., a categorical task). She argues that such results indicate that whatever type of spatial representation is used to make categorical decisions also preserves information about distance.

Several neural-network simulations presented by Kosslyn et al. (1992) suggest that Sergent's (1991) findings may not be as detrimental to the distinction between categorical and coordinate processing as she argues. In one set of simulations, networks that were split so that some of their hidden units contributed to a categorical judgment and others contributed to a coordinate judgment performed better than unsplit networks in which all hidden units contributed to both types of judgment. Such results at least indicate that there may be computational advantages for segregating the two types of judgment. In addition, Kosslyn et al. found that a neural-network model that codes only categorical information about whether a dot is above or below a line is sensitive to the distance between the dot and the line. In general, the further the dot from the line, the better the performance of the neural-network model. What this indicates is that a network devoted to the processing of categorical spatial relationships need not be immune to effects of distance. Therefore, the presence of distance effects in a categorical task does not necessarily invalidate the categorical/coordinate distinction. In another set of neural-network models, Kosslyn et al. show that they can, to some extent, mimic the luminance effects reported by Sergent. For example, adding more low-level inputs (to mimic higher contrast) eliminated the advantage of large, overlapping receptive fields for encoding coordinate information. On this basis, they conclude that the disappearance of hemispheric asymmetries at high luminance or contrast does not necessarily invalidate the hypothesis of hemispheric asymmetry for the processing of categorical versus coordinate spatial relationships. However, it may favor the revised version of the hypothesis presented by Kosslyn et al. (1992) over the original version of the hypothesis presented by Kosslyn (1987).

Relationships among Components of Vision

It is instructive to consider how the three aspects of vision reviewed here might be interrelated. The relationship between the global/local distinction and the low/high visual-spatial frequency distinction is rather clear. In general, it is the case that information about global aspects of a stimulus is carried by a lower range of visual-spatial frequencies than is information about local aspects. Thus, hemispheric asymmetry for processing of global versus local information and for utilizing low versus high spatial frequencies may well be different manifestations of the same thing. This is especially true in view of the fact that hemispheric asymmetry extends to *relatively* high and low spatial frequencies (e.g., Christman et al., 1991). However, at the present time there is insufficient evidence to know if one of these dimensions is in any sense more fundamental than the other.

There would also appear to be a relationship between the categorical/coordinate distinction and the other two, at least as hemispheric asymmetry for the processing of different spatial relationships has been envisioned by Kosslyn et al. (1992). In fact, Kosslyn et al. note the similarity between low visual-spatial frequency and visual channels with large receptive fields and the similarity between high visual-spatial frequency and visual channels with small receptive fields. The possibility that all three aspects of vision reviewed here are interrelated in this manner is extremely interesting and receives some support from an experiment reported by Cowin and Hellige (1991) on the effects of blurring on hemispheric asymmetry for processing spatial information.

Participants in the Cowin and Hellige (1991) experiment performed either the above/below task or the near/far task described earlier. In addition, the line and dot stimuli were either presented clearly or they were blurred by having the participants view them through distorting lenses. The particular method of dioptric blurring was chosen because it selectively eliminates information carried primarily by channels tuned to high visual-spatial frequencies. Dioptric blurring consistently disrupted performance of the categorical, above/below task, suggesting that the type of high-spatial-frequency information that is eliminated or made difficult to process by dioptric blurring is ordinarily important for optimizing categorical spatial judgments. Note that this conclusion is consistent with the Kosslyn et al. (1992) hypothesis. By way of contrast, the effects of dioptric blurring on performance of the coordinate, near/far task were far less consistent, suggesting that the same high-spatial-frequency information may not be relevant for optimizing coordinate spatial judgments. This is also consistent with the Kosslyn et al. hypothesis. Cowin and Hellige also found significantly fewer errors on LVF/right-hemisphere trials than on RVF/left-hemisphere trials during the first block of the near/far task. Furthermore, this LVF/right-hemisphere advantage for making the near/far coordinate judgment was also unaffected by the blurring manipulation. While research is needed to test additional aspects of the hypothesis outlined in Kosslyn's work, the results obtained by Cowin and Hellige are at least consistent with the possibility that hemispheric differences in processing spatial relationships are related to differences in dealing with low versus high spatial frequencies.

It is interesting to note that recent studies of hemispheric asymmetry for processing auditory information have also found effects of frequency—not spatial frequency, but temporal frequency. For example, in a pitch-discrimination task, Ivry and Lebby (1993) found that individuals were faster and more accurate in judging relatively low-frequency sounds when they were presented to the left ear (right hemisphere) but were faster and more accurate in judging relatively high-frequency sounds when they were presented to the right ear (left hemisphere). At the very least, this is an interesting parallel to what has been observed in vision. In fact, it is possible that there is some connection between hemispheric asymmetry for the visual components reviewed in this section and hemispheric asymmetry for components of auditory processing. One possibility considered by Ivry and Lebby is that, in both stimulus modalities, the two hemispheres process information through different asymmetric filters. Specifically, they suggest that the filter is in some sense tuned to higher spatial and temporal frequencies in the left hemisphere than in the right hemisphere. This is an interesting possibility, although the absence of hemispheric asymmetries for stimulus detection in vision places limitations on how early in visual processing such filters might operate.

An interesting parallel to the spatial-frequency hypothesis in vision is contained in a theoretical framework suggested by Guiard (1987) to account for the asymmetric division of labor from the two hands. Guiard suggests that the two hands work together in what he refers to as a kinematic chain, with the left hand (of right-handers) performing movements of low temporal and spatial frequency and the right hand building on those to perform movements of high temporal and spatial frequency. A prototypical illustration of this division of labor is handwriting, where the left hand arranges and steadies the paper (relatively few, large-scale movements) while the right hand makes frequent, smaller movements of the writing instrument (see also Hammond, 1990). It should prove useful in future studies to determine whether this division of labor between the two hands is influenced at all by the availability of visual or sensorimotor feedback and whether there is any relationship to hemispheric asymmetry in visual information processing.

Imagery

Views about hemispheric asymmetry for the generation and use of visual images have fluctuated wildly over the last decade. Earlier, the predominant view was that the right hemisphere is dominant for the use of imagery, as something of a byproduct of its hypothesized dominance for visuospatial perceptual processes. However, just as views about hemispheric asymmetry for visual perceptual processing have changed, so, too, have views about hemispheric asymmetry for imagery. For example, in a review of the literature available at the time, Ehrlichman and Barrett (1983) found little to support unequivocally the hypothesis of right-hemisphere dominance (or hemispheric asymmetry at all) for visual imagery. More recent research has been based on componential approaches to the study of visual imagery and has led to the hypothesis that it is the *left* hemisphere that is critical for the *generation* of mental images (e.g., Farah, 1984, 1986, 1989), especially for the generation of complex images that require the correct categorical arrangement of parts (e.g., Kosslyn, 1987, 1988; Kosslyn, Holtzman, Farah, and Gazzaniga, 1985). This hypothesis, too, has generated controversy (e.g., Sergent, 1989, 1990), and the issue of hemispheric asymmetry for image generation remains unresolved. Nevertheless, the componential approach represents an advance over earlier approaches.

Guided by Kosslyn's (1980) componential theory of visual imagery, Farah (1984) examined reports in the neurological literature of brain-injured patients who were described as having a loss of visual imagery. By examining the pattern of deficits and preserved abilities reported for these individuals, Farah classified as many patients as possible according to which component of visual imagery was deficient. For example, 8 patients were classified as having a deficit in the *image-generation* process, 13 cases were classified as having a deficit in *long-term visual memory,* 5 patients were classified as having a deficit in the ability to *inspect* images and percepts, and other patients were classified as having a combination of deficits or deficits in systems not related to imagery per se. From her review of lesion sites in these various groups, Farah concludes that a region in the posterior of the left hemisphere is critical for the image-generation process, whereas there was no evidence of hemispheric asymmetry for the other processes. In subsequent work, Farah (1986, 1989) argues that her visual-half-field studies with neurologically intact individuals, studies of regional cerebral metabolic activity, and a variety of other findings are consistent with the hypothesis that the left hemisphere is critical for image generation.

In a detailed review of these same empirical studies, Sergent (1990) concludes that there is very little support for this hypothesis and that, in fact, the available evidence suggests that *both* hemispheres contribute to image generation. With respect to the data from neurological patients (Farah, 1984), Sergent points out that in some cases the information about lesion location was insufficient to provide unambiguous evidence that the left hemisphere is dominant, and in other cases the behavioral tests were not sufficiently precise to single out unambiguously the image-generation process to the exclusion of the other processes. Additional methodological issues are raised with respect to relevant visual-half-field studies and studies of regional metabolic activity, and the arguments presented are sufficiently compelling to weaken the hypothesis that the left hemisphere is critical for image generation.

Whatever the eventual resolution of the issue regarding hemispheric asymmetry for image generation, it seems very unlikely that either hemisphere completely lacks the ability to generate visual images. For example, studies with split-brain patients show that tasks requiring the generation of visual images can be performed above chance level (and often equally well) by both hemispheres (e.g., Corballis and Sergent, 1988; Kosslyn et al., 1985; Sergent and Corballis, 1990). In addition, Sergent (1989) found an LVF/right-hemisphere advantage in neurologically normal individuals when they were performing a task known to involve the generation of visual images. In this task, an uppercase letter is presented on each trial and the observer must indicate whether or not the corresponding lowercase letter (which must be imagined) contains segments extending above or below the main body of the letter. Sergent argues that the LVF/right-hemisphere advantage could indicate that the right hemisphere is actually superior to the left for generating a visual image of the lowercase letter or for subsequent inspection of the image. In either case, the results argue against the idea that *only* the left hemisphere is capable of image generation.

Although Kosslyn (1987, 1988; Kosslyn et al., 1985) acknowledges that both hemispheres are capable of image generation, he hypothesizes that the left hemisphere is dominant for generating images that require the correct categorical arrangement of parts. Initial evidence for this hypothesis came from a series of experiments with a split-brain patient (J. W.), whose right hemisphere could not perform tasks that required him to make decisions about parts of imaged objects (e.g., Do a hog's ears protrude above the top of the skull?) but could perform perfectly tasks that required a decision about the overall shape or size of an imaged object (e.g., Is a book higher than it is wide? Is a hog larger or smaller than a goat?). Similar findings were reported for a second split-brain patient (V. P.), but only at early stages of practice. Support also comes from visual-half-field experiments summarized by Kosslyn (1988) in which an RVF/left-hemisphere advantage is reported for a task that is assumed to use categorical relations to arrange segments of letters within a grid. Moreover, Kosslyn (1988) argues that an additional way to form images of a multipart object is in terms of precise coordinates, and he reports an LVF/right-hemisphere advantage for a task that requires the arrangement of letter segments without the benefit of a grid (a task that is assumed to use coordinate relations to arrange segments of letters). On this basis, he concludes that both hemispheres can form images of components, but they differ in the preferred way of arranging them. At the same time, he uses the neuropsychological findings to provide converging evidence for these specific components of image generation.

The distinction regarding different types of images and different ways of arranging components is quite interesting, especially because of its obvious relationship to Kosslyn's (1987) hypothesis about hemispheric asymmetry for processing categorical versus coordinate spatial relations during perception and because of its possible relationships to the global/local and low-/high-frequency distinctions discussed earlier. However, it is important to note that the findings reported by Kosslyn and his colleagues for patients J. W. and V. P. were not replicated in a third split-brain patient (L. B.), studied by Corballis and Sergent (1988; Sergent and Corballis, 1990). In addition, hypotheses about left-hemisphere superiority for the generation of multipart images may be difficult to reconcile with the LVF/right-hemisphere advantage found when neurologically intact individuals made what would seem to be categorical decisions about parts of generated images of lowercase letters (Sergent, 1989). Of course, the fact that decisions were categorical does not rule out the possibilty that coordinate relations were used to arrange the parts of an image.

An additional aspect of the results reported by Sergent (1989) is worth considering because it suggests another factor that is likely to be important in determining hemispheric asymmetry for tasks that involve imagery. In a purely perceptual task, a lowercase letter was flashed to the LVF or RVF and observers indicated whether or not it contained a part that extended above or below the main body of the letter. On some trials the letter was presented clearly and on other trials it was blurred. One important result was that, for blurred letters, RT was faster on LVF/right-hemisphere trials than on RVF/left-hemisphere trials whereas there was no visual-field difference for clear letters. Another important result was that, for clear letters, RT was faster to letters with an extending segment than to letters without. This effect was reversed for blurred letters. As noted earlier, in the imagery task, an uppercase letter was flashed to the LVF or RVF and observers indicated whether the corresponding lowercase letter contained a part that extended beyond the main body of the letter. In addition to the LVF/right-hemisphere advantage noted earlier, it was the case that RT was faster to letters without an extending segment than to letters with an extending segment: the pattern of results found with the blurred letters in the perceptual task. This suggests that, at least for this task, the image was of lower quality than a clear percept and, specifically, contained information similar to the relatively low spatial frequencies of a blurred stimulus. Consequently, an LVF/right-hemisphere advantage may emerge because the right hemisphere's predisposition toward processing relatively low spatial frequencies extends to visual images. On this view, an RVF/left-hemisphere advantage may emerge for an imagery task to the extent that the image contains information that would be carried by relatively high spatial frequencies in a real visual stimulus.

This review makes it clear that there is a long way to go before we have a complete cognitive neuroscience of visual imagery. However, important advances have been made by taking a more componential approach to the study of imagery and to hemispheric asymmetry for imagery. Even if the components that have thus far guided the research are eventually shown to be inappropriate, their use will have provided an important step toward understanding imagery. Moreover, the additional constraints provided by neurological data will be important sources of information about the adequacy of a particular componential theory of mental imagery.

Attention

This section reviews three ways in which studies related to hemispheric asymmetry have important implications for theories of attention. The first has to do with components of visual orienting and was discussed briefly in Chapter 2. The second concerns hemispheric asymmetry for regulation of alertness, and the third considers the extent to which each hemisphere has its own limited information-processing resources that are to some extent unavailable to its partner.

Components of Visual Orienting

The human information-processing system is limited in capacity, which means that performance often suffers when an individual tries to do too many things at once or pay attention to more than one thing at a time. The ability to orient attention toward one thing rather than toward another consists of several components, and a variety of neuropsychological findings help to identify those components (e.g., Ellis and Young, 1988). With respect to hemispheric asymmetry, studies reviewed in Chapter 2 indicate that certain disorders of visual orienting (e.g., hemineglect and extinction to simultaneous stimulation) are more frequent and more severe after right-hemisphere injury than after left-hemisphere injury, leading to the hypothesis that the right hemisphere is dominant for certain aspects of spatial representation or for the orienting of attention in space. Of particular importance for understanding the mechanisms of visual orienting is the identification of component processes and their neurological substrata.

One important distinction is between *overt* shifts of visual attention (involving shifts in eye fixation) and *covert* shifts of visual attention (looking at one point and attending to another). In addition, several recent studies have used variations of a task developed by Posner and his colleagues to further identify the components of covert attention shifts and to identify the neurological correlates of some of those components (e.g., Posner et al., 1984, 1987; see also Posner, 1992).

In a typical experiment of this type, observers are asked to fixate their gaze on a centrally located point in space and then press a button as quickly as possible when they detect a target stimulus in another location. For example, the target stimulus might appear in a specific location to the right or to the left of the fixation point. On some trials a cue is presented to indicate whether the target is more likely to occur on the left side or on the right side. For neurologically intact individuals, the target is detected faster when it is presented to the cued location (valid-cue trials) than when it is presented to the uncued location (invalid-cue trials), presumably because covert attention has been directed to the cued location in anticipation of the target and must be shifted to the actual target location on invalid-cue trials.

Posner et al. (1984) show that patients with unilateral parietal-lobe injuries show an asymmetry in this type of task. Specifically, ability to shift covert attention toward the direction ipsilateral to the side of the injury (e.g., toward the right with right-hemisphere injury) is normal whereas ability to shift covert attention toward the direction contralateral to the side of the injury (e.g., toward the left with right-hemisphere injury) is not. Interestingly, this pattern of effects was obtained regardless of whether the injury was on the right or left side. The researchers are able to distinguish three components of covert shifts of attention: *disengagement* of attention from its current focus, *moving* attention to the target location, and *engagement* of attention on the target location. Furthermore, they show that the specific problem in patients with parietal-lobe injury is difficulty in *disengaging* attention from one stimulus in order to move it in a direction contralateral to the side of the injury. Problems with this component of covert visual orienting are likely to play an important role in a variety of clinical phenomena, such as hemineglect and biases in line bisection, regardless of whether the clinical manifestations are produced by unilateral injury to the right or to the left parietal lobe (Posner et al., 1987; Reuter-Lorenz and Posner, 1990).

Regulation of Alertness

An area within the frontal lobes of the right hemisphere has been found to be important for the regulation of alertness (e.g., Pardo, Fox, and Raichle, 1991; Posner, 1992; Whitehead, 1991). For example, patients with injury to the right frontal lobe are far more impaired in their ability to take advantage of a warning signal than are patients with injury to the left frontal lobe (e.g., see Pardo et al., 1991). In addition, when neurologically intact subjects are required to maintain attention continuously for longer than 10 seconds, they respond more quickly to targets presented directly to the right hemisphere (e.g., via the LVF) than to targets presented directly to the left hemisphere (Whitehead, 1991). Furthermore, recent studies using PET techniques to monitor regional cerebral blood flow have shown increased activation in the right (but not the left) frontal lobe after subjects have maintained attention for long periods (e.g., Pardo et al., 1991). It is interesting that norepinephrine (a specific neurotransmitter) seems to be particularly important in maintaining alertness and that processes dependent on norepinephrine are hypothesized to be more prevalent in the right hemisphere than in the left hemisphere (e.g., see Chapter 4 and Tucker and Williamson, 1985).

Hemisphere-Specific Priming and Interference

Research reported over the last twenty years has made it clear that effects of priming and interference are sometimes greater for one hemisphere than for the other. Much of the research involves dual-task studies that examine how performance on a task is influenced by the simultaneous performance of another task. This section provides a review of those studies and considers the implications for theories of attention.

The growth of dual-task studies related to hemispheric asymmetry is attributable in part to two complementary theoretical ideas advanced by Kinsbourne in the early 1970s (Kinsbourne, 1970, 1975; Kinsbourne and Cook, 1971). One idea dealt with selective activation and has been discussed briefly in Chapter 1. The critical assertion was that the two hemispheres could be at different levels of activation or arousal, leading to the prediction that performance would be more accurate and faster in response to stimuli presented in the visual field or ear contralateral to the more activated hemisphere. On this view, imposing some concurrent activity that selectively activated one hemisphere would create a perceptual advantage for stimuli presented directly to that hemisphere—what has been referred to as "hemisphere-specific priming." Preliminary support for this hypothesis came from studies of the effect of concurrent verbal memory (presumed to selectively activate the left hemisphere) on identification of nonverbal stimuli from the two visual fields (e.g., Bruce and Kinsbourne, 1974). The other idea dealt with selective interference. It led to the prediction that imposing a concurrent activity that demanded processing primarily from one hemisphere would interfere with the ability of that hemisphere to engage in other activity. Preliminary support for this hypothesis came from studies of the effect of concurrent verbalization (e.g., reciting a nursery rhyme) on motor activity of the two hands (e.g., Kinsbourne and Cook, 1971).

An experiment reported by Hellige and Cox (1976) demonstrates that both hemisphere-specific priming and hemisphere-specific interference can be obtained with the same experimental task and that the difficulty of the concurrent activity is a critical factor. These experiments examined the effects of concurrently holding 0, 2, 4, or 6 nouns in short-term memory on the recognition of visual stimuli presented briefly to the LVF/right hemisphere or to the RVF/left hemisphere. When the visual stimuli to be identified were complex polygon forms, the following results were obtained. On RVF/left-hemisphere trials, a relatively easy memory load of 2 or 4 nouns improved visual-recognition accuracy relative to the no-load condition, but a more difficult memory load of 6 nouns decreased visual-recognition accuracy. On LVF/right-hemisphere trials, there were no significant effects of memory load. As a result of these different patterns for the two visual fields/hemispheres, there was an LVF/right-hemisphere form-recognition advantage with 0 or 6 nouns held in memory and an RVF/left-hemisphere form-recognition advantage with 2 or 4 nouns held in memory.

Given that holding nouns in short-term memory is thought to involve primarily left-hemisphere processes, the results of this experiment suggest hemisphere-specific priming with very light loads giving way to hemisphere-specific interference as the load on the left hemisphere is increased. That is, the results on RVF/left-hemisphere trials are consistent with the well-known Yerkes-Dodson law, stating that the performance on a primary task increases with moderate increases in the level of arousal (caused by 2 or 4 nouns held in memory) but decreases as the level of arousal or information-processing load becomes too great (6 nouns held in memory). The fact that neither of these effects occurs on LVF/right-hemisphere trials suggests that both priming and interference can be restricted to one hemisphere.

Because hemisphere-specific priming very quickly gives way to hemisphere-specific interference, it is not surprising that even a small verbal memory load selectively interferes with processing stimuli presented to the RVF/left hemisphere when those visual stimuli also require verbal processing (e.g., Freidman and Polson, 1981; Freidman, Polson, Dafoe, and Gaskill, 1982; Geffen, Bradshaw, and Nettleton, 1983; Hellige and Cox, 1976; Hellige, Cox, and Litvac, 1979; Moscovitch and Klein, 1980). It is also not surprising that hemisphere-specific interference has been much easier to obtain consistently than has hemisphere-specific priming (e.g., Boles, 1979; Bouma, 1990; see also Berryman and Kennelly, 1992).

Hemisphere-specific interference is not restricted to studies using visual presentation of stimuli to one half-field or the other. Similar effects of concurrent verbal activity have been found for stimuli presented to one ear or the other during dichotic listening (e.g., Hellige and Wong, 1983; Herdman and Friedman, 1985) and in experiments that examine how concurrent tasks interfere with manual activity of the two hands (for reviews of this now extensive literature see Friedman, Polson, and Dafoe, 1988; Hellige and Kee, 1991; Kinsbourne and Hiscock, 1983).

Of what relevance is the existence of hemisphere-specific priming and interference to theories of attention? Such effects are very difficult to reconcile with the traditional idea that there is one large pool of attentional capacity or resources for which all attention-demanding activities must compete (e.g., Navon and Gopher, 1979). Instead, my colleagues and I (Hellige and Cox, 1976; Hellige et al., 1979) argued that our own early results could be accommodated reasonably well by existing theories of arousal and limited capacity if we added the proviso that each of the hemispheres could behave in a manner predicted by those theories and do so somewhat independently of each other. Moscovitch and Klein (1980) made similar suggestions on the basis of their own demonstrations of hemisphere-specific priming and interference. These ideas have been taken to their logical extreme by Friedman and Polson and their colleagues.

Quite independently of any studies having to do with hemispheric asymmetry, Navon and Gopher (1979; see also Navon, 1984) pointed out several problems with the traditional single-resource theories of attention and argued for consideration of a multiple-resource view. In an important set of theoretical and empirical articles, Friedman and Polson and their colleagues argue for a multiple-resource theory of attention in which each hemisphere has its own pool of processing resources. Furthermore, they make the strong claim that the two hemisphere-specific resource pools are completely independent of each other; that is, performance in one hemisphere can never be improved by using resources from the other. An equally strong claim is that, within each of the hemisphere-specific resource pools, the resources are completely undifferentiated; that is, all processes performed by one hemisphere compete for the same resources (although they now acknowledge that there may be some distinction between cognitive and motor processing; see Friedman et al., 1988).

The theoretical perspective taken by Friedman and Polson indicates that it is very important in dual-task studies to examine the extent to which an individual can improve performance on one task by allowing the other to suffer (that is, the extent to which there are mutual trade-offs between two tasks). According to the logic developed by Navon and Gopher (1979; also Navon, 1984), the existence of such trade-offs indicates that the tasks interfere with each other because they are competing for resources from the same limited pool. When such trade-offs are not possible (even though there is interference), the interference is said to result from "concurrence costs" or scheduling conflicts that arise from trying to coordinate two tasks. Thus, it is important to note that the conclusions reached by Friedman and Polson are based not only on the existence of hemisphere-specific interference by also on an appropriate pattern of mutual trade-offs.

An alternative account of hemisphere-specific interference has been provided by Kinsbourne and his colleagues and is based on a view of the brain as a network of parallel distributed processing (e.g., Goldman-Rakic, 1988). Central to this account is a principle referred to as *functional cerebral distance* (e.g., Kinsbourne, 1982; Kinsbourne and Hiscock, 1983). This principle maintains that two concurrent tasks interfere with each other to the extent that they involve highly interconnected regions of the cerebral cortex. The interaction can be facilitative and produce priming effects if the two tasks are compatible or be inhibiting and produce interference if the two tasks are incompatible. To the extent that the two tasks involve cortical regions that are only sparsely interconnected, the interaction between tasks decreases. With only a few exceptions, two regions within the same hemisphere are proposed to be more highly interconnected than are two regions in opposite hemispheres. In this way, hemisphere-specific priming and interference are predicted.

It is not completely clear whether the existing data are more compatible with the Friedman/Polson theory or with the Kinsbourne theory, or even whether the two theories are sufficiently precise to distinguish empirically. Potential problems with the Friedman/Polson account include the fact that it is designed to account for hemisphere-specific interference and does not account for hemisphere-specific priming in any obvious way. In addition, it seems difficult to maintain the strong view that hemisphere-specific processing resources are completely independent in view of the fact that processing in one hemisphere of split-brain patients can be influenced by the difficulty of processing being carried out in the other hemisphere (e.g., Holtzman and Gazzaniga, 1982). Potential problems with the Kinsbourne account include the lack of independent ways of determining functional cerebral distance between two tasks and whether two tasks are more compatible or more incompatible. Regardless of how these issues are eventually resolved, the existence of hemisphere-specific effects is sufficiently well documented to demand consideration when theories of attention are created or modified.

Summary and Conclusions

Even simple tasks require the coordination of a number of information-processing subsystems, components, or modules. Hemispheric asymmetry can vary from subsystem to subsystem within a task, making it important to examine asymmetry for specific subsystems or components. This requires some principled conceptualization about the subsystems involved in a task and techniques for distinguishing the subsystems empirically. As a starting point, it has proven worthwhile to consider the subsystems or components postulated by various models of perception, cognition, and action. Finding asymmetry for specific components sheds light on the nature of hemispheric asymmetry and also provides converging evidence that the decomposition of tasks into those particular components is on the right track. This componential approach to the study of hemispheric asymmetry is illustrated by studies of such diverse topics as language, vision, imagery, and attention.

The earlier notion that the left hemisphere is dominant for all aspects of language has been replaced by the view that hemispheric asymmetry differs for different components of language. The left hemisphere seems dominant for producing overt speech, phonetic decoding, using syntax, and certain (but not all) semantic processes. The right hemisphere seems dominant for using the pragmatic aspects of language, integrating information across sentences, and using context. Additional studies from cognitive neuropsychology more generally have led to a finer-grained decomposition of language into component processes and their interconnections.

The earlier notion that the right hemisphere is dominant for all aspects of visuospatial processing has also been replaced by the view that hemispheric asymmetry differs for different aspects of visual processing. For example, there is evidence of right-hemisphere dominance for processing global aspects of stimuli (e.g., the outer contour of a face) and left-hemisphere dominance for processing local aspects of stimuli (e.g., small, inner features of a face). There is also evidence that, at some level of processing beyond the sensory cortex, the left and right hemispheres are biased toward efficient use of higher and lower visual-spatial frequencies, respectively. At least three aspects of spatial frequency are relevant: (1) the absolute range of spatial frequencies contained in a stimulus, (2) the range of spatial frequencies that is most relevant for the task being performed, and (3) whether the task-relevant frequencies are high or low relative to other frequencies contained in a stimulus. With respect to localizing visual stimuli in space, it has been hypothesized that the left hemisphere makes more effective use of a categorization subsystem (which provides efficient information about the relative location of objects) whereas the right hemisphere makes more effective use of a coordinate subsystem (which provides efficient information about distance and absolute location). All three of these aspects of vision seem to be interrelated. For example, information about global versus local aspects of a visual stimulus is carried by relatively low versus relatively high ranges of spatial frequency, respectively. Furthermore, computer simulation models suggest that categorical aspects of spatial location are computed better by networks with relatively small, nonoverlapping receptive fields (similar to relatively high spatial frequency) and that coordinate aspects of spatial location are computed better by networks with relatively large, overlapping receptive fields (similar to relatively low spatial frequency). Such differences in the visual modality may also be related to hemispheric asymmetry for processing relatively high versus relatively low sound frequencies and to hand asymmetry for making movements of different temporal and spatial frequencies.

Views about hemispheric asymmetry for visual imagery have also been influenced by a componential approach. The earlier view that the right hemisphere is dominant for visual imagery has been replaced by the hypothesis of left-hemisphere dominance for a particular component process: the *generation* of visual images. This hypothesis has generated a great deal of controversy and will undoubtedly lead to additional empirical tests. At the present time, it seems very unlikely that either hemisphere completely lacks the ability to generate images. On the basis of an even finer-grained componential analysis, it has been suggested that the left hemisphere is dominant for the generation of multipart images and especially for the generation of correct categorical relations among the parts of an image, whereas it has been suggested that the right hemisphere is dominant for arranging the parts of an image according to precise coordinates. This provides an interesting parallel to hemispheric asymmetry for the perceptual processing of visual input.

Studies of patients with hemineglect and extinction to simultaneous stimulation (usually after injury to the right hemisphere) have helped to identify components of visual orienting of attention. For example, covert shifts of attention require disengagement of attention from its current focus, moving attention to the new location, and engagement of attention at that location. In addition, patients with unilateral parietal-lobe injury have a particular difficulty in disengaging attention from a stimulus in order to move it in a direction contralateral to the injury. It is interesting that this seems to be true regardless of the side of injury. At the same time, the frontal lobe of the right hemisphere seems to be particularly important for the regulation of alertness.

A variety of dual-task studies have demonstrated that priming and interference are sometimes greater for one hemisphere than for the other, with the demonstration of hemisphere-specific interference being particularly clear. Such hemisphere-specific effects are difficult to reconcile with the traditional idea that there is one single pool of attentional capacity or processing resources for which all attention-demanding activities must compete. Instead, the results suggest that the information-processing resources of the two hemispheres are at least somewhat independent of each other. Although there is disagreement about the best way to conceptualize this independence, its existence must be considered by any complete theory of attention.

This variety of theoretical and empirical work serves to illustrate the value of a componential approach to the study of hemispheric asymmetry. Instead of trying to force all instances of hemispheric asymmetry into a single dichotomy, we must incorporate the study of hemispheric asymmetry into the larger enterprise of trying to understand how the brain produces cognition. To the extent that reliable hemispheric asymmetries emerge in specific components of processing, we learn something important about the nature of hemispheric asymmetry. At the same time, the fact that two components produce different patterns of hemispheric asymmetry reinforces the idea that they are in some sense separable. Of course, there is absolutely no *a priori* guarantee that all (or even any) of the information-processing components derived from purely cognitive models will map onto regions of the brain in any understandable way at all. This is not particularly problematic if the goal is to develop theories and models of mental activity with little or no particular regard for the brain. Indeed, this has been the case for much of cognitive psychology. However, from the perspective of *cognitive neuroscience,* understanding cognition is not sufficient. Instead, the goal is to understand how the brain produces cognition. From that perspective, it is critical to discover which types of componential analyses are reinforced by neuropsychological data and which are not. Appropriate studies of hemispheric asymmetry can be an important part of that enterprise.

4

Biological Asymmetries in the Human Brain

At first glance the human brain appears symmetrical, leading one to wonder how it is that behavioral asymmetries emerge from biologically identical hemispheres. Despite the generally symmetrical appearance of the two hemispheres, however, a number of biological asymmetries have been documented during the last hundred years. The present chapter reviews some of the better-established biological asymmetries and considers possible correlations among biological and behavioral asymmetries. The focus is on asymmetries in humans, but where it is particularly relevant, there is some discussion of asymmetries in other species. Asymmetries in other species are discussed in more detail in Chapter 5.

The chapter begins with a discussion of *anatomical* asymmetries in the human cortex, followed by a discussion of *pharmacological* and *chemical* asymmetries. For each of these types of asymmetry, the pattern of results for right-handed individuals will be contrasted with the pattern for non-right-handed individuals whenever possible. After discussing the biological asymmetries in the two hemispheres I turn to a consideration of the connection between them, the corpus callosum, and its possible relationship to asymmetries. An important part of this discussion concerns individual variation in callosal connectivity and in biological asymmetry. In the final section of the chapter I examine the relationship between biological and behavioral asymmetries.

Anatomical Asymmetries

A variety of techniques have been used to identify anatomical asymmetries in the human brain. The techniques include such things as the measurement of important anatomical landmarks during autopsy of the brains of deceased individuals, similar measurements in living individuals using magnetic resonance imaging (MRI) and other radiological techniques (see Chapter 1), and the measurement of endocranial markings on the inside of the skull. In addition, in recent years it has become possible to observe asymmetries in areas defined by cellular architecture or cytoarchitectonic structure. This includes the measurement of such things as cell size, cell density, cell type, cell organization, and so forth. The use of these various techniques permits the determination of certain biological asymmetries for a single individual. In addition, the typical asymmetries for a population can be estimated by compiling results across a group of individuals from that population. In some cases, it has been possible to compile results separately for right-handed and non-right-handed individuals. This comparison is interesting in view of the fact that certain behavioral asymmetries are often more systematic for the right-handed population than for the non-right-handed population (see Chapter 7). The remainder of this section reviews some of the anatomical asymmetries that have been discovered.

In the majority of human brains, the frontal region is wider and protrudes farther forward in the right hemisphere than in the left hemisphere. At the same time, the occipital region is wider and protrudes farther rearward in the left hemisphere than in the right hemisphere. These asymmetries, illustrated in Figure 4.1, are sometimes described as giving the brain a kind of "counterclockwise torque" (e.g., Bradshaw and Nettleton, 1983). As reviewed in Chapters 5 and 9, a similar counterclockwise torque also characterizes the brains of certain nonhuman primates.

In view of the well-established hemispheric asymmetry for certain aspects of language, it is not surprising that the search for anatomical asymmetries has been particularly focused on areas of the cortex believed to be important for language (e.g., certain temporal and parietal areas). In fact, several anatomical asymmetries in those areas have been documented for right-handed individuals, with the asymmetries tending to be smaller in non-right-handed individuals (for an extensive review, see Geschwind and Galaburda, 1987).

One of the most often studied asymmetries of this type involves a cortical landmark known as the lateral or sylvian fissure. This deep groove on the lateral surfaces of the cerebral hemispheres marks the boundary between the frontal and parietal lobes (which lie above the fissure) and the temporal lobe (which lies below the fissure). For approximately a century it has been known that the left and right sylvian fissures are typically asymmetric at the posterior end (see Figure 4.2). Specifically, in most brains the right-hemisphere fissure curls upward more than the left-hemisphere fissure (e.g., Cunningham, 1892; Eberstaller, 1884). This particular asymmetry has been verified in more recent studies and is much stronger among right-handed individuals than among non-right-handed individuals. For example, Hochberg and LeMay (1975) studied the brains of 100 right-handers and 28 non-right-handers and found that for right-handed individuals the height of the posterior end of the sylvian fissure was higher in the right

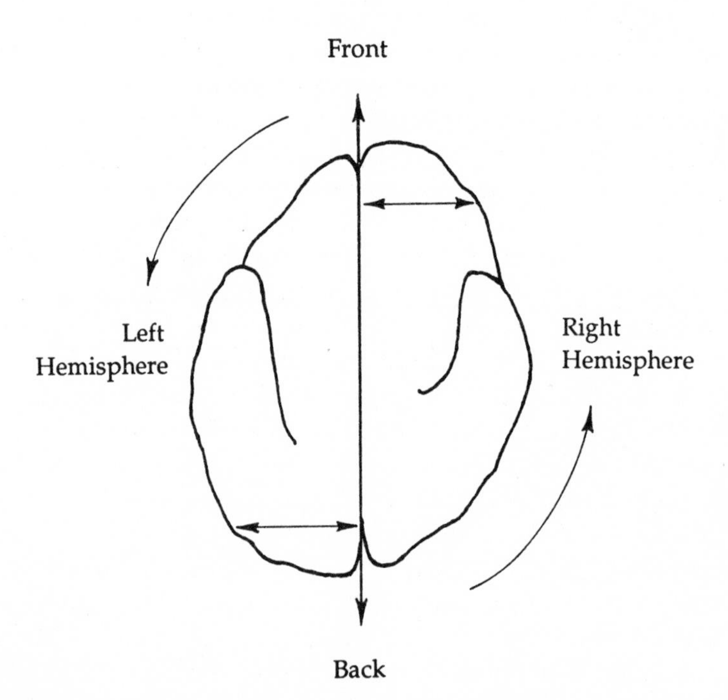

Figure 4.1. The counterclockwise torque of the human brain. Note the wider right frontal and left occipital areas.

hemisphere in 67 percent of the brains studied, equal in the two hemispheres in 26 percent of the brains studied, and higher in the left hemisphere in 7 percent of the brains studied. The corresponding percentages for non-right-handers were 22 percent, 71 percent, and 7 percent, respectively. Similar distributions have been reported for the brains of adults and children (e.g., Geschwind and Galaburda, 1987; Hochberg and LeMay, 1975).

It is also the case that the sylvian fissure is typically longer in the left hemisphere than in the right hemisphere, leading to the suggestion that the temporal and parietal opercula are larger in the left hemisphere (e.g., Geschwind and Galaburda, 1987). Additional evidence for this suggestion comes from the finding that a cortical area known as the planum temporale (which, in the left hemisphere, is an extension of Wernicke's area, known to be important for certain aspects of language) is usually larger in the left hemisphere. For example, Geschwind and Levitsky (1968) examined the brains of 100 right-handers and reported that the length of the lateral edge of the planum temporale was longer in the left hemisphere than in the right in 65 percent of the brains examined, equal in the two hemispheres in 24 percent of the brains examined, and longer in the right hemisphere in 11 percent

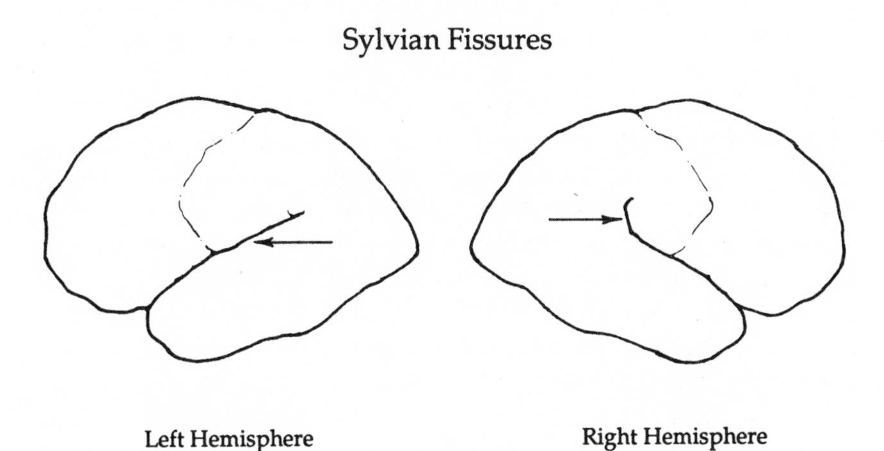

Figure 4.2. The sylvian fissure is typically longer and straighter in the left hemisphere than in the right hemisphere.

of the brains examined. Furthermore, the asymmetry in favor of the left hemisphere was sometimes very dramatic, with the left planum being ten times larger than the right planum in some brains. Such dramatic differences in the opposite direction are very rare.

Asymmetries in the length of the sylvian fissures have also been reported in the brains of children and infants, with the distribution being similar to that reported for adults (see Geschwind and Galaburda, 1987). Furthermore, asymmetries in the size of the planum temporale are also found in children and adolescents (e.g., Hynd, Semrud-Clikeman, Lorys, Novey, and Eliopulos, 1990; Larsen, Hoien, Lundberg, and Odegaard, 1990). Of additional interest is the fact that this anatomical asymmetry is much less likely in children and adolescents who have developmental dyslexia. For example, Hynd et al. used MRI scans to study the brains of 10 children with dyslexia, 10 children with attentional deficit disorder/hyperactivity (ADD/H), and 10 age- and sex-matched control children. They report that 70 percent of the control children and ADD/H children had a larger left planum whereas only 10 percent of the dyslexic children did. In fact, 90 percent of the dyslexic children showed a reverse pattern of asymmetry, which could be attributed to a smaller left planum than in the other groups and to the right planum being about the same size as in the other groups. In an independent MRI study, Larsen et al. studied 19 dyslexic adolescents and 19 control adolescents. Approximately 70 percent of the control subjects had a larger left planum whereas only 30 percent of the dyslexic subjects had a larger left planum. The remainder of the subjects in both groups were classified as having symmetric plana. Implications for the neurological bases of dyslexia will be discussed in Chapter 7. For present purposes, it is important to note that the percentage of asymmetric brains reported for the control groups is similar to that reported for adults.

The sylvian fissures tend to leave their impression on the inner surface of the skull. Therefore, it is possible to learn about asymmetries of the sylvian fissure by examining the skull. From casts of the inside of fossil skulls, it has been determined that several of the anatomical asymmetries found in present-day adults and children were also present in our evolutionary ancestors (for review see Geschwind and Galaburda, 1987; see also Chapter 9 for discussion of the evolutionary implications).

In addition to gross anatomical landmarks, cytoarchitectonic features such as cell density and type may also be used to identify boundaries between cortical areas. Asymmetries have been demonstrated in several cytoarchitectonic areas that are believed to be involved in language. For example, an area known as Tpt, located on and around the planum temporale, was found to be larger in the left hemisphere than in the right hemisphere for three of the four brains of right-handers studied by Galaburda, Sanides, and Geschwind (1978). Furthermore, there was a perfect rank-order correlation between Tpt asymmetry and asymmetry of the planum temporale, indicating that gross asymmetries of the planum are reflected at the level of cell architecture.

In a study using the same 100 brains examined by Geschwind and Levitsky (1968), Galaburda, Rosen, and Sherman (1990) report that there is a negative correlation between the *total size* of the planum temporale (i.e., left-hemisphere plus right-hemisphere area) and *asymmetry* of the planum temporale (i.e., left-hemisphere minus right-hemisphere area). This correlation results from the fact that there is little individual variation in the size of the left planum whereas there is considerable variation in the size of the right planum. Viewed another way, in asymmetric brains the planum on one side (usually the right side in right-handed individuals) is reduced in size relative to the symmetric brain, whereas the planum on the other side is about the same size as the planum of each hemisphere in the symmetric brain. A similar negative correlation between total size and asymmetry has been reported in the visual cortex of the rat (Galaburda et al., 1990), suggesting that such relationships may be common. Galaburda et al. note that there are three ways by which a cortical area can be reduced in size: by decreasing the number of neurons, by decreasing the distance between neurons (i.e., increasing the packing density), and by a combination of both of these factors. Although definitive studies remain to be carried out with humans, recent studies with other species suggest that the number of neurons is the critical factor.

In addition to the anatomical asymmetries already discussed, it has been shown that an area known as the pars opercularis of the frontal lobe (in an area where injury leads to Broca's aphasia) is typically larger in the left hemisphere than in the right hemisphere (see Geschwind and Galaburda, 1987). The same is true of a cytoarchitectonic area of the parietal lobe known as area PG, which is also thought to be involved in language (Eidelberg and Galaburda, 1984). As might be expected, there was a significant correlation between asymmetry in area PG and asymmetry in the planum temporale. In the same study, a different architectonic area of the parietal lobe (PEG) that was hypothesized to be important for visuospatial function was found to be larger in the right hemisphere than in the left hemisphere. It is interesting to note that there was no correlation between asymmetry in area PEG and asymmetries in the areas thought to be important for language. This leads to interesting questions about which biological asymmetries are correlated and which are independent and whether there is a relationship between this independence and the independence of various behavioral asymmetries as discussed in Chapter 2. Of course, convincing data about independence versus association will require the examination of much larger sample sizes than has heretofore been the case.

Scheibel, Fried, Paul, Forsythe, Tomiyasu, Wechsler, Kao, and Slotnik (1985) found that the extent of higher-order dendritic branching was greater in certain speech areas of the left hemisphere (e.g., Broca's area) than in homologous areas of the right hemisphere. At the same time, the lower-order dendritic branches were longer in the right hemisphere than in the left hemisphere. Scheibel et al. speculate that these patterns reflect different sequences of growth for the two hemispheres and for different areas within each hemisphere. In particular, they suggest that the greater length of lower-level dendritic branches on the right side occurs because the right hemisphere develops more quickly than the left during the first year or so of life (see also Chapter 8). They also suggest that development of the left hemisphere eventually catches and surpasses the right (at least in speech areas), accounting for the greater extent of higher-order dendritic branching in the left-hemisphere speech areas. Although their sample size was small, it is interesting that Scheibel et al. found fewer asymmetries of this sort in the brains of non-right-handed individuals.

Pharmacological and Chemical Asymmetries

Certain drugs have asymmetric effects when administered systemically, and some neurotransmitters are more abundant and more effective on one side of the brain compared to the other. The clearest demonstrations of these types of asymmetry—pharmacological and chemical, respectively—have come from research with nonhuman species. Of particular note is the work of Glick and his colleagues with rats (e.g., Carlson and Glick, 1989; Glick, Carlson, Drew, and Shapiro, 1987). For example, in rats the administration of drugs that stimulate dopamine pathways produces an asymmetric change in motor function such that animals tend to circle predominantly in one direction. Furthermore, the direction in which a particular animal tends to circle is determined by left/right asymmetry of dopamine content and of sensitivity to dopamine in the nigrostriatal pathway of the rat brain. This work will be discussed in some detail in Chapter 5. For the moment, it is important to note that similar pharmacological and chemical asymmetries have also been found in the human brain.

In an extensive review, Tucker and Williamson (1984; also Tucker, 1987) conclude that the distribution of two important types of neurotransmitter processes is asymmetric in the human brain. Specifically, they argue that processes dependent on dopamine are more prevalent in the left hemisphere than in the right hemisphere, whereas processes dependent on norepinephrine are more prevalent in the right hemisphere than in the left hemisphere. For example, Glick, Ross, and Hough (1982), who re-examined data reported by Rossor, Garret, and Iversen (1980) on neurotransmitter levels in different brain areas at the time of autopsy argue that levels of dopamine and other chemical substances known to contribute to dopaminergic processes are higher in the left than in the right globus pallidus. The globus pallidus is one of the basal ganglia (a set of five subcortical nuclei) and functions as an excitatory structure of the extrapyramidal motor system. Tucker and Williamson also note that in a PET scan of a living brain, there was a higher concentration of dopamine terminals in the left than in the right basal ganglion (Wagner et al., 1983). Given the rich interconnections between the basal ganglia and the cerebral cortex, it is reasonable to suppose that chemical asymmetries at this subcortical level have implications for asymmetry at the cortical level. In addition, Amaducci, Sorbi, Albanese, and Gainotti (1981) reported greater activity of choline acetyltransferase (which is associated with higher levels of dopamine) in the left hemisphere than in the right hemisphere in an area known as Brodmann Area 22. In view of the anatomical asymmetries discussed earlier, it is interesting that Brodmann Area 22 contains cytoarchitectonic area Tpt, which tends to be larger in the left hemisphere than in the right hemisphere. By way of contrast, Oke, Keller, Mefford, and Adams (1978) reported that concentration of norepinephrine was higher in the right than in the left side of certain regions of the thalamus that are important for somatosensory function (see Tucker and Williamson, 1984, for additional discussion and converging evidence for their assertion that processes dependent on norepinephrine are more prevalent in the right hemisphere of the human brain).

Although considerably more research is needed to increase our knowledge about pharmacological and chemical asymmetries, it is useful to consider potential consequences of such asymmetries. One of the most interesting and ambitious hypotheses has been advanced by Tucker and Williamson (1984; Tucker, 1987). In fact, they argue that the asymmetry in neurotransmitter pathways outlined earlier leads to far-reaching systemic differences between the cerebral hemispheres.

Tucker and Williamson (1984) distinguish between two major regulatory systems that they term the *activation* and *arousal* systems (after Pribram and McGuinness, 1975). Furthermore, they argue that the activation system is mediated by pathways that depend on dopamine whereas the arousal system is mediated by pathways that depend on norepinephrine. Consequently, to the extent that these two types of neurotransmitters are represented asymmetrically in the human brain, we might expect behavioral and cognitive asymmetries that derive from activation versus arousal regulatory systems. Tucker and Williamson make exactly this argument.

According to Tucker and Williamson (1984), the major function of the activation system is to maintain readiness for action. On the hypothesis that the left hemisphere is organized around such an activation system, it makes sense that the left hemisphere would be superior for complex motor functions (e.g., hand use and sequencing of movements), including the programming of the articulatory movements used during speaking. Tucker and Williamson go on to suggest that once the left hemisphere became organized around the activation system, the properties of that system came to influence not only motor activity but also perception, cognition, and emotion carried out by the left hemisphere. For example, they argue that the left hemisphere developed a bias toward analytic processing from the tonic activation system around which it became organized during the course of evolution because of the redundancy bias that such a regulatory system is thought to involve (e.g., Pribram and McGuinness, 1975; Tucker, 1987).

In contrast, the major function of the arousal system is hypothesized to be the production of phasic responses to stimulus input and orientation to new stimulus input. On the hypothesis that the right hemisphere is organized around such an arousal system, it makes sense that the right hemisphere would be superior for integrating bilateral perceptual information, for the distribution of attentional orienting across space and for maintaining alertness (see Chapter 3). Tucker and Williamson (1984) suggest that once the right hemisphere became organized around the arousal system, in a manner complementary to the left hemisphere's organization around the activation system, the properties of the arousal system came to influence many aspects of perception, cognition, action and emotion carried out by the right hemisphere. Thus, they argue, the right hemisphere's predisposition to holistic processing and attention emerged from the novelty selection bias of the phasic arousal and orienting regulatory system.

The theory of hemispheric asymmetry advanced by Tucker and Williamson (1984) requires a great deal of additional testing before it can be judged. To be sure, certain aspects can be questioned. For example, there is reason to question the analytic/holistic distinction between left- and right-hemisphere processing (see Chapter 2)—a distinction that is generally accepted and used by Tucker and Williamson. However, it is exciting to see a theory that attempts to derive hemispheric asymmetry at the cognitive, emotional, and behavioral levels from a biological asymmetry that is judged to be fundamental. Even if this specific theory falls short, it at least illustrates the potential importance of pharmacological and chemical asymmetries in the human brain.

Callosal Connectivity

The major fiber tract that connects the left and right cerebral hemispheres is the *corpus callosum,* shown in midsagittal cross-section in Figure 4.3. As suggested by the figure, there are no gross anatomical landmarks within the corpus callosum that allow it to be partitioned into distinct anatomical regions. However, there is a good deal of evidence that different regions of the corpus callosum contain fibers originating in different cortical areas. For example, the anterior portions of the corpus callosum (the rostrum and genu) contain fibers that originate from anterior regions of the cortex (e.g., premotor and prefrontal regions), the middle portion of the corpus callosum (midtrunk or midbody) contains fibers that originate from motor and somatosensory areas of the cortex, and the posterior fifth or so of the corpus callosum contains fibers that originate from the temporal, post-parietal, and peri-striate areas of the cortex. Because individual variations may not be the same for all of these regions of the corpus callosum, investigators have attempted to subdivide the corpus callosum into segments (e.g., fifths) and examine variation in the various sub-areas that result. Particular emphasis has been placed on the investigation of anatomical differences related to handedness and biological sex.

An illustration of the nature of studies of corpus callosum anatomy is provided by Witelson (1985). She examined the area of portions of the corpus callosum from photographs that come from the postmortem examinations of individuals with known hand preference. The brains came from cognitively normal individuals who were seriously ill with non-neurological forms of cancer, who underwent neurological testing while alive, and who agreed to autopsies in the event of death. One of the behavioral measures was a 12-item handedness questionnaire. Each of the individuals was classified as 100 percent right-handed (the consistent right-handers) or as mixed-handed (less than 100 percent preference for either hand.) There were no completely consistent left-handers in her sample. The data reported came from 27 consistent

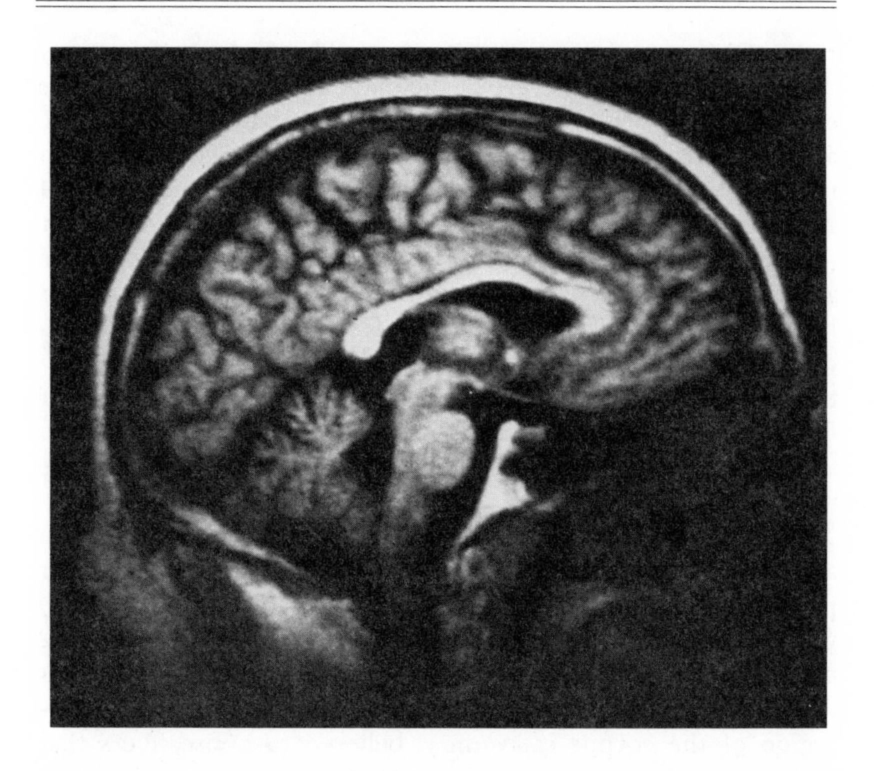

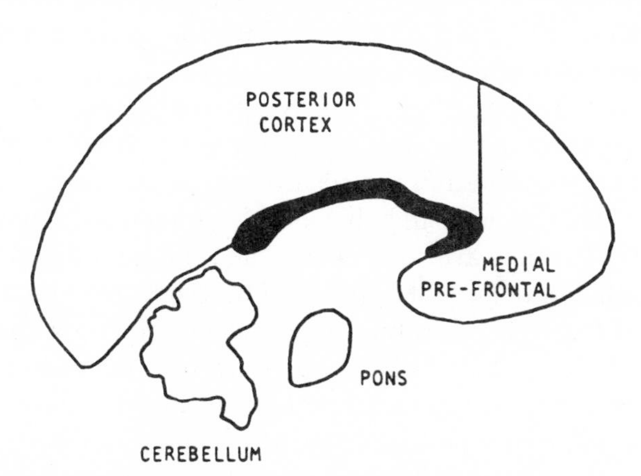

Figure 4.3. Cross-section of the human corpus callosum illustrated by photograph from an MRI scan in the upper portion of the figure and by diagram in the lower portion. In the lower portion, the corpus callosum is shaded. (Photograph provided by Adrian Raine)

right-handers and 15 mixed-handers. The finding of primary importance was that the size of the midsagittal area of the corpus callosum varied with hand preference. Specifically, the midsagittal area of the corpus callosum was larger by an average of 11 percent in the brains of mixed-handed individuals than in the brains of consistently right-handed individuals. This effect was present for measures of the total area of the corpus callosum, the area of the anterior half of the corpus callosum, and the area of the posterior half of the corpus callosum. These differences between the handedness groups continued to be observed even when Witelson controlled for cerebrum weight and sex. The area of the posterior fifth of the corpus callosum (which corresponds roughly to an area known as the splenium) did not differ significantly for the right- and mixed-handed groups.

In a subsequent analysis of these brains and others, Witelson and Kigar (1987) and Witelson (1989) found that the posterior part of the trunk of the corpus callosum (especially the isthmus) was particularly large in the mixed-handed group, averaging 19 percent larger than in the right-handed group. Witelson and Kigar argue that this finding is particularly important because this region of the corpus callosum is believed to house fibers that connect the parietal and temporal lobes, which contain areas known to be involved in language, learned motor sequences, and visuospatial functions—functions for which hemispheric asymmetry is frequently reported.

As Witelson (1985, 1989; Witelson and Kigar, 1987) notes, hand preference is related to hemispheric asymmetry for cognitive functions, although the relationship is far from perfect. The general finding is that behavioral asymmetries that are typical for right-handers are smaller (on the average) in mixed-handers and left-handers (see Chapter 7). Therefore, her findings on the size of the corpus callosum suggest, albeit indirectly, that callosal morphology may be related to individual differences in *behavioral* asymmetry. Witelson notes that the corpus callosum plays an important role in integrating information across the hemispheres, and she suggests that a tendency away from hemispheric asymmetry in left- and mixed-handers may be associated with greater anatomical connectivity between the hemispheres. Whether the tendency toward hemispheric asymmetry results from less callosal connectivity early in development or, as an alternative, callosal size depends on hemispheric asymmetry is unclear. In fact, it is not clear whether the two things are *causally* related at all. In any event, Witelson notes that recent findings in the neurobiological literature indicate that naturally occurring regressive events (e.g., death of neurons, the elimination of axon collaterals, etc.) play an important role in development of the nervous system. She wonders whether these regressive events might be factors in individual differences in brain morphology and whether right-handers may be individuals with more extensive early elimination of certain neural components.

Witelson and Kigar (1987) also consider the rather unique findings for the region of the corpus callosum called the splenium (approximately the posterior one-fifth). Despite large individual variation in the shape of this area, they found no size difference between the two handedness groups. They suggest that this feature of the splemium may be related to differences between it and the rest of the corpus callosum in terms of anatomy and function. For example, the splenium undergoes the greatest relative increase in size after birth and is the first region of the callosum to begin myelination. In functional terms, the splenium is thought to transmit mainly sensory information, in contrast to the higher-level processed information that may be transmitted by more anterior regions of the corpus callosum.

Witelson (1985, 1989) also compared the area of the corpus callosum for males versus females. The most striking finding is that the difference between the handedness groups in the area of the isthmus interacted with biological sex. Specifically, the handedness effect noted earlier was present for males but not for females.

Earlier in the present chapter it was noted that certain *biological* asymmetries are more prevalent among right-handers than among non-right-handers. The combination of this finding with Witelson's (1985, 1989) finding that the corpus callosum is larger in non-right-handers suggests a relationship between biological asymmetry and callosal connectivity. Specifically, the suggestion is that increasing biological asymmetry is accompanied by less callosal connectivity. This is a very interesting possibility, and it merits more direct investigation by examining the size of the corpus callosum and various biological asymmetries for the same brains to see whether there is any correlation.

Thus far, direct examinations of this type have not been reported for humans. However, suggestive data have been reported by Galaburda et al. (1990) for callosal connectivity and asymmetry in the rat brain. The visual cortices of rats are often asymmetric, with the amount of asymmetry varying across individual rats. In addition, the two visual cortices in rats (area 17) are connected via the corpus callosum. These factors suggest that the visual cortices and corpus callosum in rats provide a suitable model system for examining the relationship between callosal connectivity and biological asymmetry. Galaburda et al. report that more symmetric brain areas had a greater percentage of callosal terminations than did more asymmetric brain areas, with the correlation between a measure of asymmetry and the percentage of callosal terminations being −.899. Galaburda et al. also note that the decrease in the percentage of callosal fibers in asymmetric brains is even larger than the proportionate decrease that would be expected in view of the fact that asymmetric areas are generally associated with fewer neurons than are symmetric areas (see the earlier discussion of anatomical asymmetries). This leads Galaburda et al. (1990, p. 537) to suggest that

> some of the neurons that are not lost in the brains with asymmetric areas withdraw their callosal projections and re-route them within the ipsilateral hemisphere during development. This set of events would lead to a situation whereby brains with symmetric areas are organized in a relatively interhemispheric fashion, while those with asymmetric areas are relatively more intrahemispherically connected.

This is a very interesting hypothesis that merits additional examination in both humans and in nonhuman species.

The handedness and sex differences found so consistently in the postmortem examinations reported by Witelson (1985, 1989) have not always been replicated in MRI studies that examine the size of the corpus callosum in living subjects (for reviews of these failures to replicate see Clarke, 1990; Habib, Gayraud, Oliva, Regis, Salamon, and Khalil, 1991; Witelson, 1989). The reasons for this are not completely clear. In could be that the postmortem findings are more accurate than the MRI findings because of uncertainties inherent in the imaging technique. Alternatively, it could be that the postmortem findings are less accurate than the MRI findings because of postmortem deformation. It could also be that much of the inconsistency is related to differences in the methods used to determine handedness groups. For example, many of the MRI studies used writing hand to divide subjects into groups whereas Witelson (1985, 1989) used a multiple-item questionnaire that permitted a distinction between subjects who were consistently right-handed and those who were not. From this perspective, it is interesting that two recent MRI studies that have classified subjects into handedness groups in a manner similar to that used by Witelson have also provided at least a partial replication of her findings (Clarke, 1990; Habib et al., 1991).

In an extensive MRI and behavioral study of 60 graduate students (15 subjects in each of four groups defined by handedness and sex), Clarke (1990) found no overall effect of handedness or sex on measures of the size of the corpus callosum. However, when he looked at the area of the isthmus normalized for callosum size, Clarke did replicate the handedness-by-sex interaction reported by Witelson (1989) for this same normalized measure. Habib et al. (1991) obtained MRI and handedness data from 53 neurologically intact individuals and found that the corpus callosum was significantly larger in nonconsistent right-handers than in consistent right-handers, especially in the anterior half. Furthermore, this handedness effect was much larger for males than for females, and the interaction between handedness and sex was particularly dramatic in the area of the isthmus. In general, these results confirm many of Witelson's (1985, 1989) findings and may help to explain the discrepant results from other MRI studies.

Biology and Behavior

It is obviously important to determine whether there are relationships between biological and behavioral asymmetries and whether there are behavioral consequences of individual differences in callosal connectivity. Despite the importance of knowing about such relationships, there is little in the way of relevant data on which to base hypotheses, and much of the data that do exist are indirect. For example, it has already been noted that certain anatomical asymmetries tend to be larger and more frequent in the brains of right-handers than in the brains of non-right-handers. When this observation is combined with the fact that many behavioral asymmetries also tend to be larger and more frequent in the brains of right-handers than in the brains of non-right-handers, a relationship between biological and behavioral asymmetries is suggested. However, from this suggested relationship alone it is impossible to determine whether the individual who shows a large behavioral asymmetry for some particular task is the same individual with a large biological asymmetry of some specific type. Thus, noting a correlation between handedness and biological asymmetry is limited in the conclusions that it permits.

A preferred strategy would be to collect data on a variety of behavioral and biological asymmetries from the same individuals. Only in this way can we provide strong tests of various hypotheses about the correlation between the two kinds of asymmetry. One way to collect such data is to administer relevant behavioral tasks to individuals who agree to have their brains autopsied after death. The other way is to obtain behavioral data from individuals who agree to have biological aspects of their brains studied *in vivo,* using techniques such as magnetic resonance imaging (MRI). The few studies that have been conducted in these ways offer some indication that biology and behavior are related.

As noted earlier, Witelson (1983; Witelson and Kigar, 1987) has been able to obtain measures of behavioral asymmetry from individuals who were seriously ill with various non-neurological forms of cancer and who had agreed to have their brains autopsied after death. For some of these individuals, behavioral measures included hand preference as indicated on a questionnaire, finger-tapping rate, ear asymmetry for identifying consonant-vowel syllables (see Chapter 1), and ear asymmetry for identifying melodies. Of 12 cases reported in Witelson (1983), nine were strongly right-handed. Of these nine cases, seven individuals had a markedly larger planum temporale in the left hemisphere than in the right hemisphere. Of these seven, four individuals had performed all of the behavioral tasks listed above and all four showed a pattern of asymmetry that is typical of right-handers: they tapped faster with the right than with the left hand and showed a right-ear advantage with syllables and a left-ear advantage with melodies. One of the other two right-handed individuals had nearly identical left and right plana and was also the only right-handed individual with almost identical scores from both ears on the dichotic-listening tasks. The right-hand advantage on the finger-tapping task was also smaller than is typical for right-handers. The other right-handed individual had a larger right planum and also showed little ear asymmetry on the dichotic-listening tests and no hand difference in tapping rate. Thus, there appears to be some connection between asymmetric size of the left and right plana and certain behavioral asymmetries, although the small sample size demands a cautious interpretation.

Three of the cases reported by Witelson (1983) were left-handed. One case had a larger left planum, but the dichotic-listening results for this individual were uninterpretable because of a left-hemisphere lesion. However, on the basis of other observations, Witelson believed the individual to be left-hemisphere dominant for language. The second case had a left planum that was only slightly larger than the right planum and showed a "slight" right-ear advantage on the syllable-identification task and greater tapping rate with the left hand than with the right hand. The third left-hander had equal plana on the two sides, but no dichotic tests were administered. Clearly, additional data of this type are needed before strong conclusions can emerge.

Studies of brains *in vivo* are also providing an opportunity to study the relationship between size of the corpus callosum and various measures of behavior. For example, Clarke (1990) administered a battery of behavioral laterality tasks to the 60 subjects in his MRI study of corpus callosum anatomy. He found that the size of the corpus callosum was not related to behavioral measures of the ability to transfer information from one hemisphere to the other or to ear differences in dichotic listening, arguing against the hypothesis that a larger corpus callosum allows more efficient interhemispheric communication. In addition, the size of the corpus callosum correlated *positively* with performance asymmetries for visual and tactile discrimination tasks. That is, contrary to Witelson's (1985, 1989) suggestion, a larger corpus callosum was associated with greater functional asymmetry. This pattern of results leads Clarke to consider the possibility that a larger corpus callosum provides a larger *inhibitory barrier* between the cerebral hemispheres.

Additional data on the relationship of callosal size to behavior is provided by Raine et al. (1990), who examined structural and functional characteristics of the corpus callosum in schizophrenics, psychiatric controls, and normal controls. They used MRI scans to obtain measures of the thickness of anterior, posterior, and middle regions of the corpus callosum and had the individuals perform tasks that either measured hemispheric asymmetry or required the two hemispheres to share information: (1) identifying syllables presented dichotically, (2) recognizing pitch contours presented dichotically, (3) repeating either on the same (uncrossed) or opposite (crossed) hand a sequence of finger touches administered by the experimenter, (4) palpating a random shape with one hand and then making a same/different judgment when a second shape was presented to either the same (uncrossed) hand or opposite (crossed), hand and (5) solving block-design problems (see Chapter 1) with either the left or right hand. For tasks 3 and 4, the difference between crossed and uncrossed conditions provides a measure of the efficiency of interhemispheric transfer.

In the normal control subjects, males had a thicker anterior and posterior corpus callosum than females—a finding that has sometimes been reported by others (although it is not always statistically significant; e.g., Witelson, 1985) and that is likely caused by the fact that the brains of males are uniformly larger than the brains of females. In both the schizophrenic and psychiatric control patients, this sex difference was reversed, despite the fact that the overall area of male brains continued to be larger than the overall area of female brains. Raine et al. (1990) note that similar findings have been reported for schizophrenics in earlier studies and they provide an instructive review of those studies as well as studies that have produced discrepant results.

For present purposes, it is important to note that the structural differences among groups were not paralleled by differences on any of the behavioral tasks. Note that for the tasks measuring interhemispheric transfer of information and dichotic listening, this replicates the results reported by Clarke (1990). Although it would be premature to reach any unambiguous conclusion from these experiments, they provide no evidence for the assertion that callosal size correlates negatively with either behavioral asymmetry or correlates at all with behavioral measures of interhemispheric connectivity. Of course, we do not know whether other aspects of callosal morphology would be more predictive of behavioral measures.

Summary and Conclusions

Despite the generally symmetrical appearance of the two cerebral hemispheres, a number of biological asymmetries have now been documented in humans. For example, the wider right frontal and left occipital areas have been said to give the brain a counterclockwise torque. In addition, the sylvian fissure is typically longer and the height of the posterior end of the fissure is lower in the left hemisphere than in the right hemisphere. The planum temporale is also typically larger in the left hemisphere than in the right hemisphere. These anatomical asymmetries are present in the brains of children as well as adults, and there is evidence from the examination of fossil skulls that similar asymmetries were characteristic of our evolutionary ancestors. Recent studies suggest that these gross asymmetries are also reflected at the level of cell architecture. It is noteworthy that these anatomical asymmetries tend to be larger and more frequent in the brains of right-handers than in the brains of non-right-handers, especially in view of the fact that many behavioral asymmetries also tend to be larger and more frequent among right-handers than among non-right-handers.

There is also some indication that certain systemically administered drugs have asymmetric effects, perhaps because some neurotransmitters are more abundant and more effective on one side of the brain than on the other. A particularly interesting hypothesis is that processes more dependent on the neurotransmitters dopamine and norepinephrine are more prevalent in the left and right hemispheres, respectively. In fact, it has even been suggested that the cognitive, emotional, and behavioral asymmetries discussed in earlier chapters result from this chemical asymmetry. These specific ideas require a great deal of additional testing before they can be judged, but in view of studies with other species it is reasonable to suppose that pharmacological and chemical asymmetries in the brains of humans have consequences for behavior.

The major fiber tract that connects the two cerebral hemispheres is the corpus callosum. Recent studies have shown considerable individual variation in the size of portions of the corpus callosum, although the data on how this might relate to individual variation in hemispheric asymmetry are equivocal. There is some indication that portions of the corpus callosum are larger in non-right-handers than in right-handers, with this handedness effect being larger for males than for females. There is also some indication in studies with rats that there is a strong positive correlation between a measure of left/right anatomical asymmetry and the percentage of callosal fibers, raising the possibility that increasing asymmetry is accompanied by less callosal connectivity.

Given the existence of both behavioral and biological asymmetries, it is important to determine whether the two are related and whether there are behavioral correlates of individual differences in the size of the corpus callosum. The fact that both behavioral and biological asymmetries are larger and more frequent for right-handers than for non-right-handers suggests that some relationships exist. In addition, some preliminary data suggest relationships within right-handers between asymmetries of the planum temporale and various behavioral asymmetries. In contrast, there is little evidence for the hypothesis that behavioral asymmetry is diminished in brains with relatively large corpora callosa or that a large corpus callosum necessarily leads to more efficient interhemispheric communication. In fact, there is some evidence that a larger corpus callosum provides a more effective inhibitory barrier between the hemispheres. However, strong conclusions about the relationship between behavioral asymmetries and various biological factors are prevented by the facts that the database is sparse and that the biological measures are still relatively gross.

Although the five themes outlined in Chapter 1 were formulated in terms of functional hemispheric asymmetry, it is interesting to note that they may also be used to characterize biological asymmetry. As reviewed in the present chapter, there is clear evidence of biological asymmetry in the brains of humans (theme 1) and at least some preliminary indication that biological asymmetry is related to callosal connectivity (theme 2). We shall see in the next chapter that many of the biological asymmetries found in the brains of humans are also found in other species (theme 3). Furthermore, there is clear individual variation in the pattern of biological asymmetry and in biological aspects of callosal connectivity (theme 4). Finally, we shall see in Chapters 8 and 9 that at least some of these biological asymmetries are present as the human fetus develops *in utero* and were also present in our evolutionary ancestors (theme 5).

5

Behavioral and Brain Asymmetries in Nonhuman Species

In recent years there has been increasing interest in the study of behavioral and brain asymmetries in nonhuman species. At one time, left/right asymmetries in nonhuman species were thought to be rare and unrelated to hemispheric asymmetry in humans. In part, this bias derived from the emphasis on hemispheric asymmetry for language and the belief that language processes are unique to humans. However, the growing realization that behavioral asymmetry in humans is not restricted to language led to renewed interest in asymmetry in other species and to a renewed search for animal models of behavioral and brain asymmetry. The resulting literature, much of it published in the last ten years, has made it clear that behavioral and biological asymmetries are ubiquitous in nonhumans.

At the outset it is important to consider a distinction made in biology between *analogous* and *homologous* systems. Analogous systems are similar in function but not in origin or structure. For example, the wing of a bird and the wing of a bat have similar functions but have distinct structures and distinct phylogenetic and ontogenetic origins. Homologous systems share a common structure and common origins. They often have similar functions (like analogous systems), although their functions may not be identical. For example, the wing of a bat and the foreleg of a mouse are homologous even though their functions overlap only partially. Suppose that two species both show a right-sided bias for manipulating small objects (e.g., a right-hand bias in both humans and in one of the nonhuman primate species). To the extent that these asymmetries are similar in function, they are clearly analogous, but they are homologous only to the extent that they are similar in origin and structure. The distinction between analogous and homologous systems is important because an asymmetry found in one species is likely to be a better model system for a corresponding asymmetry in another species to the extent that the two are homologous (rather than merely analogous). At the present time it is possible to identify several asymmetries in nonhumans that are clearly analogous to certain asymmetries seen in humans. However, it is much more difficult to determine whether they are also homologous and, in most cases, additional research is needed to resolve this question.

The strategy taken in this chapter is to review asymmetries in nonhumans that seem at least analogous to the asymmetries seen in humans. Where there has been speculation about whether a certain asymmetry is also homologous to an asymmetry found in humans, that speculation and debate surrounding it will be noted. For convenience of exposition, the review is organized into the general types of behavioral asymmetry that were used to review human asymmetries in Chapter 2: (1) asymmetric motor performance, (2) asymmetries in the production and perception of vocalizations, (3) asymmetries for visuospatial processes, and (4) asymmetry of motivation and emotion. This organization is not meant to imply that all of the asymmetries seen in nonhumans are homologous to those found in humans. Rather, it is intended to facilitate consideration of the parallels between humans and nonhumans. For each of these general types of asymmetry, studies are at least crudely grouped by species (primates, rodents, birds, etc.).

To the extent that studies of biological asymmetry have accompanied studies of behavioral asymmetry, they will be considered together. Additional biological asymmetries will be discussed in a separate section. Some of the asymmetries observed in nonhuman species are different for males and females and, of those, many are influenced by the level of hormones, such as testosterone. Consequently, the effects of sex and hormones will be noted explicitly throughout this review.

Motor Performance

As was noted in Chapter 2, the most obvious behavioral asymmetry in humans is handedness. It is instructive to consider two important characteristics of handedness in humans. The first is that an individual typically prefers to use the same hand for a wide variety of tasks, and this consistency across tasks is especially characteristic of right-handers. Of course, there is variation within the population in the strength of handedness, but when one hand is strongly preferred for one task (e.g., right-hand preference for writing), that same hand is typically preferred for many additional tasks (e.g., lighting a match; but see Chapter 7). The second important characteristic is that the human population is strongly biased toward right-handedness: approximately 90 percent of the population is classified as right-handed.

There are many examples of motor asymmetry in nonhuman species, some of which are quite dramatic at the level of an individual member of the species. Although these asymmetries bear some resemblance to handedness in humans, none of them exhibits *both* of the important characteristics of human handedness. For example, in certain species of nonhuman primates any population preference for one hand or the other differs across tasks and does not have the consistency that is typical of individual humans. By way of contrast, for many species of rats the paw preference shown by an individual may be strong and consistent across tasks but there is little or no evidence for a population bias toward one paw or the other. That is, there are approximately equal numbers of left- and right-pawed individuals. Despite these differences from human handedness, some of the motor asymmetries in other species seem analogous to some of the asymmetries in humans, and some may even be homologous to asymmetries in humans.

Primates

The conventional point of view has been that nonhuman primates do not exhibit population-level asymmetries of hand preference. This is certainly true in the sense that hand preference in these species fails to exhibit both of the characteristics of human handedness discussed earlier. But in a reconsideration of the studies on primate handedness available at the time, MacNeilage, Studdert-Kennedy, and Lindblom (1987) argued that there are, in fact, several statistically significant population-level asymmetries. The direction of the asymmetries varies from task to task, however, which means that the population-level asymmetries are obscured if a global measure of handedness is based on collapsing a variety of tasks into one single measure of hand preference. The generalization reached by MacNeilage et al. is that there is a left-hand preference for reaching (e.g., for a piece of food) and a right-hand preference for manipulation (e.g., unfastening a hasp with two fingers to open a container). These generalizations are based on the authors' reconsideration of previously published data obtained from eight different species of nonhuman primates.

MacNeilage et al. (1987) hypothesize that this pattern of primate handedness evolved as a result of adaptations to feeding and that these two opposite hand preferences are precursors to hemispheric asymmetries in humans. Specifically, they hypothesize that the asymmetries evolved in the following way. The first asymmetry to emerge was the left-hand/right-hemisphere dominance for performing visually guided movements such as reaching: the left-hand dominance for these movements remained in monkeys but not in humans. MacNeilage et al. suggest that this preference may have been accompanied initially by dominance of the right arm and hand for postural support. With the emergence of an upright posture, the need for postural support from the upper limb diminished, freeing the right arm and hand to become dominant for other things (see Chapter 9). The specific suggestion is that with upright posture and the development of an opposable thumb the right hand/left hemisphere became dominant for manipulation and bimanual coordination—a population-level asymmetry that they argue persists in both monkeys and humans.

The thesis advanced by MacNeilage et al. (1987) is both interesting and controversial for the radical suggestion that certain patterns of handedness in primates are homologous to handedness in humans and for the hypotheses about the evolution of hemispheric asymmetry in humans (for additional discussion of hypotheses about evolution, see Chapter 9). Much of the controversy is illustrated in the commentaries that accompany the article, where a good deal of skepticism is expressed about both the data on primate handedness and about the suggestion that primate handedness and human handedness are homologous. Additional data published more recently leave little doubt that the majority of individuals within a number of prosimian species prefer to use their left hand for reaching, though the extent of that asymmetry varies with age and sex (e.g., Ward, 1991) and also with the novelty and spatiotemporal scale of the movements (e.g., Fagot and Vauclair, 1991; see also Westergaard, 1991). With respect to the latter factors, Fagot and Vauclair argue that manual asymmetry is often found for high-level tasks, such as making a somatosensory discrimination, but not for low-level tasks, such as simple food reaching. For example, the left-hand preference for food reaching in prosimians is restricted to tasks that require the animals to maintain an erect posture—a requirement that Fagot and Vauclair suggest introduces a number of spatiotemporal constraints that are not present with a nonerect posture. Fagot and Vauclair suggest that manual asymmetry for high-level tasks is related to hemispheric asymmetry and possibly to human handedness, whereas asymmetry for low-level tasks is not. Interestingly, the direction of hand preference for high-level tasks is not consistent across species. For example, rhesus monkeys show a left-hand preference for making a somatosensory discrimination whereas chimpanzees (which are more closely related to humans) show right-hand dominance.

More data are needed to affirm or deny the specific hypotheses advanced by MacNeilage et al. (1987) and by Fagot and Vauclair (1991). In addition, it may prove worthwhile to look more closely at the strength of handedness for different tasks in humans to see whether there might be subtle task-by-hand interactions of the sort found in other primates. For example, with respect to the planning and execution of sequential movement, Summers and Sharp (1979) have already argued that the right hemisphere is more involved in moving the limbs and fingers to exact locations whereas the left hemisphere is more involved in producing the correct sequential order, and Haaland and Harrington (1989) suggest that the left hemisphere is dominant for controlling preprogrammed, open-loop movements but not for controlling closed-loop movements (see Chapter 2). In addition, Steenhuis and Bryden (1989) have found that both right- and left-handed humans show stronger hand preferences for what they term "skilled" activities (e.g., writing) than for less-skilled activities (e.g., picking up objects).

Rodents

Interesting motor asymmetries are also found in rats and mice. Two types of asymmetry are considered here: paw preference and preferred direction of circling. For both of these behaviors, individual animals often show a reliable preference toward one side or the other, with the direction and strength of the preference varying from animal to animal. Although a given animal often prefers the same paw for more than one task, there is little in the way of a dramatic population-level asymmetry such as that found in humans. That is, the number of left- and right-pawed animals is approximately equal. Nevertheless, these motor asymmetries in rats and mice have provided ways of testing hypotheses about the genetic transmission of motor asymmetry and about the biological mechanisms responsible for motor asymmetry.

As an illustration of paw preference in mice, consider one of the paradigms used by Collins and his colleagues (e.g., Collins, 1985). A food-deprived mouse is placed into a testing chamber in which a food particle is available in a cylindrical feeding tube attached perpendicular to one wall. In order to obtain the food, the mouse must stand on its hind legs and reach into the tube with one paw. A typical testing session might record the paw used on 50 trials of this sort. In this way, both the direction and magnitude of paw preference can be obtained, examined across testing sessions, and so forth. At least in the species studied by Collins and his colleagues, it is the case that the paw preference shown by an individual mouse is reliable and enduring and is not task-specific.

In several experiments, Collins (1985) has found that in mice it is not possible to breed selectively for the *direction* of paw preference. That is, two "right-pawed" mice are no more likely to produce "right-pawed" offspring than are two "left-pawed" mice. In contrast, it has been possible to breed selectively for the *magnitude* or degree of asymmetry. That is, the offspring of two strongly asymmetric mice are more likely to be strongly asymmetric than are the offspring of two weakly asymmetric mice. These results indicate the potential independence of the direction and degree of asymmetry and raise interesting questions about the genetic transmission of other asymmetries in both mice and other species, including humans. In fact, on the basis of a three-generation study of handedness in humans, Bryden (1987; see also Bryden and Steenhuis, 1991) suggests that the degree of hand preference is more heritable than the direction of hand preference (see also Chapter 7). This may occur because the direction of asymmetry is influenced by a variety of early (perhaps prenatal) environmental influences (see Chapter 8).

Additional work suggests that the two strains of mice bred by Collins (High Asymmetry and Low Asymmetry) differ in the extent of biological brain asymmetry: the brains of the High Asymmetry strain are larger and more asymmetric than the brains of the Low Asymmetry strain (see Collins, 1985). It is also interesting to note that, in the original strain of mice studied by Collins (before selective breeding), the degree of asymmetry for paw preference was larger for females than for males. In addition, there is some evidence that the High and Low Asymmetry strains differ in characteristics related to sexual dimorphism. For example, the levels of circulating testosterone were higher in adult males from the High Asymmetry strain than from the Low Asymmetry strain. Such results are consistent with the hypothesis that behavioral and biological asymmetries are related in potentially complex ways to biological sex and to sex-related hormones at various stages of development (for speculation about the nature of some of these influences, see Chapter 7, as well as Geschwind and Galaburda, 1987).

In an extensive series of investigations spanning the last fifteen years or so, Stanley Glick and his colleagues have studied the tendency of rats to circle predominantly in one direction or the other (for reviews see Carlson and Glick, 1989; Glick and Shapiro, 1985; Glick, Carlson, Drew, and Shapiro, 1987). For example, when neurologically intact rats are tested with amphetamine stimulation, they typically rotate predominantly in one direction. Although in some strains there may be a population bias toward circling to the right, this bias is small relative to the wide individual variation in the direction and magnitude of preferred rotation. The preferred direction of rotation has been shown to be related to asymmetries in the dopamine (DA) content and sensitivity to DA in the nigrostriatal pathway of the rat brain. Originally it was thought that all rats circled predominantly in the direction opposite to the side of the brain containing more DA or more highly activated postsynaptic DA receptors. However, recent studies suggest a more complex picture of two populations of rats: those that circle consistently away from the side with greater DA innervation and those that circle consistently toward the side with greater DA innervation (e.g., Glick et al., 1987; Shapiro, Camarota, and Glick, 1989). In addition, there appear to be a variety of other biochemical asymmetries as well as sex differences and rat-strain differences in the magnitude and direction of the asymmetries.

Of particular interest are relationships between asymmetries in rotation and other aspects of behavior. One important dimension is the strength of rotation bias. For example, only rats with a strong rotation bias toward one side were able to learn in an operant conditioning study that reinforced rats for making 360-degree turns or for making a left-right discrimination (for review, see Glick and Shapiro, 1985). This has led to the suggestion that, unless an animal has a sufficiently strong side preference, it has difficulty learning an association between a spatially biased movement and reward. Another important dimension may be the direction of preferred rotation. For example, rats find it easier to learn to circle for reward in their preferred direction than in their nonpreferred direction. In addition, (1) right-rotators have been reported to be more active and have stronger side preferences than left-rotators, (2) bilateral lesions of the frontal cortex decrease side preferences in right-rotators and increase side preferences in left-rotators, and (3) administration of cocaine induced more rotating in right- than in left-rotating female Sprague-Dawley rats, but the effects were exactly opposite for right- and left-rotating males.

The foregoing is not meant to be an exhaustive review of what are now many studies of rotational behavior in rats. However, even this brief sampling indicates that reliable rotation biases exist in rats and that these biases are related to neurochemical asymmetries in the brain, to sex, and to other behaviors—especially behaviors that demand left-right discriminations.

In addition, recent studies have examined the contribution of genetic and environmental factors to the development of rotation biases. One type of study involved breeding male-female pairs with the same or opposite rotational biases. In these studies, male offspring tended to have the same direction of rotation bias as the male parent and opposite bias from the female parent. Female offspring tended to have a rotation bias opposite that of the female parent only if the litter from which they came had more males than females. From this and from a variety of other observations, it has been suggested that the direction of rotation bias depends in part on genetic factors and in part on the levels of fetal testosterone (see Carlson and Glick, 1989). Another type of study has attempted to breed for the strength of rotation; that is, to produce strains of strongly and weakly rotating rats. A strain difference was found after eight generations, but only in females. Note that the result for females is similar to the results for both male and female mice reported by Collins (1985) in his experiment of breeding for strength of paw preference. It is interesting that a left-sided rotation bias emerged for both males and females in the weakly rotating strain, for this is consistent with the possibility of a relationship between side of rotation and strength of rotation.

It is difficult to know how these rotation asymmetries in rats might relate to asymmetries in other species, including humans. However, it is interesting to note that reliable rotation biases have also been obtained in cats (Glick, Weaver, and Meibach, 1981), prosimians (e.g., Ward, 1991), and humans (Bracha, Seitz, Otemaa, and Glick, 1987; Gospe, Mora, and Glick, 1990; see also Chapter 7) and that the effects in humans are related to sidedness for other behaviors and to sex. In humans, rotation biases are measured by having subjects wear a device that counts the number of full (360-degree) and quarter (90-degree) turns as they go about their daily routines. For adult humans, men who were consistently right-sided for hand, foot, and eye dominance rotated more to the right than to the left whereas the relationship was reversed for women. It is also interesting to note that hemispheric asymmetries in DA innervation have been proposed in humans and that such neurochemical asymmetries have been suggested to underlie functional asymmetries in humans (see Chapter 4 and Tucker and Williamson, 1984). In fact, Bracha (1987) has found evidence that humans (like some rats) prefer to circle in a direction opposite the side of the brain with greater DA activity. Whether the asymmetries in rats are homologous to what seem similar asymmetries in humans is not clear at the present time.

Other Species

Individual members of other species also show interesting motor asymmetries. For example, connoisseurs of a famous Chinese dish, braised bear paw, insist on left paws because they tend to be softer and more succulent—presumably because the bears lick them more often than right paws (*Newsweek,* July 29, 1991, p. 35). A variety of other motor asymmetries and their possible relationship to human handedness are reviewed by Bradshaw (1989), Geschwind and Galaburda (1987), and Glick and Shapiro (1985). An illustrative example comes from the fact that several species of parrot are known to be left-footed at the population level for food manipulation (Rogers, 1980, 1986). Note that although this preference seems opposite right-handedness in humans and in nonhuman primates for manipulation, one could argue (as did Bradshaw, 1989) that in parrots it is the beak that is used for finer manipulation and that, consequently, it is the beak that may be analogous (or even homologous) to the right hand of humans and other primates. From this point of view, the left foot is analogous (or possibly homologous) to the left hand, for it reaches for an object and positions it for the beak to manipulate. Regardless of the merit of this argument, it illustrates the difficulty of making precise cross-species comparisons.

The Production and Perception of Vocalizations

The most widely cited cognitive asymmetries in humans involve left-hemisphere dominance for several aspects of language. There is growing evidence that some of those asymmetries in humans have parallels in other species. In particular, asymmetries in the production or perception of vocalizations have been demonstrated in certain primate species, in some species of song birds, and in at least one species of canine.

Primates

There is evidence that the left hemisphere is dominant for the perception of communicatively relevant species-specific vocalizations in Japanese macaques *(Macaca fuscata)*. For example, Peterson, Beecher, Zoloth, Moody, and Stebbins (1978) trained five

Japanese macaques and five other Old World monkeys to discriminate among vocalizations produced by Japanese macaques in the field ("coo" sounds). The coo sounds are brief tonal sounds composed of a fundamental-frequency band and an extended series of harmonics. In the field, coo sounds occur primarily during social, contact-seeking behavior. Two communicatively relevant subtypes of coos can be identified acoustically by the relative temporal location of the "peak" fundamental frequency (early versus late). Coos with the peak occurring late are produced primarily by estrous females soliciting males, whereas coos with the peak occurring early are used by males and females in a variety of circumstances referred to by Petersen et al. as contact-seeking. One task required the animals to discriminate stimuli using the temporal location of the peak fundamental frequency. A second task required the animals to discriminate an orthogonal feature of the same stimuli (pitch of the stimuli, relatively high versus relatively low) that is not communicatively relevant to Japanese macaques. Using an adaptation of the dichotic-listening procedure that has been used to study hemispheric asymmetry in humans (see Chapter 1), Petersen et al. found that all five Japanese macaques discriminated the temporal location of the peak fundamental frequency better when the stimuli were presented to the right ear (left cerebral hemisphere) than to the left ear (right cerebral hemisphere). By way of contrast, the Japanese macaques showed either a left-ear advantage or no ear difference for the pitch-discrimination task. That is, for the same set of stimuli, a right-ear advantage was obtained only when the feature to be discriminated was relevant for communication. The other Old World monkeys, for whom neither peak-location nor pitch information was communicatively relevant, showed no consistent ear advantage for either discrimination task.

The results obtained with Japanese macaques are similar to the findings with humans that a right-ear advantage is obtained for the perception of linguistic stimuli, such as words and nonsense syllables, but not for nonlinguistic sounds. These results lead Petersen et al. (1978) to conclude that Japanese macaques engage the left hemisphere more than the right for the analysis of communicatively significant speech sounds and that this is analogous to hemispheric asymmetry for speech perception in humans.

In additional studies of Japanese macaques, Heffner and Heffner (1984, 1986) set out to determine the neuroanatomical basis of the right-ear advantage reported by Petersen et al. (1978). To do so, they used a conditioned avoidance task to teach monkeys to discriminate between the communicatively relevant subtypes of coos ("peak-discrimination") used in the previous study. Discrimination training was carried out before and after unilateral and bilateral ablation of the temporal cortex, with the area of ablation including the primary and secondary auditory cortex. Eleven animals were trained to make the critical discrimination prior to any ablations. Five of these animals then received unilateral lesions of the left cerebral cortex and five received unilateral lesions of the right cerebral cortex. The other (control) animal received a unilateral lesion of the left superior temporal gyrus that spared part of the auditory cortex.

Unilateral ablation of the left hemisphere resulted in an initial impairment in the ability to discriminate the temporal location of the peak fundamental frequency; some level of initial impairment was present for all five animals. Animals in this group required from 5 to 10 additional discrimination-training sessions to reach their preoperative level of discrimination. In contrast, unilateral ablation of the right hemisphere produced no measurable effect whatsoever on postoperative performance. After the animals reached normal, preoperative levels of discrimination performance, similar lesions were made in the opposite hemisphere of three monkeys with initial left-hemisphere ablation and two monkeys with initial right-hemisphere ablation. Bilateral lesions of these brain regions made it impossible for these animals to perform the discrimination. Even as long as nine months after surgery, none of these animals could discriminate even a single pair of coos. Neither unilateral left-hemisphere nor subsequent bilateral lesions in the control animal (lesions that spared a portion of the auditory cortex) had any effect on discrimination performance.

As noted by Heffner and Heffner (1984, 1986), this pattern of deficits after unilateral and bilateral lesions indicates hemispheric asymmetry for the discrimination of communicatively relevant coo vocalizations. The fact that unilateral ablation of areas of the left hemisphere produced an initial impairment whereas unilateral ablation of corresponding areas of the right hemisphere had no effect suggests that the left hemisphere ordinarily plays a dominant role in making the discrimination. Animals with unilateral left-hemisphere ablation can re-learn the discrimination, however, and subsequent ablation of the right hemisphere makes them completely unable to make the discrimination. These findings indicate that the right hemisphere is not completely without ability to make the discrimination, but whether it can do so when the left hemisphere is intact or whether it can do so as efficiently as the left hemisphere remains unclear. It is also unclear whether the asymmetry reported by Heffner and Heffner is restricted to discrimination of communicatively relevant features (their study used only "peak" discrimination), but the absence of a right-ear advantage for a feature that is not communicatively relevant (Petersen et al., 1978) makes it reasonable to expect such a restriction.

The relevance of the communication value of a stimulus is also illustrated by Hopkins, Morris, and Savage-Rumbaugh (1991) in a visual-half-field study of hemispheric differences in priming in two language-trained chimpanzees *(Pan troglodytes)*. In their experiments, Hopkins et al. required the two subjects to respond as quickly as possible to the onset of a small patch of light presented at the fixation point. On some trials, this response stimulus was preceded by a visual warning stimulus flashed briefly to either the LVF/right hemisphere or RVF/left hemisphere. Examples of the warning stimuli used in one of their experiments are shown in Figure 5.1. The primary dependent variable was the difference in reaction time to the response stimulus when a warning stimulus was present versus when no warning stimulus was given (i.e., the priming effect caused by the warning stimulus). For these two language-trained chimpanzees, the warning stimulus had a larger effect when presented to the RVF/left hemisphere than to the LVF/right hemisphere when the priming stimulus was meaningful—that is, when the stimulus was a symbol that the chimpanzees could reliably comprehend. No such visual-field differences were found when the warning stimuli were similar symbols that were either unfamiliar or familiar but without meaning. Hopkins et al. compare these results to studies with humans that suggest that the amount of priming produced by each hemisphere depends on hemispheric asymmetry for the processing of the warning stimuli, and they suggest that these results for language-trained chimpanzees are similar to the RVF/left-hemisphere dominance seen for processing language-related symbols in humans. Given these similarities, they propose that "basic phylogenetic neuropsychological systems related to activation and priming processes may link nonhuman primate and human studies of lateralization" (p. 46).

Although many aspects of language are associated with left-hemisphere dominance in humans, there are exceptions. For example, adult humans discriminate a variety of speech sounds according to the phonetic labels assigned to each sound, a phenomenon known as categorical perception. That is, adults will readily discriminate the voicing contrast that separates */ba/* from */pa/*, but they perform at chance levels when trying to discriminate between two different */ba/*s that differ acoustically just as much as */ba/* differs from */pa/*. Interestingly, this type of categorical perception can also be seen in the auditory evoked response (AER) measured

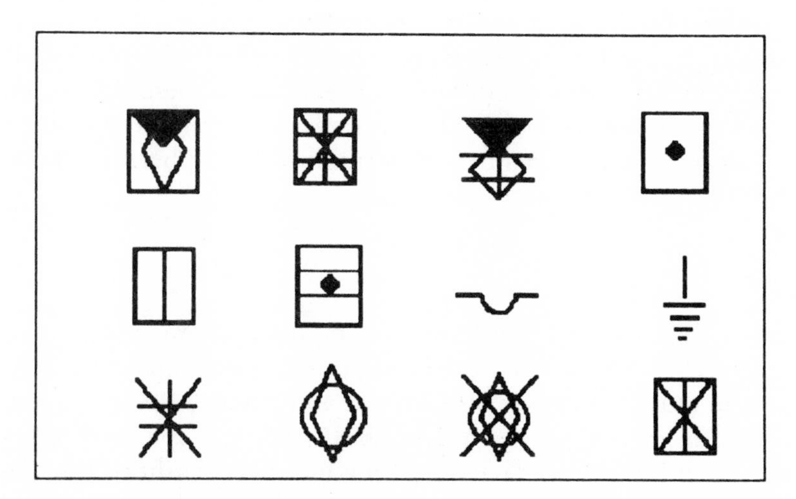

Figure 5.1. The 12 lateralized warning stimuli used in the studies of language-trained chimpanzees reported by Hopkins et al. (1991). [Reprinted from W. D. Hopkins, R. D. Morris and E. S. Savage-Rumbaugh, "Evidence for Asymmetrical Hemispheric Priming Using Known and Unknown Warning Stimuli in Two Language-Trained Chimpanzees *(Pan troglodytes),*" *Journal of Experimental Psychology: General,* 120 (1991):46–56. Copyright 1991 by the American Psychological Association. Reprinted by permission.]

from certain areas of the scalp (see Chapter 1 for a discussion of evoked potentials). In both human adults and infants, certain components of the AER more often respond categorically over the right hemisphere than over the left hemisphere (for reviews see V. Molfese, D. Molfese, and Parsons, 1983; D. Molfese and Burger-Judisch, 1991). Morse, D. Molfese, Laughlin, Linnville, and Wetzel (1987) found that a late component of the AER recorded from the right temporal region of rhesus monkeys *(Macaca mulatta)* also discriminated between stimuli in a categorical manner. The authors note the similarity of these results to the results of humans and suggest that the cortical mechanisms associated with categorical perception (including asymmetry in those mechanisms) may be similar across human and nonhuman primates.

Rodents

Mouse pups emit ultrasonic calls to gain attention from their mothers. These calls are perceived categorically (like human speech), and Ehret (1987) has shown that they are preferentially perceived by the right ear/left hemisphere of the mothers. Interestingly, no ear/hemisphere differences are obtained from female mice who have been conditioned to respond to these signals but who have never had pups. Thus, the communication value of the signals may be important for determining hemispheric asymmetry.

Birds

Among the first and best-known examples of asymmetry in nonhumans for the production of vocalization were reports by Nottebohm (1970, 1979; see also Arnold and Bottjer, 1985) of asymmetries in certain species of song birds. For example, sectioning of the left hypoglossus nerve of an adult male chaffinch leads to permanent loss of many components of bird song whereas sectioning of the right hypoglossus nerves leads to the loss of only a few components. If the left-sided damage occurs before song is first developed, then song may develop normally, indicating that the right side is not completely unable to produce song. Although the central mechanisms responsible for this asymmetry are not yet understood completely (e.g., Bottjer and Arnold, 1985), the behavioral asymmetry is very striking, and the parallels to left-hemisphere dominance for human speech are intriguing.

Other Species

Most reports of asymmetries in the production and perception of vocalizations have come from studies of primates and birds, but there may be related asymmetries in other species. For example, Adams, Molfese, and Betz (1987) recorded AERs from the left and right temporal and parietal scalp regions of 15-week-old border collies while the puppies listened to a series of consonant-vowel syllables (e.g., */ba/*, */pa/*). Components of the AER recorded over the right hemisphere discriminated between the sounds in a categorical manner, with the critical components occurring between 50 and 180 milliseconds after speech onset. Note that this asymmetry is similar to that obtained with the same paradigm in rhesus monkeys and in human adults and infants. This is consistent with the possibility that some of the basic brain mechanisms that subserve aspects of human speech perception are also present in nonprimates.

Visuospatial Processes

In humans there are well-established hemispheric asymmetries for a variety of visuospatial processes (see Chapter 2). The traditional view has been that the right hemisphere of humans is dominant for visuospatial processing, but recent studies from a more computational perspective suggest that the direction of hemispheric dominance depends on specific viewing conditions and task demands (see Chapter 3). Evidence has also accumulated for the existence of asymmetries for visuospatial processes in other species, with the direction of asymmetry sometimes (but not always) similar to that reported for humans. Of course, it is rarely the case that the same stimuli are used for humans and other species and it is virtually never the case that humans are tested with the same behavioral paradigms (e.g., operant conditioning) that are used with other species. Therefore, differences in the direction of certain asymmetries across species must be treated with caution. Nevertheless, it is important to acknowledge that functional asymmetries related to spatial processing do exist.

Primates

Hopkins and Morris (1989) examined visual-field asymmetries for processing the location of a short line contained within a geometric form in the same two language-trained chimpanzees that were tested in the study by Hopkins et al. (1991) reviewed earlier. Recall that for these two chimpanzees Hopkins et al. found RVF/left-hemisphere dominance for processing language-related symbols and no asymmetry for processing meaningless symbols. For one of the chimpanzees, Hopkins and Morris found significantly faster reaction times for the line-location task when the stimuli were presented tachistoscopically to the LVF/right hemisphere, regardless of which hand was used to respond. The other chimpanzee showed a similarly significant LVF/right-hemisphere advantage when using the left hand (controlled by the right hemisphere) to respond. A trend in the same direction was found when the right hand was used to respond, but in this case the visual field difference was not statistically significant. These same two chimpanzees plus a third language-trained chimpanzee also show an LVF/right-hemisphere advantage for processing meaningless forms of the sort that produce a similar asymmetry in humans (Hopkins, Washburn, and Rumbaugh, 1990).

The results reported by Hopkins and Morris (1989; Hopkins et al., 1990) are important for several reasons. Chimpanzees are very similar to humans phylogenetically and neuroanatomically and it is interesting that the LVF/right-hemisphere advantages found by Hopkins and colleagues are similar to asymmetries found in humans for tasks that demand processing of location (at least location in a coordinate system) or recognition of nonsense forms. The fact that these same chimpanzees produce opposite visual-field asymmetries for processing language-related symbols and line location rules out interpretations of either effect in terms of a general dominance or superiority of one hemisphere, regardless of task demand. This, too, is reminiscent of the results for humans. Of course, it must be kept in mind that the chimpanzees tested in these studies have undergone extensive language training, and it is unknown whether any of their asymmetries are a product of that training.

In their study of form recognition, Hopkins et al. (1990) found an RVF advantage for two rhesus monkeys *(Macaca mulatta)*, an asymmetry exactly opposite that found for chimpanzees. Although the sample sizes are very small, this finding reinforces the possibility that different primate species may have developed analogous asymmetries in different directions. Although this possibility requires a great deal of additional investigation, other research also makes it clear that functional asymmetries for visuospatial processing are not restricted to the great apes.

Hamilton and Vermeire (1982) examined hemispheric asymmetry for learning a discrimination based on the comparison of sequentially presented visual stimuli in 12 split-brain rhesus monkeys *(M. mulatta)*. Each of the monkeys underwent midsagittal division of the corpus callosum, hippocampal and anterior commissures, and the optic chiasm. As a result, input from each eye was restricted to only one cerebral hemisphere. By covering one eye, it is possible to train only one hemisphere at a time and to compare the performance of the two hemispheres. Although Hamilton and Vermeire found no overall advantage for one hemisphere or the other, they reported a significant correlation (r = .768) between hand dominance and hemispheric dominance. That is, the same hemisphere tended to be dominant for both hand use and for the sequential-learning task. That the sequential nature of the visual task was important is suggested by the fact that similar asymmetries were not found for a control discrimination that required simultaneous rather than successive matching. Hamilton and Vermeire suggest that because of its sequential nature, the sequential-matching task would produce a left-hemisphere advantage in right-handed humans. While it is true that in humans there is some evidence of left-hemisphere dominance for remembering and producing a sequence of movements (see Chapter 2), it is difficult to make any strong statement about human asymmetry for the specific task used by Hamilton and Vermeire.

Jason, Cowey, and Weiskrantz (1984) studied the performance of 9 rhesus monkeys *(M. mulatta)* on a task for which there is some evidence of right-hemisphere dominance in humans. For the monkeys, the task required a discrimination between two squares, one containing a dot exactly in the center and another containing a dot displaced upward from the center. After the monkeys were trained to discriminate squares when the dot was either centered or displaced 26.5 mm upward, thresholds were determined to discover the minimum displacement upward that would continue to lead to correct responding. For each monkey, the threshold was determined preoperatively and then four animals received a left-sided unilateral occipital lobectomy combined with a splenial transection (the left-hemisphere lesion group) and five animals (the right-hemisphere lesion group) received corresponding surgery on the right side. After surgery, thresholds were again determined for all animals. The result of primary importance is that all monkeys with left-hemisphere lesions performed worse than all monkeys with right-hemisphere lesions (i.e., there was no overlap in the performance of the two groups). In contrast to the results reported by Hamilton and Vermeire (1982), functional asymmetry was independent of handedness.

In discussing the direction of the asymmetry in their rhesus monkeys, Jason et al. (1984) suggest that asymmetries in opposite directions for humans and monkeys may indicate that the asymmetries are analogous rather than homologous; that is, the asymmetries may have evolved independently, but perhaps for similar reasons. It is also possible that if humans were tested with the exact stimuli and procedures used with the monkeys they would also show left-hemisphere dominance. This possibility is particularly interesting in view of recent suggestions that the left hemisphere of humans is dominant for judging categorical spatial relations like up/down (see Chapter 3).

An important issue in the study of functional asymmetries in humans is whether various asymmetries are statistically independent of each other or whether they are correlated (see Chapter 2). What evidence there is indicates that many asymmetries are independent of each other. It is interesting that a similar finding has been reported by Hamilton and Vermeire (1988, 1991). In an investigation of 25 split-brain macaques, they found that 18 of 25 showed faster learning when visual stimuli were restricted to the right hemisphere when the stimuli were photographs of monkey faces. In contrast, 22 of 25 showed faster learning when the stimuli were restricted to the left hemisphere when the stimuli were lines differing in orientation by 15 degrees. Despite the fact that both tasks produced significant asymmetries, the two asymmetries were independent. This can be seen in Table 5.1, which shows the number of monkeys falling into each cell of a 2-by-2 contingency table defined by the hemisphere showing superiority for each of the two tasks. The numbers in parentheses are the expected frequencies under the assumption that the two asymmetries are independent. The obtained frequencies are very close to the expected values and there was no hint of a correlation between left-right difference scores for the two tasks ($r = 0.01$). Thus, although the majority of monkeys (17) are lateralized in opposite directions for face and line stimuli, the two asymmetries are statistically independent of each other. This pattern of independence is generally similar to what has been reported for humans, suggesting that functional asymmetries can develop independently in both species.

It is interesting to compare the direction of asymmetries reported by Hamilton and Vermeire (1988, 1991) with similar asymmetries found with humans. As Hamilton and Vermeire note, the right hemisphere of humans tends to be dominant for a variety of tasks that involve face recognition (but not all; see Chapters 2 and 3). As a result, they argue, the right-hemisphere dominance for face processing in their monkeys is similar to that found in humans. Hamilton and Vermeire also note that inverting photographs of faces eliminates the right-hemisphere advantage in both monkeys and humans, further suggesting similar mechanisms in

Table 5.1. Number of split-brain monkeys showing right- and left-hemisphere advantages for two tasks

	Line-orientation task	
	RH	LH
Face-recognition task		
RH	2 (2.2)	16 (15.8)
LH	1 (0.8)	6 (6.2)

Note: Numbers in parentheses are the frequencies expected if the two asymmetries are distributed independently of each other.

Source: Hamilton and Vermeire (1988, 1991).

the two species. In monkeys there was no asymmetry for learning to discriminate geometric figures, a result that Hamilton and Vermeire also see as consistent with data obtained from humans. Where Hamilton and Vermeire see an inconsistency is in the left-hemisphere advantage for discriminating line orientation in monkeys. While it is certainly true that there have been reports of a right-hemisphere advantage for judging line orientation in humans, the tasks used with humans differ sufficiently from the tasks used with monkeys to make a direct comparison difficult. In view of the fact that hemispheric asymmetry for visuospatial processing in humans depends on a variety of perceptual and task factors, it would be worthwhile to obtain data from humans using exactly the discrimination required of the monkeys. Furthermore, Mehta, Newcombe, and Damasio (1987) have shown in humans that at least some judgments of line orientation (as well as other visuospatial judgments) are more impaired by injury to the left than to the right hemisphere.

Rodents

It has been suggested that in certain species of rats there is hemispheric asymmetry for "spatial behavior" (e.g., Denenberg and Yutzey, 1985). Although the behaviors tested do not really fall under the heading of "*visuo*spatial," it is nevertheless instructive to consider the kind of spatial behaviors for which there is some indication of asymmetry. Denenberg and his colleagues have shown that rats who have been subjected to early handling show different amounts of activity in an open field depending on which of the two hemispheres has been ablated. However, the direction of the asymmetry depended on whether or not the rats had also been exposed to an enriched environment. For handled and enriched rats, activity was higher with the right hemisphere intact than with the left hemisphere intact. For handled rats that were not enriched, the asymmetry was in the opposite direction. Interestingly, there were no asymmetries for nonhandled rats. As Denenberg and Yutzey note, the amount of exploration in the open field is likely related to both spatial and affective factors, so an interpretation in terms of either type of factor is problematic. Additional studies of affective behavior will be discussed later. With respect to spatial components of behavior, Denenberg and

Yutzey report that when handled rats are released into an open field, they most often go leftward, suggesting to them that the right hemisphere is more involved in spatial exploration. Once again, there were no asymmetries of this sort for nonhandled rats.

Why behavioral asymmetries in rats depend on early experience (handling and enrichment) is unclear. At the very least, however, such effects indicate that early experiences are not having the same effect on both sides of the developing brain. This suggests that the pattern of asymmetry that eventually emerges is determined by the interaction of a number of biological and environmental factors, a point to which I will return in Chapter 8.

Birds

The visual system of birds makes them particularly appropriate for studying asymmetry for visuospatial processing. This is because each eye projects to only the contralateral side of the brain. Birds do not have a corpus callosum; instead, they have a number of smaller commissures that are not thought to permit much crossing of visual information (e.g., Gunturkun, 1985). As a consequence of these anatomical facts, when one eye is covered by a patch, visual information is restricted to the side of the brain contralateral to the uncovered eye and the behavior of the bird is restricted by the ability of that side of the brain to process the relevant visual information. Thus, functional asymmetries can be discovered by comparing the performance of left-eye-occluded and right-eye-occluded conditions. Of course, it is also possible to lesion one side of the brain or the other and examine how performance is affected by right- versus left-side ablations.

Gunturkun (1985) trained adult homing pigeons *(Columba livia)* binocularly on two different tasks that involved the discrimination of two successively presented visual patterns. After reaching a criterion rate of correct responding, each bird was tested with (1) the right eye occluded, (2) the left eye occluded, and (3) with neither eye occluded. Both the total number of responses and the percentage discrimination scores were higher when the right eye was uncovered (left-hemisphere stimulation) than when the left eye was uncovered (right-hemisphere stimulation). The author suggests that either the left hemisphere is dominant for this task or there is a bias for the left hemisphere to be more involved than the right hemisphere in learning the task with both eyes opened. In fact, he notes that the initial level of performance for a few birds was so low when only the left eye was uncovered that it seems likely that for those birds the right hemisphere learned very little with both eyes open.

In a follow-up study von Fersen and Gunturkun (1990) trained pigeons to discriminate 100 different visual patterns from an additional 625 similar stimuli (see Figure 5.2). During the extensive training sessions, neither eye was occluded. A retention test was administered approximately seven months later and was conducted using the same viewing conditions used by Gunturkun (1985). The percentage of correct responses was significantly higher with both eyes uncovered than with either eye covered and performance was significantly better with the right eye uncovered than with the left eye uncovered. The authors argue that asymmetry in the pigeon's visual system depends in part on the superiority of the left hemisphere for forming and retaining visual memories.

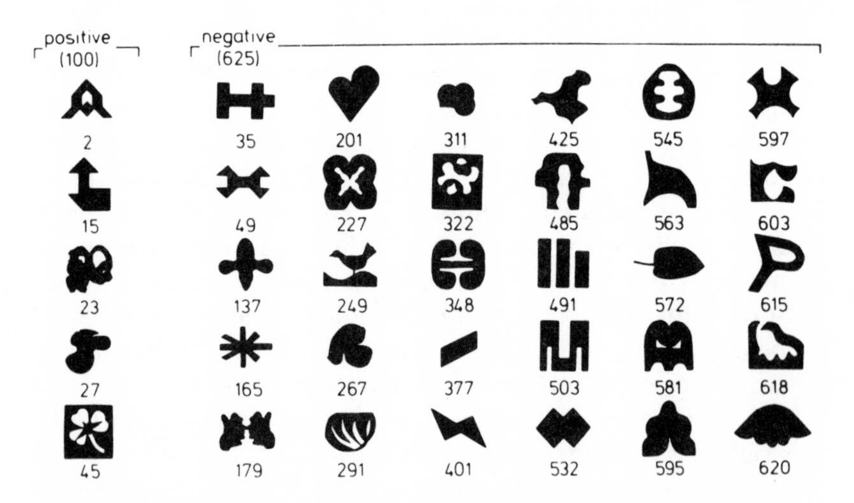

Figure 5.2. Examples of rewarded (positive) and nonrewarded (negative) stimuli in the study with pigeons reported by von Fersen and Gunturkun (1990). In the study there were 100 different positive stimuli and 625 different negative stimuli. [Reprinted from L. von Fersen and O. Gunturkun, "Visual Memory Lateralization in Pigeons," *Neuropsychologia,* 28 (1990):1–7. Copyright 1990 by Pergamon Press plc. Reprinted by permission.]

Rogers and colleagues (e.g., Arnold and Bottjer, 1985; Rogers, 1980, 1986) have also reported that the left hemisphere is dominant for visual-pattern discrimination in domestic chicks. For example, domestic chicks were tested in the second week of life in the following way for visual-discrimination learning. Food-deprived chicks were placed on a cage floor scattered with chick mash. Small inedible pebbles were glued to the floor. These pebbles resembled the grains of mash in size, shape, and color but not in brightness and texture. Chicks typically learn rapidly to discriminate the pebbles from the grains. In order to test for asymmetry in this task, some of the chicks had received intracranial injections on the second day of life of either a protein-synthesis inhibitor or glutamate, both of which cause long-lasting changes in brain function. Learning was normal if only the right hemisphere had been injected with either of these two drugs. However, learning was disrupted if the drugs had been injected into only the left hemisphere or into both hemispheres, leading Rogers to suggest that the neural circuits involved in this task depend on a normally functioning left hemisphere. Similar conclusions have come from studies using monocular testing of untreated male chicks. That is, in the second week of life, chicks learn the visual-discrimination task more rapidly if only the right eye (left hemisphere) receives stimulation than if only the left eye (right hemisphere) receives stimulation.

Rogers (1986) has speculated about the developmental origins of a variety of asymmetries in chicks and suggests that the left-hemisphere dominance may be related to the fact that after day 17 of incubation, the left eye of the chick embryo is covered by the wing and body so that the right hemisphere receives little or no light stimulation. In contrast, the right eye is next to the air sac and receives more light stimulation. Consistent with this possibility is the report by Rogers that chicks incubated in darkness from day 17 on show little or no asymmetry for a variety of behavioral tasks that are otherwise asymmetric.

Dolphins

The visual system of dolphins also provides an interesting opportunity to study asymmetries in visual processing. Unlike most mammals, each eye of the dolphin projects to only the contralateral cerebral hemisphere, so that what each eye sees goes first to the contralateral cerebral hemisphere. Hemispheric asymmetries have recently been tested by presenting stimuli to only one eye at a time and comparing reaction times in the two viewing conditions.

Morrel-Samuels, Herman, and Bever (1989) used this procedure to examine asymmetries in the perception of sign-language gestures in a female dolphin who had previously demonstrated good comprehension of the gesture language used in the research. The gestures used in this study were divided into simple gestures (e.g. "Hoop Under") and syntactically complex strings or "sentences" of gestures (e.g., "Hoop Ball Fetch"). Responses to simple gestures were faster when the gestures were viewed by the left eye/right hemisphere whereas responses to the complex gesture commands were faster when the gestures were viewed by the right eye/left hemisphere. In a follow-up study, in which 10 simple gestures were shown 7 times each, an initial left-eye/right-hemisphere advantage gave way to a right-eye/left-hemisphere advantage with repeated presentations. The authors suggest that the left hemisphere of this dolphin is dominant for tasks that are similar to complex language comprehension. Morrel-Samuels, Herman, and Pack (1990) reported a left-eye/right-hemisphere superiority for visual object recognition in this same female dolphin, which indicates that dolphin asymmetries are not restricted to stimuli with propositional content. Given certain parallels between the results of dolphins and humans, the authors suggest that the emergence of hemispheric asymmetry in humans is not likely to be contingent on such things as bipedalism, tool use, or language production per se. While the findings reported in these studies are fascinating and the potential implications are far-reaching, the results must be treated with caution until more than one subject has been studied.

Motivation and Emotion

In humans, there is some evidence that the right hemisphere is typically more involved than the left in the production and perception of emotion (see Chapter 2). Asymmetries related to motivation and emotion have also been reported in other species—in particular, in certain species of rats and in chicks.

Rats

Denenberg and Yutzey (1985) summarize a variety of research showing asymmetry for conditioned taste aversion and mouse killing in rats. As in studies of spatial behavior discussed earlier, asymmetries depended on whether rats were handled or not handled as infants. Conditioned taste aversion was used to determine whether one hemisphere or the other was better at remembering learned fear. Intact rats were subjected to typical conditioned taste-aversion training. Briefly, they were allowed to drink a novel sweetened milk solution and then injected with lithium chloride to make them ill. Four weeks later, different groups of rats received a left or right neocortical lesion, a sham operation, or no surgery at all. As in earlier studies, there were no asymmetric effects for nonhandled rats. However, the following results were obtained for handled rats. Rats with only an intact right hemisphere consumed the least amount of sweetened milk, indicating the greatest amount of conditioned fear. Rats with only an intact left hemisphere consumed more, and rats with both hemispheres intact even more. It is interesting that when both hemispheres were intact, rats showed less conditioned fear than when only the right hemisphere was intact; Denenberg and Yutzey suggest that in the completely intact brain the less fearful left hemisphere inhibits the more fearful right hemisphere.

Similar paradigms have been used to study asymmetry for mouse killing, a well-known behavior of laboratory rats. Again, there were no asymmetries for nonhandled rats. For handled rats, animals with only the right hemisphere intact showed a greater incidence of mouse killing than either animals with only the left hemisphere intact or animals with both hemispheres intact (with the latter two groups not differing). The equality of the latter two groups leads Denenberg and Yutzey (1985) to speculate that in the completely intact brain, the less emotional left hemisphere inhibits the more emotional right hemisphere. This speculation is supported by the fact that the incidence of mouse killing is much higher in split-brain rats (eliminating the opportunity for the left hemisphere to inhibit behavior controlled by the right hemisphere) than in rats with intact brains.

Chicks

Rogers and colleagues (e.g., summarized by Rogers, 1986) have used the unilateral drug treatment paradigm described earlier to examine asymmetry for attack behavior and copulation performance in chicks. To examine attack behavior, the fingers of the experimenter's hand are arched and moved rapidly at a chick's beak level, with the chick's responses scored in a standard way. For example, a chick might avert its gaze (scored as a zero) or engage in a variety of leaping, pecking, sparring, and so forth (up to a maximum score of 10). Copulation behavior was elicited by thrusting an outstretched hand (palm down) at the chick's chest level and holding it still. A maximum copulation score of 10 involved mounting the hand, pelvic thrusting, and so forth. An elevation in attack and copulation responses was found when only the left-hemisphere had been injected with a protein-synthesis inhibitor. No such elevation was found when only the right hemisphere or both hemispheres had been injected.

Rogers (1986) notes that the elevation of attack and copulation is found only when an imbalance is generated by drug injection into the left hemisphere and refers to this as an asymmetry that is dependent on "hemispheric coupling." On the basis of these results and results of additional experiments that involve monocular viewing, Rogers suggests that the right hemisphere may activate these behaviors while the left hemisphere suppresses them. Note the similarity to conclusions reached by Denenberg and Yutzey (1985) in their studies of conditioned taste aversion and mouse killing in rats.

As noted earlier, Rogers (1986) has considered how environmental factors can contribute to the development of asymmetry. For example, the typical asymmetries in attack behavior and copulation are reduced or eliminated if chicks are incubated in darkness from day 17 on. Rogers also suggests that the administration of testosterone can influence and may even reverse some of these asymmetries. Furthermore, the effects of testosterone appear to be different for males and females.

Additional Evidence of Biological Asymmetry

Chapter 4 reviewed evidence for a variety of biological asymmetries in the brains of humans. As noted throughout the present chapter, some of the functional asymmetries found in other species are known to be related to certain biological asymmetries. For example, the rotational bias of rats is related to asymmetric DA content and sensitivity in the nigrostriatal pathway. In addition to those biological asymmetries that have already been noted, there are a variety of others that have been reasonably well established, although their relation to function remains to be determined. For example, LeMay (1985) summarizes a variety of morphological asymmetries in the brains and skulls of nonhuman primates. She notes that, in general, the asymmetries are similar to but less frequent than some of those found in humans. For example, in both humans and apes the brain has a kind of counterclockwise torque. That is, the front portion of the right hemisphere tends to be wider and farther forward than the front portion of the left hemisphere, whereas the posterior portion of the left hemisphere tends to be wider and protrude farther toward the rear than the posterior portion of the right hemisphere. The brains of Old and New World monkeys show similar but less striking asymmetries. Of particular interest is the fact that the sylvian fissure is more frequently longer on the left than on the right side in both humans and chimpanzees, although not in rhesus monkeys (e.g., Yeni-Komshian and Benson, 1976).

There is also evidence in a variety of nonhuman species for asymmetries of brain size or cortical thickness. The finding most often reported is that the right side of the brain is larger than the left. This has been shown for brain weight or surface size in species of rats, mice, cats and rabbits (e.g., Diamond, 1985; Kolb, Sutherland, Nonneman, and Whishaw, 1982). In more detailed measurements of the brains of rats, Kolb et al. found that the neocortex was thicker on the right side than on the left side, but there were no asymmetries in the cerebellum, hippocampus, thalamus, or brain stem. Diamond has found that the asymmetries in rats are more marked in males than in females and more marked early in life rather than later in life. In fact, some of the asymmetries may be reversed in males and females and at least some of the asymmetries are influenced by sex steroid hormones (e.g., Diamond, 1985; Stewart and Kolb, 1988).

Recent studies in rats have also examined the influence of sex and early experience on the size and ultrastructure of the corpus callosum. Although direct comparisons with human data are impossible, certain effects with rats provide interesting parallels and contrasts to the results from humans that were summarized in Chapter 4. For example, Berrebi, Fitch, Ralphe, J. Denenberg, Friedrich, and V. Denenberg (1988) examined the size of the corpus callosum in handled and nonhandled rats at 110 and 215 days of age. Their most striking finding was that the corpus callosum was larger in males than in females, even when adjusted for the larger overall brain weight of males. This was true at both ages. Interestingly, at 110 days of age (but not at 215 days of age) there was a sex-by-handling interaction such that handled males had the largest callosa whereas handled females had the smallest callosa. These sex differences in the size of the corpus callosum of rats are reminiscent of the handedness-by-sex interactions reported in humans (see Chapter 4). However, it is impossible to compare any of the effects directly.

In a study of both size and ultrastructure, Juraska and Kopcik (1988) examined the corpora callosa of male and female nonhandled rats who were reared in either a complex (enriched) or isolated environment. With respect to callosum size, Juraska and Kopcik found no differences between males and females but they did find that the posterior third of the corpus callosum was larger for rats raised in a complex environment than for rats raised in an isolated environment. In addition to measuring callosum size, Juraska and Kopcik examined an area of the callosum known as the splenium with electron microscopy. This ultrastructural analysis revealed several differences between males and females. For example, regardless of the rearing environment, females had more unmyelinated axons than males, and rats of both sexes from the complex environment had larger and more unmyelinated axons than rats from the isolated environment. In addition, for rats reared in the complex environment, females had a greater number of unmyelinated axons than males but the unmyelinated axons were larger for males than for females. The functional importance of these sex differences remains to be determined, but their existence suggests the need for similar examinations of ultrastructure in humans.

Summary and Conclusions

The studies reviewed in this chapter illustrate the ubiquity of behavioral and brain asymmetries in nonhuman species (consistent with theme 3 outlined in Chapter 1). At least some of those asymmetries bear a striking resemblance to asymmetries seen in humans and provide interesting parallels to the other themes regarding human asymmetry outlined in Chapter 1. Of course, it should be kept in mind that much of the search for asymmetries in other species has been guided by what is already known about humans and there is the possibility that this biases the search in a way that maximizes the chance that only similar asymmetries will be discovered. Nevertheless, several of the parallels between asymmetries in humans and those in other species are too noteworthy to ignore.

Research has demonstrated a variety of motor-performance asymmetries in a number of species. For example, it has been hypothesized that there is a left-hand preference for reaching in several species of primates, accompanied by a right-hand preference for manipulation. However, this pattern of effects seems to depend on variables such as sex and age and whether the tasks that are used require low- versus high-level manual activity. Rotation biases have been demonstrated in a number of species, and the biases in rats are clearly related to asymmetries in DA innervation. In some ways, but not in others, these motor-performance asymmetries parallel human handedness. The fact that individual members of other species differ reliably from each other in the direction and magnitude of various motor-performance asymmetries is consistent with the view that individuals can differ reliably in patterns of brain asymmetry and that such individual differences can have behavioral consequences (theme 4 in Chapter 1).

There is also a noteworthy parallel between left-hemisphere dominance in humans for aspects of language and asymmetries in other species for the production and perception of vocalizations. For example, there is evidence that in Japanese macaques the left hemisphere is dominant for the discrimination of communicatively relevant species-specific vocalizations but not for the discrimination of vocalizations using a cue that is not communicatively relevant. In language-trained chimpanzees, an RVF/left-hemisphere advantage for processing visual warning stimuli is similarly restricted to communicatively relevant symbols. Some evidence of asymmetry for the production of vocalizations comes from the left-sided dominance for the control of song in some species of song birds.

Asymmetries have also been reported for a variety of visuospatial processes. For example, language-trained chimpanzees show an LVF/right-hemisphere advantage for processing the location of a short line within a geometric figure and for identifying meaningless forms of the sort found to produce a similar asymmetry in humans. Rhesus monkeys have been shown to have right-hemisphere dominance for recognizing monkey faces and left-hemisphere dominance for processing the orientation of lines and for the categorical location of a dot within a square. Furthermore, asymmetry for processing faces is statistically independent of asymmetry for processing line orientations, a pattern similar to the statistical independence of various asymmetries in humans. It has also been argued that the right hemisphere of handled rats is more involved than the left hemisphere in spatial exploration. The fact that variables such as handling influence asymmetry is consistent with the view that, across the life span of an individual, functional asymmetries are shaped by the complex interaction of both biological and environmental factors (theme 5 in Chapter 1). Evidence that this sort of interaction begins very early in development comes from the observation that the emergence of left-hemisphere dominance for visual-pattern discrimination in chicks depends on whether or not light strikes only the right eye during a particular period of incubation.

Research with rats and chicks indicates asymmetry for a variety of emotional behaviors. For example, in handled rats the right hemisphere seems to be more emotional, while the left hemisphere normally serves to inhibit emotional activity. It is interesting that a similar kind of right-hemisphere activation and left-hemisphere inhibition of emotional behaviors has been suggested for domestic chicks. These results illustrate some of the ways in which the two hemispheres interact with each other to provide a unified system for controlling behavior (theme 2 in Chapter 1).

With respect to other aspects of brain-behavior relationships (e.g., sensory processes, memory), great advances in understanding have come about through the study of animal models. In fact, the contribution of animal studies has been so great that it seems impossible that so much could have been learned without them. Until very recently, there was little or no emphasis on the use of animal models to study the mechanisms of hemispheric asymmetry. In part, this bias was based on the belief that hemispheric asymmetry depended on the development of language and was restricted to humans. In fact, it has occasionally been argued that many of the unique properties of the human mind (e.g., superior intellect, creative abilities, conscious thought, and so forth) have come about because of the unique hemispheric asymmetry that characterizes the human brain. Findings such as those reviewed in this chapter indicate that behavioral and brain asymmetries do exist in other species and that some of those asymmetries have excellent potential as animal models. Furthermore, at the very least the emergence of so many asymmetries in other species indicates it is possible for behavioral and cognitive asymmetries to develop independently of language.

6

Varieties of Interhemispheric Interaction

As we move around the world looking at objects, touching them, hearing sounds, and so forth, most of the information is taken in by both cerebral hemispheres. In addition, both hemispheres are usually able to generate some appropriate behavioral response. This being the case, it should come as no surprise that both hemispheres seem to be involved in one way or another in almost everything we do or think about doing. For example, measures of electrophysiological activity and of regional metabolism indicate that virtually all tasks activate many areas of both hemispheres (see Chapter 1). At the same time, we have seen that the two hemispheres have different information-processing abilities and propensities, with the differences sometimes being very striking. Such differences create an opportunity for conflicts in perception, cognition, emotion, and action—an opportunity that, unfortunately, is seldom fulfilled. The present chapter considers how it is that these two differing hemispheres interact with each other to form an integrated information-processing system.

It is instructive to consider the different kinds of mechanisms that could prevent conflict between the two cerebral hemispheres and increase the efficiency with which the whole brain operates —various mechanisms of interhemispheric interaction. In order to illustrate the range of possibilities, it is useful to think about some of the ways that two individuals interact with each other. There are times when the best way to minimize conflict and increase efficiency is to isolate the two individuals from each other. For example, if one student is studying for a final exam in American history and another is practicing for a tuba recital it is probably better if they are not even in the same room (or even in the same building). Working in isolation tends to be useful when the individuals are engaged in activities that are either completely separate or mutually inconsistent. When two individuals must cooperate to solve a common problem, however, efficiency can usually be increased if they collaborate in a way that allows each individual to take the lead for his or her areas of greatest expertise. For example, if an architect and a carpenter are working together to build a house, it would seem most efficient for the architect to take the lead in designing the space, drawing a set of plans, obtain the building permits, and so forth and for the carpenter to study the plans, order the materials needed to implement the plan, frame the house according to the plans, and so forth. Note that for this type of collaboration to work efficiently, there must be clear communication of *relevant* information back and forth between the two individuals. That is, they cannot be isolated completely from each other. At the same time, the communication of *irrelevant* information is likely to interfere with their performance and each may still need a certain degree of isolation from the other in order to perform certain tasks in a maximally efficient manner. Thus, the sharing of information and the need for some isolation from each other are not mutually exclusive. Another way in which individuals sometimes interact with each other is for one to dominate over the other. This often occurs when two individuals both have sufficient ability to perform a task, but they each want to go about it in different (sometimes contradictory) ways. While it would be a mistake to make too much of the analogy between interactions between two people and interactions between the cerebral hemispheres, the analogy does serve to illustrate the fact that there can be several types of interaction and to suggest the kinds of interaction that one might expect.

Despite the obvious importance of understanding the manner in which the two hemispheres interact to produce a unified behavioral response, it is only in the last few years that the topic of interhemispheric interaction has begun to receive the attention that it deserves. Several experimental paradigms have been developed to study interhemispheric interaction, each with somewhat different questions in mind. One conclusion that emerges from these studies is that there is probably no such thing as *the* mechanism of interhemispheric interaction. Instead, there are likely to be several varieties of interhemispheric interaction, with such things as the information-processing demands of a task determining how the hemispheres interact while performing that task. The present chapter reviews these varieties of interhemispheric interaction, discusses the task factors that seem to be important, and considers potential biological mechanisms of interhemispheric interaction.

The chapter begins with a discussion of cooperation between the hemispheres for tasks that require processing components or subsystems for which different hemispheres are dominant or for tasks that require the coordinated activity of both hemispheres. Included here is a discussion of the biological mechanisms responsible for interhemispheric cooperation, with a focus on the potential roles of the corpus callosum and of subcortical structures. This is followed by discussion of the benefits and costs associated with the integration of information across the two hemispheres. The third section of the chapter discusses a distinction that has been made between hemispheric ability and hemispheric dominance and considers the extent to which one hemisphere might assume control of processing as an additional way of resolving potential conflicts.

Cooperation between the Hemispheres

The Need for Cooperation

For many tasks, the two hemispheres are dominant for different task-relevant processing components or subsystems (see Chapters 2 and 3). What happens in the intact brain when such a task is performed under conditions that provide both hemispheres with access to the relevant stimuli? A likely answer to this question is that the two hemispheres coordinate their activities, so that each takes the lead for those components of processing that it handles best (much as the architect and the carpenter in our earlier analogy). For example, suppose you are listening to a radio drama about a mild-mannered college professor who has been thrust unwittingly into the center of a diabolical plot to kidnap the pope. As discussed in earlier chapters, the left hemisphere is usually superior to the right for processing phonetic, syntactic, and certain semantic aspects of language, whereas the right hemisphere is usually superior to the left for processing intonational and pragmatic aspects of language. All of these processes (and a host of others) are relevant for your understanding and appreciation of the radio drama. There is good reason to suppose that in the intact brain each of these aspects of language and communication is handled primarily by the hemisphere that is dominant for it.

It would be possible for the two hemispheres to process the same stimulus information in parallel with little or no interhemispheric interaction. In order for the hemispheres to coordinate their activities, however, it is necessary for them to share various types of information with each other. For example, both phonetic information and intonation are relevant for reaching a correct conclusion about what a character in a radio drama is likely to do next. Therefore, there must be some mechanism that allows these two types of information to be integrated. To the extent that different hemispheres take the lead in extracting one of these two types of information, the integration of the two types of information requires some transfer of information between the hemispheres. In the absence of evidence to the contrary, it is often assumed that in the intact brain virtually all information that is available to one hemisphere as a result of its own processing becomes available very quickly to the other hemisphere, although there may be some degradation of information as it is transferred from one hemisphere to the other. Certainly, this notion has a great deal of intuitive appeal and would seem necessary to allow for efficient processing in a brain composed of two functionally asymmetric cerebral hemispheres. At the same time, it may still be useful to insulate certain hemisphere-specific processes from each other in order for them to proceed efficiently in parallel. For example, it is possible that the on-line extraction of phonetic information may be carried out most efficiently if it is insulated from the concurrent on-line processing of prosody and intonation.

The Corpus Callosum

The largest fiber tract that connects the two cerebral hemispheres is the corpus callosum, with at least 200 million nerve fibers.

Consequently, it should come as no surprise that the corpus callosum plays important roles in coordinating the activity of the two hemispheres, although there is room for disagreement about exactly what those roles are. Before considering the roles that have been suggested, let us review briefly what is known about the topographical structure of the corpus callosum.

As noted in Chapter 4, different regions of the corpus callosum contain fibers originating in different cortical areas. In fact, studies in both monkeys and humans indicate that many callosal fibers exhibit a homotopic arrangement with respect to the cerebral cortex; that is, many of the fibers connect homologous regions of the two hemispheres (see Cook, 1984, 1986). Although there are no gross anatomical landmarks that divide the corpus callosum into discrete regions, it is generally the case that the anterior portions of the corpus callosum contain fibers that connect premotor and frontal regions of the two hemispheres, the middle portions of the corpus callosum contain fibers that connect motor and somatosensory regions of the two hemispheres, and the posterior portions of the corpus callosum contain fibers that connect temporal, post-parietal, and peri-striate regions of the two hemispheres (see Chapter 4). While many of the callosal fibers connect homologous regions of the two hemispheres, there are also many fibers that originate in a specific region of one hemisphere and terminate in a completely different region of the opposite hemisphere. The presence of such heterotopic fibers allows for the possibility that neural activity in one hemisphere has generalized effects on the neural activity in the opposite hemisphere.

Given its topographic structure, it is not surprising that the corpus callosum plays an important role in the transfer of information from one hemisphere to the other. This has been demonstrated in a variety of studies comparing intact and split-brain monkeys (for review see Dimond, 1972) and in the work with split-brain humans (see Chapter 1). For example, recall that the left hemisphere of split-brain humans cannot identify an object flashed visually to the LVF/right-hemisphere or touched only by the left hand. In fact, these failures of interhemispheric cooperation in split-brain patients indicate the limitations of information transfer from one hemisphere to the other via various subcortical fiber tracts. It is also the case that patients with injury to only a portion of the corpus callosum may show selective disruption of the ability to transfer certain types of information but not others and that difficulty in transferring a specific type of information depends on exactly what area of the corpus callosum has been damaged. For example, Bentin, Sahar, and Moscovitch (1984) found poor transfer of haptic information (i.e., information obtained from touching nonsymbolic patterns with the fingers) in three patients with injury to a specific region in the anterior part of the trunk of the corpus callosum, although the same patients could name familiar objects and letters after exploring them with either hand. That is, the callosal lesions of these particular patients selectively affected the transfer of "pure touch" information about complex, meaningless patterns. By way of contrast, no such deficit was found for patients whose callosal injury was either more anterior or more posterior.

The idea that one important function of the corpus callosum is to permit information to be transferred from one hemisphere to the other is undisputed, but there is not as much agreement on the way in which this takes place. One issue has to do with the extent to which interhemispheric connections across the corpus callosum are excitatory or inhibitory. In discussing this issue, we must keep in mind the difference between excitation versus inhibition at a neural level and excitation versus inhibition at a functional level. Unfortunately, this distinction is not always made clear in discussions of whether the corpus callosum is primarily excitatory or inhibitory.

At a neural level, excitation occurs when an increase in the firing rate of one neuron (e.g., a callosal fiber) causes an *increase* in the firing rate of the neurons onto which it synapses (e.g., fibers in the receiving hemisphere), whereas inhibition occurs when an increase in the firing rate of one neuron causes a *decrease* in the firing rate of the neurons onto which it synapses. The terms *excitation* and *inhibition* are more difficult to define at a functional level, but with respect to the corpus callosum the terms sometimes refer to whether processing that involves specific regions of one hemisphere tends to activate or suppress processing in similar regions of the other hemisphere. A somewhat different use of the terms has been to refer to the corpus callosum as an "inhibitory barrier" that blocks the flow of at least some types of information from one hemisphere to the other, thereby allowing each hemisphere to work efficiently in some degree of peaceful isolation from its partner. Although there may be some connection between excitation versus inhibition at the neural level and these various notions of excitation versus inhibition at functional levels, the connection is not mandatory. That is, the activation or suppression of something as abstract as "processing" at a functional level could be accomplished by neurons with either excitatory or inhibitory properties. Consequently, we must be cautious about interpreting evidence for excitation versus inhibition at one level in terms of excitation versus inhibition at the other level.

An extremely interesting hypothesis has been advanced by Cook (1984, 1986), who sees evidence for what he calls homotopic callosal inhibition. While it is not always clear whether he is talking about neural or functional levels, it is clear that his hypothesis is ultimately about function whereas the strongest empirical data come from the neural level. The critical idea is that the corpus callosum connects homotopic areas of the two hemispheres in an inhibitory fashion, perhaps involving both callosal neurons and interneurons. At the finest-grain level, activation of a single column of cortical neurons in one hemisphere leads to inhibition of the homotopic column of cortical neurons in the opposite hemisphere. He supplements this idea with the hypothesis that adjacent areas within a hemisphere are connected in a mutually inhibitory manner, in something of a center-surround fashion. As a result, when a small cortical area in one hemisphere is inhibited (e.g., by callosal fibers activated by the homologous area in the opposite hemisphere), the immediately surrounding area becomes more active. Thus, the corpus callosum tends to produce mirror-image patterns of activation and inhibition in the two hemispheres. Using these principles, Cook attempts to explain hemispheric asymmetry by suggesting that homotopic callosal inhibition not only leads to transfer of information from one hemisphere to the other but also leads the two hemispheres to become dominant for complementary functions. The evidence for various aspects of Cook's hypothesis is uneven. The existence of homotopic connections is well established and there is good evidence that many of those connections are inhibitory, at least at the neural level. However, it is not so clear that the notion of hemispheric dominance for complementary functions is necessarily predicted by inhibition at either the neural or functional level. For example, two hemispheres with completely mirror-image patterns of activation and inhibition could just as easily become completely redundant with respect to the information that could be encoded.

The corpus callosum has been proposed by Kinsbourne (1975) to have a different sort of inhibitory role. On this view, the left and right hemispheres are always in a state of mutual inhibitory balance with each other. That is, activation of regions in one hemisphere tends to inhibit the general level of activity throughout the opposite hemisphere. Together with various subcortical commissures, the corpus callosum serves to regulate the state of asymmetric activation or arousal. Note that this is different from the very specific form of homotopic inhibition proposed by Cook (1984, 1986). In Chapter 1, I noted how this view could explain such things as left/right perceptual asymmetries in the intact brain. Briefly, when a task (e.g., CVC recognition) activates one hemisphere (e.g., the left hemisphere) more than the other, attention is directed more easily to the side of space contralateral to the more activated hemisphere (e.g, the RVF), producing a performance asymmetry. The notion of two hemispheres that are in a mutually inhibitory relationship to each other is also consistent with the hypothesis that in rats and chicks the less emotional left hemisphere normally inhibits the more emotional right hemisphere (see Chapter 5).

Related to the idea that an important functional role of the corpus callosum is inhibitory is the hypothesis that the corpus callosum serves as an "inhibitory barrier" between the hemispheres, preventing maladaptive cross-talk between the processes for which each hemisphere is dominant. This notion of an inhibitory barrier is an important component of the "functional cerebral distance" principle advanced by Kinsbourne and his colleagues (e.g., Kinsbourne, 1982; Kinsbourne and Hiscock, 1983) to account for hemisphere-specific priming and interference (see Chapter 3).

At the present time, it is impossible to reduce the corpus callosum to any one of the single biological or functional roles considered here. Instead, the corpus callosum is likely to be involved in a number of aspects of normal interhemispheric cooperation.

Whether it is by excitation or inhibition, the corpus callosum plays an important role in the transfer of at least certain types of *information* from one hemisphere to the other. In addition, it may serve to reduce maladaptive cross-talk between mutually inconsistent *processes* by functioning as something of a barrier between the two hemispheres. It remains to be determined how best to reconcile these differing viewpoints about the roles of the corpus callosum. One possibility is that, with respect to interhemispheric transfer, there is an important distinction between the information that results from hemisphere-specific processes and the processes themselves. To return to an analogy used earlier, for a carpenter to frame a house according to an architect's plans, the carpenter must have a copy of those plans (the result of the architect's processing activity). However, the carpenter need not have any knowledge of the step-by-step procedures used by the architect to produce those plans. In fact, if the architect were required to communicate each processing step to the carpenter along the way, it might well increase the time taken to produce the final plans. So, too, it may be that the hemispheres work together most efficiently by sharing the results of hemisphere-specific processes while at the same time insulating those processes from each other. The corpus callosum is undoubtedly an important component in the biological substrate for this sort of interhemispheric interaction.

Subcortical Structures

The examination of split-brain patients, whose hemispheres can no longer communicate through the corpus callosum, provides an opportunity to study the contribution of other structures to interhemispheric communication. In some patients, the anterior commissures are also severed so that interaction is restricted to subcortical structures. On the basis of investigation of such patients, it has been argued that connotative and contextual information can be transmitted through subcortical structures but information about the name or identity of a stimulus cannot (e.g., Myers and Sperry, 1985). Although there is merit in this position, more recent studies suggest far more subcortical transmission than previously acknowledged.

As noted by Sergent (1983), "one startling fact about commissurotomized [split-brain] patients is that, despite having two independent and different cognitive processors, they behave as unified individuals and seldom display signs of hesitation, confusion or dissociation in their day-to-day activities" (p. 800). In a series of clever experiments using several such patients, Sergent (1983, 1986, 1990, in press) provides evidence that this is because the intact subcortical structures play an important role in both the transmission of information between hemispheres and in coordinating the activities of the two hemispheres. The following examples illustrate the range of information that can and cannot be transmitted between the hemispheres of split-brain patients.

All of the patients studied by Sergent exhibit the typical "disconnection syndrome," which has come to be one of the hallmarks of successfully severing the corpus callosum. That is, these patients cannot indicate whether two stimuli presented one to each visual field (hemisphere) are the same or different and they cannot speak the name of an object shown only to the LVF/right hemisphere. Despite this, there is much that they can do that involves integrating information across the two hemispheres. For example, Sergent (1983) presented two letters simultaneously (one to each visual field/hemisphere) to split-brain patient J. W. When the patient was asked to indicate whether the two letters were identical, performance was not significantly different from chance—a result that is consistent with the typical disconnection syndrome. When the patient was asked to indicate whether either of the two letters was a vowel, however, he displayed no overt signs of hesitation and his performance was virtually perfect. This was true regardless of which hand was responding and even when one letter was a vowel and the other was not (so that each hemisphere would come to a different decision). Thus, each hemisphere was able to receive information about the decision made by the other hemisphere and the two decisions could be reconciled to produce a single response, even though there did not appear to be transfer of letter identity from one hemisphere to the other (see also Lambert, 1991).

In subsequent experiments with other split-brain patients, Sergent (1986) found that patients could indicate whether either of two colored stimuli (one presented to each hemisphere) was a predetermined color (e.g., red), even when the two stimuli did not match (e.g., one red and the other green). Once again, however, the patients were unable to indicate whether the two stimuli on a trial were the same color. In a similar vein, Sergent (1990) reported that split-brain patients presented with two digits (one to each hemisphere) could not indicate whether the two digits were the same or different. However, they could perform other tasks that required the integration of information about the two digits. For example, they performed above chance when asked to judge whether each digit was odd or even and indicate whether or not both digits came from the same category. Patients were also highly accurate in indicating which of the two digits was higher in numeric value. Additional experiments required the same patients to report verbally (i.e., from the left hemisphere) various types of information about faces projected to only the LVF/right hemisphere. Patients were unable to verbally describe visual characteristics of the faces and they could not identify familiar faces, but they were able to provide certain types of semantic information about the face that was presented (indicating the emotion that was displayed on the face, indicating the occupation of the person pictured, etc.).

From these studies Sergent argues that the disconnection syndrome does not extend to all forms of knowledge that can be derived about a given object. Whereas information about stimulus identity is not transferred in a way that allows the split-brain patient to indicate whether two stimuli are identical when each is presented to a different hemisphere, a variety of other information about the stimuli is transferred. Among these other types of information are the categories to which a stimulus belongs (odd/even, happy/sad , U.S. President, etc.) and certain aspects of information about location in space (e.g., see later discussion of Sergent, in press). Sergent also argues that subcortical structures play an essential role in coordinating the activity of the two hemispheres, permitting a single unified response to be produced by the dual-brain system.

Of course, when the corpus callosum is intact it may mediate some of the functions attributed to subcortical structures in split-brain patients. Nevertheless, it is now clear that any consideration of the biological mechanisms of interhemispheric communication must include the roles taken by subcortical structures as well as the roles taken by the corpus callosum.

Benefits and Costs of Interhemispheric Cooperation

Although the two hemispheres are capable of sharing many types of information in the intact brain, ranging from very early perceptual input to complex decisions, this does not mean that cooperation at all levels necessarily takes place all of the time. When a stimulus is presented directly to only one hemisphere, it is sometimes the case that the hemisphere that receives the information directly carries out all of the processing (see Chapter 1 for discussion of this "direct-access" model of perceptual asymmetry). For example, from a series of lexical-decision experiments that presented letter strings to one visual field or the other, Hardyck (1991) argued that the only thing the hemispheres shared at a functional level was the final decision about whether or not the stimulus was a word or a nonword. That is, despite the fact that the corpus callosum was intact, when a letter string was presented to one visual field, the hemisphere that received the input directly appeared to do all of the processing—including making the critical word-nonword decision (for additional evidence that lexical-decision tasks are performed in this direct-access fashion, see Zaidel, 1983).

It is necessary to understand the factors that determine when it is more efficient in the intact brain for the hemispheres to operate independently and when it is more efficient for the hemispheres to operate collaboratively. That is, it is important to know to what extent the likely benefits of collaboration are offset by the possible costs of transferring information and otherwise coordinating the activities of the two hemispheres. With this in mind, the present section reviews studies that examine the benefits and costs of distributing information-processing across both hemispheres.

In Chapter 3 we have seen that two unrelated concurrent tasks often interfere with each other more when they both demand resources from the same hemisphere than when they can be processed by different hemispheres. This raises the possibility that even a single task might be performed more efficiently when the various information processing components can be distributed across the two hemispheres. On the other hand, there may well be some cost associated with the need to transfer information across the hemispheres or the need to coordinate the activities of the two hemispheres, and this cost may be sufficiently detrimental to offset the benefits of cooperation. In fact, studies have shown this to be the case and have suggested that whether distributing processing across both hemispheres is beneficial or detrimental depends on a variety of factors—especially on how demanding a task is of processing resources. Specifically, as the processing requirements of a task become more demanding, performance is enhanced by distributing processing across both hemispheres.

Task Difficulty

In an investigation of the benefits of interhemispheric cooperation, Green (1984) reported the results of several choice reaction-time experiments that varied the visual field to which stimuli were presented and also the hand used to make the response. For example, in one experiment subjects indicated whether or not two uppercase letters flashed to the LVF/right hemisphere or RVF/left hemisphere were physically identical. Subjects made their response by pressing the index finger of one hand if the letters matched and the middle finger of the same hand if the letters did not match. Reaction times were faster when stimuli were presented to the visual field contralateral to the side of the response hand (see the left panels of Figure 6.1). Similar interactions between visual field and response hand were found in choice reaction-time tasks involving the comparison of letter names and the comparison of schematic drawings of faces.

Recall that each visual field projects directly to the contralateral hemisphere and that the voluntary finger movements of each hand are programmed by the contralateral hemisphere. Thus, reaction times were slower when stimuli had to be processed initially by the same hemisphere that had to program the response than when at least some components of stimulus and response processing could be distributed across the two hemispheres. However, the advantage of this distribution of processing across both hemispheres seems to be related to the overall demand for processing resources because no such advantage was found for any of the tasks when a simpler go/no go response was substituted for the choice response (see the right panels of Figure 6.1). A similar conclusion about the importance of task difficulty comes from studies that examine the consequences for information processing when the two hemispheres are required to share information about relevant stimuli in order to perform a task.

Consider a task that requires observers to indicate whether two visually presented letters are physically identical. If both letters

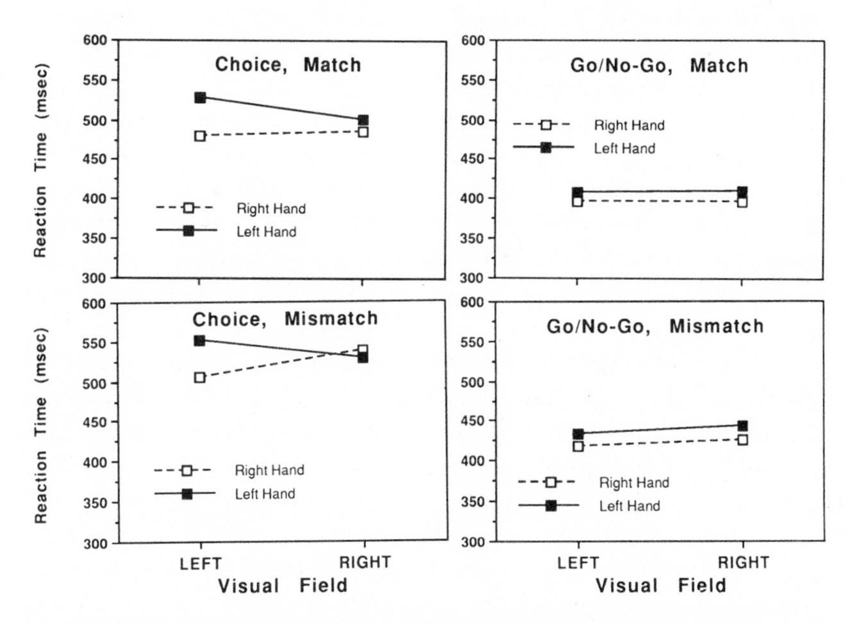

Figure 6.1. Reaction time needed to indicate whether two uppercase letters were physically identical when the letter pairs were flashed to the left or right visual field and when the left or right hand was used to respond. The two panels on the left side of the figure come from an experiment (the Choice experiment) in which observers made one of two choice responses on each trial. The two panels on the right side of the figure come from an experiment (the Go/No-Go experiment) in which the observer responded to only one type of letter pair during the experiment. The upper panels show the results from trials where the two letters matched (the Match trials) and the lower panels show the results from trials where the two letters did not match (the Mismatch trials). [Adapted from J. Green, "Effects of Intrahemispheric Interference on Reaction Times to Lateral Stimuli," *Journal of Experimental Psychology: Human Perception and Performance,* 10 (1984):292–306. Copyright 1984 by the American Psychological Association. Adapted by permission.]

are presented to the same visual field (the *within-hemisphere* condition), then the task does not *demand* any interhemispheric sharing of information or collaboration. However, if one letter is presented to each visual field (the *between-hemisphere* condition), then some collaboration between the hemispheres is absolutely *required* in order to perform the task. As a result, the comparison of within-hemisphere and between-hemisphere conditions provides information about what has been referred to as the cross-hemispheric integration of information.

One indication that processing demands are relevant in such experiments comes from the fact that a between-hemisphere advantage has been reported for a variety of stimuli when the two stimuli to be compared are presented simultaneously but a within-hemisphere advantage has been reported when the two stimuli are presented successively (for review see Banich and Belger, 1990; Hellige, 1987; Sereno and Kosslyn, 1991). For example, Hellige (1987) describes an experiment in which two letters were sometimes presented to the same visual field (within-hemisphere trials) and sometimes to opposite visual fields (between-hemisphere trials) and the observer was required to indicate as quickly as possible whether or not the letters were same or different. There was a between-hemisphere advantage when the letters were presented simultaneously or within 100 msec of each other, and this between-hemisphere advantage was eliminated or reversed to a within-hemisphere advantage when the letters were separated by an interval of 500 msec. When the two letters are presented with a sufficiently long interval between them (e.g., 500 msec), perceptual processing of the first letter is likely to be completed before presentation of the second letter (Hellige, 1987). In contrast, when the two letters are presented simultaneously (or nearly so) the perceptual processing of the two letters must overlap in time—creating more demand for processing resources. In this sense, the fact that a between-hemisphere advantage occurs with simultaneous but not successive presentation of the two stimuli is consistent with the conclusion that performance is enhanced by distributing processing across both hemispheres to the extent that the processing requirements of a task become sufficiently demanding.

Even clearer demonstrations of the importance of processing

demands come from experiments reported by Banich and Belger (1990). Their studies used displays similar to those shown in Figure 6.2. For the task illustrated in Figure 6.2, observers were presented on each trial with a single digit at the fixation point and with three uppercase letters. The two top letters were always in opposite visual fields and different from each other. The bottom letter was always in one visual field or the other and it could match either or neither of the top letters. The observers were required on each trial to press a button as quickly as possible if the bottom letter matched either of the top two letters and to refrain from pressing if the bottom letter did not match either of the upper

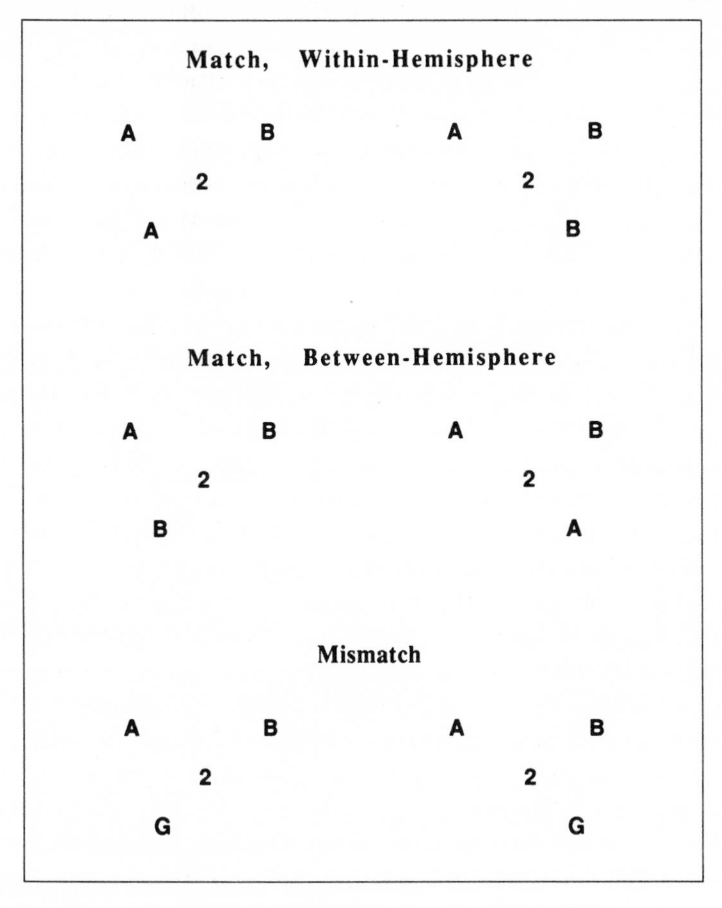

Figure 6.2. Stimulus displays similar to those that have been used to study costs and benefits of interhemispheric cooperation.

letters. After making this response, observers identified the fixation digit (this task was included to encourage appropriate eye fixation). Note that for half of the matching trials, the two letters that matched were both presented to the same visual field/hemisphere (e.g., the within-hemisphere examples illustrated in Figure 6.2). For the remaining match trials, the two letters that matched were presented to opposite visual fields/hemispheres (e.g., the between-hemisphere examples illustrated in Figure 6.2).

For the physical-identity letter-matching task just described, reaction times on matching trials were faster for within-hemisphere trials than for between-hemisphere trials. For another task, the two top letters were always uppercase but the bottom letter was always lowercase and observers were required to indicate whether the bottom letter matched either of the top letters in name. Not surprisingly, it took longer to make name-identity matches than to make physical-identity matches—suggesting that the processing demands are greater for the name-identity task than for the physical-identity task. In addition, the pattern of within- versus between-hemisphere effects was exactly opposite that found for the physical-identity task. That is, for the name-identity task, reaction times on matching trials were faster for between-hemisphere trials than for within-hemisphere trials (for replication and extension of this effect, see Eviatar, Hellige, and Zaidel, 1992). In subsequent experiments, Banich and Belger found that a task that required physical matching of single digits produced a within-hemisphere advantage similar to that found with physical matching of letters. In contrast, more demanding tasks with digits (requiring observers to indicate whether the sum of the bottom digit and either of the top digits was greater than or equal to 10 or indicating whether the value of the bottom digit was less than either of the top digits) produced between-hemisphere advantages similar to that found with matching letters on the basis of name. Taken together, these experiments indicate that when task requirements are demanding, performance can be enhanced by distributing processing across the two hemispheres. When task requirements are sufficiently light, however, the costs associated with this type of cross-hemispheric integration may outweigh the benefits so much that performance is actually harmed by the requirement to distribute processing across the two hemispheres.

Practice

There may be additional factors that determine whether the requirement to distribute processing across both hemispheres is beneficial or harmful. One such factor is level of practice with the experimental task. For example, Liederman, Merola, and Martinez (1985) either presented word pairs to a single visual field (within-hemisphere condition) or projected one word to each visual field (between-hemisphere condition). Observers responded as quickly as possible whenever the words came from the same semantic category. At the beginning of each of their experiments, the reaction time for the between-hemisphere condition was as fast as the better of the two within-hemisphere conditions (i.e., the RVF/left-hemisphere condition for this linguistic task). In this sense, there was no cost associated with dividing the input between the two hemispheres. However, performance improved with practice for the two within-hemisphere conditions but not for the between-hemisphere condition. As a consequence, by the end of the experiment the reaction time for the between-hemisphere condition was slower than for the better of the two within-hemisphere conditions. That is, by the end of the experiment there was a significant cost associated with dividing the input between the two hemispheres. The interpretation of this sort of practice effect is not completely clear. However, as a task is practiced more and more it is likely to become less demanding of processing resources. In view of the conclusions reached earlier about the importance of resource demands for determining the cost associated with cross-hemispheric integration, it is interesting that in the experiments reported by Liederman et al. such costs became apparent only as the tasks became less demanding because of increased practice. This is at least consistent with an interpretation in terms of the demand for processing resources.

Mutually Exclusive Processes

Another factor that may determine whether the requirement to distribute processing across both hemispheres is beneficial or harmful is the extent to which each hemisphere is required to engage in more than one kind of processing. For example, Liederman, Merola, and Hoffman (1986; see also Liederman, 1986) had observers attempt to identify four simultaneously presented letters. In their displays, two letters were upright and two were inverted. When the two upright letters were presented to one hemisphere and the two inverted letters were presented to the other hemisphere, performance was superior to the conditions in which all four letters were presented to one hemisphere. That is, there was a between-hemisphere advantage. However, if one upright and one inverted letter were presented to each hemisphere, performance was approximately the same as in the within-hemisphere conditions. On the basis of this pattern of results, Liederman and her colleagues suggest that dividing stimulus input between the two hemispheres is advantageous if the hemispheres are engaged in mutually exclusive perceptual processes (e.g., the processing of upright versus inverted letters). When this is not the case (e.g., when each hemisphere must process both an upright letter and an inverted letter), then no between-hemisphere advantage occurs. This is an interesting hypothesis that merits additional investigation.

Hemispheric Ability, Hemispheric Dominance, and Metacontrol

Ordinarily, objects in the environment are experienced in a way that does not bias processing in favor of one hemisphere or the other. For example, when inspecting an object visually, we usually turn our eyes and head so that an image of the object is presented at the center of our visual field. The preceding sections of this chapter indicated how the two hemispheres might collaborate with each other: perhaps each hemisphere may take the lead for different components of information processing. A somewhat different way in which the hemispheres might interact in such situations is for one or the other hemisphere to assume control of processing more generally, rather than for just some subset of processing components. Although this seems relatively inefficient at first glance, efficiency can be very difficult to determine when there is some cost associated with transferring information from one hemisphere to the other. In fact, the research findings reviewed in this section are consistent with this possibility as an additional variety of interhemispheric interaction.

Studies of Split-Brain Patients

On the basis of their studies with four split-brain patients, Levy and Trevarthen (1976) distinguished between *hemispheric ability* and *hemispheric dominance* and also introduced a new concept referred to as *metacontrol. Hemispheric ability* refers to how well each hemisphere can perform a particular task and, as indicated by previous chapters in this book, differences in hemispheric ability have been the subject of a great deal of research. *Hemispheric dominance* refers to the degree to which each hemisphere tends to assume control of information processing and behavior when given a chance to do so. The term *metacontrol* is used to refer to the neural mechanisms that determine the extent to which each hemisphere attempts to assume control of processing. If one hemisphere were to assume control of a particular task, it would make sense that it would be the hemisphere with the greater ability to perform the task. Levy and Trevarthen found that this was frequently the case in the split-brain patients, but not always. That is, ability differences between the hemispheres were not the only determinants of hemispheric dominance as they defined it.

An illustrative example of the dissociation between hemispheric ability and hemispheric dominance in split-brain patients comes from a study reported by Levy, Trevarthen, and Sperry (1972). One of their tasks presented a pattern of three vertically oriented *X*s and squares, with each hemisphere being shown a different permutation of symbols. For example, the left hemisphere might see (from top to bottom) "*X, X*, square" while the right hemisphere sees "square, square, *X*" at the same time. On each trial, the patient chose a permutation from a set of eight possibilities. In a free-response situation (where patients could point to the response with either hand), the right hemisphere proved to be strongly dominant for this task, in the sense that metacontrol mechanisms favored the right hemisphere and whatever mode of processing it was using. However, when the left hemisphere was forced to assume control of responding by requiring the patients to give a verbal description of the stimulus sequence, the level of performance achieved by the left hemisphere was superior to that achieved by the right hemisphere under the free-response condition. As noted by Levy and Trevarthen (1976, p. 300):

> Thus, there was a *negative* correlation on this test between dominance, on the one hand, and ability, on the other. This result strongly suggests that a capacity difference between the two sides of the brain is not the sole determinant of hemispheric dominance; that, in fact, a hemisphere assumes control of processing as a result of set or expectation as to the nature of processing requirements *prior* to actual information processing, and that it remains in control even if its performance, for whatever reason, is considerably worse than that which could have been produced by the opposite side of the brain.

The biological mechanisms responsible for this type of metacontrol remain to be determined. However, the fact that it can occur in the brains of callosotomy patients indicates that it need not be mediated by the corpus callosum. Instead, in these patients the mechanisms would seem to reside in the hemispheres themselves and in the coordination that can occur subcortically.

Studies of Neurologically Intact Individuals

It is important to consider the possibility that mechanisms of metacontrol also contribute to interhemispheric interaction in the intact brain. When the brain is intact, it is very unlikely that either hemisphere is ever completely uninvolved in ongoing processing. Consequently, any effects of metacontrol mechanisms are likely to be more subtle than those that can be seen in split-brain patients. In this section, I review the results of several studies that illustrate some of those subtle effects. These studies take advantage of the fact that, for a variety of tasks, both hemispheres have reasonable competence, but they go about processing in qualitatively different ways (see Chapters 2 and 3). The existence of such tasks raises interesting questions about what happens when the same stimulus information is available to both hemispheres in a manner that does not bias processing in favor of one hemisphere or the other. The possibility raised by the studies of Levy et al. (1972) and Levy and Trevarthen (1976) is that one hemisphere will tend to dominate processing and that the dominant hemisphere may not always be the one that is associated with better performance.

The redundant BILATERAL paradigm. A major problem in extending the investigation of metacontrol to the intact brain is finding a way to determine whether one hemisphere in some sense dominates processing when both hemispheres have equivalent access to stimulus input and response output. Recent studies with visual stimuli have attempted to do this in the following way (see Hellige, 1987, 1991). A task is chosen for which both hemispheres have some competence but for which the processing differs qualitatively as a function of which visual field/hemisphere receives the stimulus input. On different trials, stimulus information is presented to the LVF/right hemisphere or the RVF/left hemisphere or exactly the same stimulus information is presented to both visual fields (and hemispheres) simultaneously (redundant BILATERAL trials). In this way, it is possible to compare the qualitative nature of processing when both hemispheres have an equal chance to control processing and the qualitative nature of processing when each of the single hemispheres receives the stimulus input. To the extent that the mode of processing on redundant BILATERAL trials matches the mode of processing on one of the unilateral trials (but not the other), there is some evidence that metacontrol mechanisms also operate in the intact brain. Note that this sort of result does not mean that all processing necessarily takes place in one hemisphere. Instead, such a result suggests that one hemisphere dominates the selection of what might be termed an information-processing strategy.

One operational definition of a *qualitative* difference in the mode of processing on LVF/right-hemisphere and RVF/left-hemisphere trials is that there is an interaction between visual field (hemisphere) and some manipulated task variable (e.g., Hellige, 1983, 1987)—what Zaidel (1983) refers to as a processing dissociation. Given this operational definition and the logic outlined above, the question is whether the effects of the manipulated task variable are the same on BILATERAL trials as on one of the unilateral trials. Several recent experiments using this logic suggest that the concept of metacontrol extends to interhemispheric interaction in the intact brain, at least insofar as the mode of processing favored by one hemisphere is sometimes chosen over the mode of processing favored by the other hemisphere.

Comparison of cartoon faces. Hellige, Jonsson, and Michimata (1988) had subjects indicate whether two successively presented cartoon faces were identical or differed in one of four variable features (hair, eyes, mouth, or jaw). On each trial, the first face was presented at the observer's fixation point and the second face was presented to the LVF/right-hemisphere or to the RVF/left-hemisphere, or the same probe face was presented simultaneously to both LVF and RVF locations (BILATERAL presentation). In previous experiments, Sergent (1982, 1984) had reported that when the two faces on a trial differed, reaction time to detect the difference depended on which single feature differed, with the pattern of feature location effects depending on whether stimuli were presented to the LVF/right hemisphere or the RVF/left hemisphere. That is, there was a processing dissociation as defined above, suggesting that the two hemispheres processed the cartoon faces in qualitatively different ways (see Sergent, 1982, 1984, for interpretations of this qualitative difference). We also found different patterns of feature location effects on LVF/right-hemisphere and RVF/left-hemisphere trials. For example, reaction time was faster if only the eyes of the two faces differed than if only the mouth of the two faces differed, with this effect being larger on RVF/left-hemisphere trials than on LVF/right-hemisphere trials. In fact, when only the eyes differed, reaction time was faster on RVF/left-hemisphere trials than on LVF/right-hemisphere trials, but when only the mouth differed reaction time was faster on LVF/right-hemisphere trials than on RVF/left-hemisphere trials. Of particular importance was the finding that the pattern of feature-location effects obtained on BILATERAL trials was identical to the pattern obtained on RVF/left-hemisphere trials. For example, on BILATERAL trials the reaction-time difference between the eyes-different and mouth-different conditions was equal to that obtained on RVF/left-hemisphere trials. These results are consistent with the possibility that metacontrol mechanisms bias processing in favor of the left hemisphere under conditions that allow both hemispheres an equal opportunity to influence processing.

Same/different letter matching. Hellige and Michimata (1989b) reported similar results for an experiment that required observers to indicate whether two simultaneously presented uppercase letters were physically identical. The letter-pair on each trial was again presented to the LVF/right-hemisphere or the RVF/left-hemisphere, or the same letter pair was presented to both LVF and RVF locations (BILATERAL trials). Consistent with the results of certain earlier experiments with lateralized stimuli, there was an interaction between hemisphere (LVF/right hemisphere versus RVF/left hemisphere) and letter similarity (same versus different). As shown in Figure 6.3, on RVF/left-hemisphere trials reaction time was faster when the two letters were the same than when they were different. This effect was absent on LVF/right-hemisphere trials. This processing dissociation suggests qualitatively different modes of processing by the two hemispheres, as discussed by Bagnara, Boles, Simion, and Umilta (1983). As shown in Figure 6.3, the pattern of results obtained on BILATERAL trials was once again similar to the pattern obtained on RVF/left-

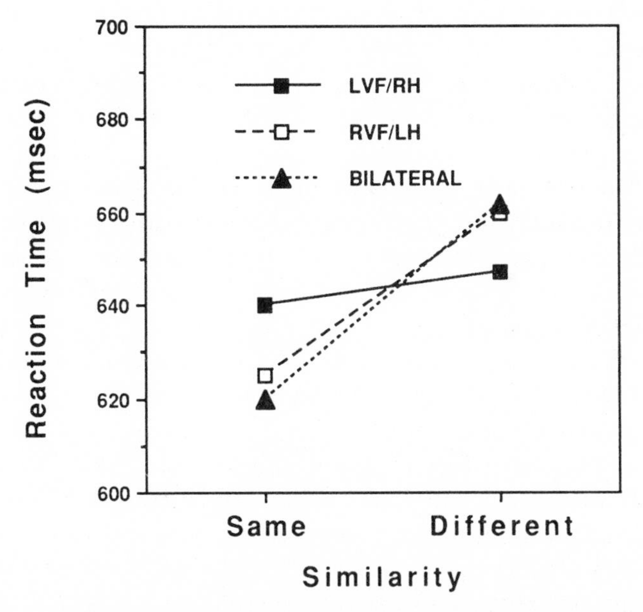

Figure 6.3. Reaction time needed to indicate whether two uppercase letters were the same or different. The results are shown for each of three visual-field conditions: LVF/right hemisphere (LVF/RH), RVF/left hemisphere (RVF/LH), and BILATERAL (the same letter pair was presented to both visual fields at the same time). [Adapted from J. B. Hellige and C. Michimata, "Visual Laterality for Letter Comparison: Effects of Stimulus Factors, Response Factors and Metacontrol," *Bulletin of the Psychonomic Society,* 27 (1989):441–444. Copyright 1989 by the Psychonomic Society. Adapted by permission.)

hemisphere trials and different from the pattern obtained on LVF/right-hemisphere trials, consistent with a bias toward left-hemisphere processing on BILATERAL trials.

Processing of spatial relationships. In an experiment reviewed for other reasons in Chapter 3, Hellige and Michimata (1989a) required observers to perform two tasks that required the processing of different types of spatial relationships between a horizontal line and a dot. The above/below task required observers to indicate whether the dot was above or below the line and the near/far task required observers to indicate whether the dot was within 3 cm of the line. Stimuli on each trial were presented to the LVF/right hemisphere, to the RVF/left hemisphere, or to both LVF and RVF locations (BILATERAL trials). In an analysis restricted to the unilateral trials, there was a hemisphere-by-task interaction, suggesting qualitatively different modes of processing on LVF/right-hemisphere and RVF/left-hemisphere trials. One way to describe the interaction is to note that reaction time was faster for the above/below task than for the near/far task and this difference was significantly larger on RVF/left-hemisphere trials than on LVF/right-hemisphere trials (see Chapter 3 for an interpretation of this hemispheric difference in processing categorical versus coordinate aspects of spatial relationships). For present purposes, it is important to note that on BILATERAL trials the reaction-time difference between the two tasks was identical to that found on RVF/left-hemisphere trials and significantly larger than that found on LVF/right-hemisphere trials. Once again, the qualitative pattern found on BILATERAL trials was consistent with a meta-control bias toward the left hemisphere.

The experiments reported by Hellige et al. (1988) and by Hellige and Michimata (1989a,b) make very different information-processing demands and yet certain aspects of the results are very similar. Of most note is the fact that in each of these experiments the qualitative pattern of effects obtained on BILATERAL trials was identical to the qualitative pattern obtained on RVF/left-hemisphere trials and both were different from the qualitative pattern obtained on LVF/right-hemisphere trials. From this it would be tempting to conclude that metacontrol mechanisms are biased in favor of the left hemisphere when viewing conditions permit either hemisphere to guide the mode of processing, regardless of the specific demands of the task. However, additional experiments to be reviewed next indicate that such a broad conclusion is inappropriate. They also provide the clearest indication to date that the dissociation between hemispheric ability and hemispheric dominance reported by Levy and Trevarthen (1976) for split-brain patients also extends in a more subtle way to the intact brain.

Identification of nonsense syllables. Hellige, Taylor, and Eng (1989) required observers to identify consonant-vowel-consonant (CVC) nonsense syllables presented briefly and followed by a masking stimulus. As illustrated in Figure 6.4, on each trial the CVC was presented to the LVF/right hemisphere, or the RVF/left hemisphere, or the same CVC was presented to both LVF and RVF locations (redundant BILATERAL trials). This task was patterned after an experiment reported by Levy, Heller, Banich, and Burton (1983a), which produced evidence of qualitatively different modes of CVC processing on LVF/right-hemisphere and RVF/left-hemisphere trials. The goals of the experiments reported by Hellige et al. were to replicate the evidence for this qualitative difference and to determine whether the qualitative mode of processing on BILATERAL trials matched that of one hemisphere or the other.

Levy et al. (1983a) found evidence for qualitatively different modes of CVC processing on LVF/right-hemisphere and RVF/left-hemisphere trials by examining the nature of the errors that were made when stimuli were directed to each hemisphere. To do this, errors were classified into one of three types and the normalized frequency of the different types was compared. A first-letter error (FE) occurred when the first letter of the CVC was missed but the third letter was correct, with the correctness of the middle letter being irrelevant (e.g., *CAG* or *CEG* as responses to *DAG*). A last-letter error (LE) occurred when the last letter of the CVC was missed but the first letter was correct, with the correctness of the middle letter again being irrelevant (e.g., *DAC* or *DEC* as responses to *DAG*). All other types of errors were categorized as other errors (OEs). As expected for a verbal-identification task, Levy et al. reported a large RVF/left-hemisphere advantage (also see Chapter 1). Given that the overall error rate is so much lower on RVF/left-hemisphere trials than on LVF/right-hemisphere trials, merely comparing the *number* of FEs, LEs, and

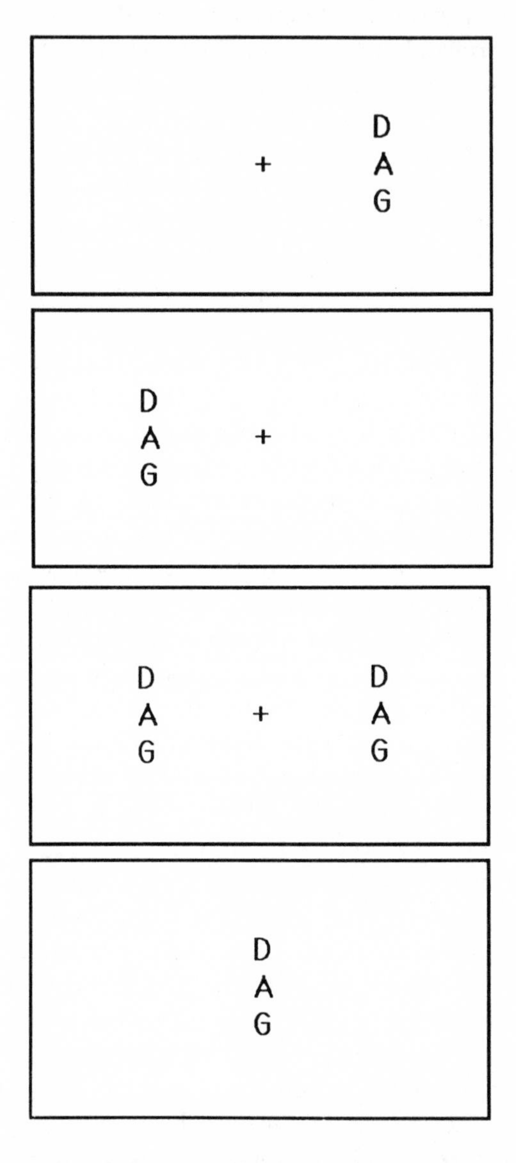

Figure 6.4. Examples of the same stimulus *(DAG)* presented in four different visual-field conditions. From top to bottom: RVF/left hemisphere; LVF/right hemisphere; BILATERAL; CENTER.

OEs for the two visual fields/hemispheres makes it difficult to compare the *qualitative* aspects of performance. This is so because the number of different types of errors confounds qualitative aspects of performance with overall accuracy level. Because of this, Levy et al. divided the number of FEs, LEs, and OEs for each visual-field condition by the total number of errors for that condition. The resulting *normalized* error scores provide an indication of the qualitative pattern of errors in each condition that is independent of the overall level of accuracy in that condition.

Using these normalized error scores, Levy et al. (1983a) found that on LVF/right-hemisphere trials observers failed to identify the last letter far more often than they failed to identify the first letter. That is, there were many more normalized LEs than normalized FEs. They attribute this to the fact that the right hemisphere lacks phonetic processing ability (see Chapters 2 and 3; also Levy and Kueck, 1986) and, as a consequence, treats the CVC stimulus as three individual letters processed sequentially. On RVF/left-hemisphere trials, the difference between the percentage of normalized FEs and LEs was much smaller. Levy et al. argued that this occurred because the left hemisphere supplements or replaces a letter-by-letter analysis of the CVC stimulus with a phonetic or linguistic mode of processing that distributes attention more evenly across all letter positions. (Additional evidence that is consistent with this interpretation is provided by Ellis, Young, and Anderson, 1988, and by Reuter-Lorenz and Baynes, 1992.)

In each of two experiments, Hellige et al. (1989) replicated the effects reported by Levy et al. (1983a). That is, the overall error rate was significantly lower on RVF/left-hemisphere trials than on LVF/right-hemisphere trials. In addition, in an analysis restricted to LVF/right-hemisphere and RVF/left-hemisphere trials, there was a significant interaction between visual field and error type for the normalized FE, LE, and OE scores. The upper panel of Figure 6.5 pictures this interaction for one of the experiments. As the figure shows, the difference between normalized FEs and LEs was much larger on LVF/right-hemisphere trials than on RVF/left-hemisphere trials.

The results reviewed so far are consistent with the assertion that the two hemispheres are predisposed toward qualitatively different ways of processing the CVC stimuli. Furthermore, the

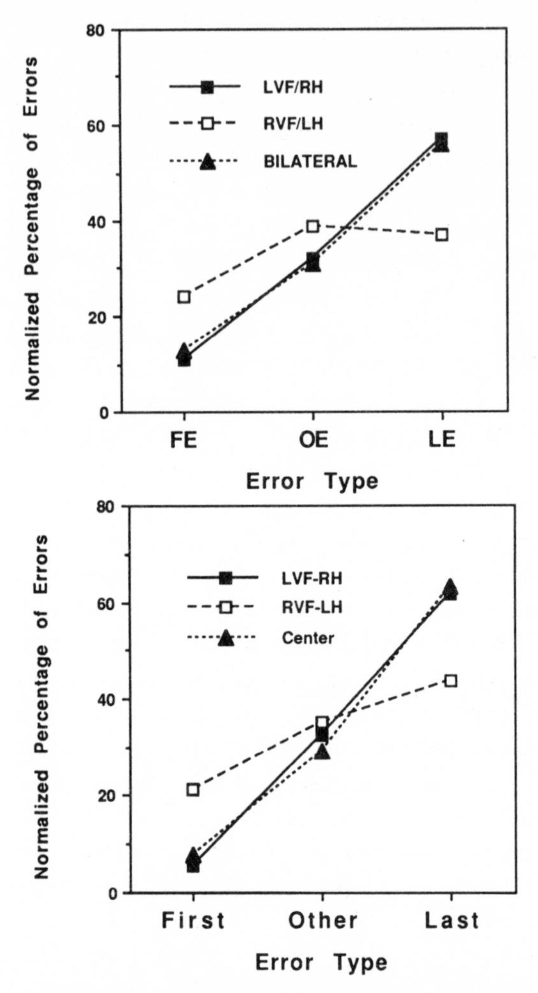

Figure 6.5. Normalized percentage of different types of errors (First, Other, Last) for different visual-field conditions. The upper panel shows results for LVF/right-hemisphere (LVF/RH) trials, RVF/left-hemisphere (RVF/LH) trials, and BILATERAL trials from an experiment reported by Hellige et al. (1989). [Adapted from J. B. Hellige, A. K. Taylor, and T. L. Eng, "Interhemispheric Interaction When Both Hemispheres Have Access to the Same Stimulus Information," *Journal of Experimental Psychology: Human Perception and Performance,* 15 (1989):711–722. Copyright 1989 by the American Psychological Association. Adapted by permission.] The lower panel shows results for LVF/RH trials, RVF/LH trials, and CENTER trials from an experiment reported by Hellige et al. (in press). [Adapted from J. B. Hellige, E. L. Cowin, and T. L. Eng, "Recognition of CVC Syllables from LVF, RVF and Central Locations: Hemispheric Differences and Interhemispheric Interaction," *Journal of Cognitive Neuroscience* (in press). Copyright by MIT Press. Adapted by permission.]

fact that the overall percentage of errors is lower on RVF/left-hemisphere trials than on LVF/right-hemisphere trials indicates that the mode of processing favored by the left hemisphere leads to superior performance, at least with stimuli presented briefly and off-center. In view of this, it would be reasonable to expect that the brain would spontaneously engage in the mode of processing associated with superior performance when the viewing conditions allow both hemispheres equivalent access to the stimulus. From this perspective, the qualitative error pattern obtained on redundant BILATERAL trials is predicted to match the pattern obtained on RVF/left-hemisphere trials.

Despite the intuitive appeal of the prediction just outlined, the results actually obtained by Hellige et al. (1989) were quite different. As shown in the upper panel of Figure 6.5, the qualitative error pattern actually obtained on redundant BILATERAL trials matched the pattern obtained on LVF/right-hemisphere trials. This counterintuitive result is not an artifact of the BILATERAL mode of stimulus presentation, because Hellige, Cowin, and Eng (in press) found very similar results when the redundant BILATERAL trials were replaced with the presentation of a single CVC to the center of the viewing field (see Figure 6.4 for an example of the CENTER presentation of a CVC). That is, Hellige et al. (in press) used CENTER presentation as another way of creating a viewing condition that does not favor one hemisphere or the other. As shown in the lower panel of Figure 6.5, we replicated the qualitative error differences between LVF/right-hemisphere and RVF/left-hemisphere trials and found that the qualitative error pattern obtained on CENTER trials matched the pattern obtained on LVF/right-hemisphere trials. That is, the redundant BILATERAL and CENTER methods of stimulus presentation led to very similar results.

These results indicate that, in the intact brain, there is sometimes a dissociation between hemispheric superiority for a task (e.g., the RVF/left-hemisphere superiority for recognizing CVC stimuli) and which hemisphere's preferred mode of processing is favored when conditions do not bias processing toward either hemisphere (e.g., the mode of CVC processing used on LVF/right-hemisphere trials). Note that this more subtle effect on the mode of processing is similar to the more dramatic results reviewed earlier for split-brain patients. This similarity suggests that the same neural mechanisms that serve to establish metacontrol in split-brain patients may also serve to influence the mode of processing used by individuals whose brains are intact.

Hellige et al. (1989, in press) consider several possible reasons for this counterintuitive pattern of results. For example, the brain may be biased in favor of a mode of processing that can be used by both hemispheres, unless stimulus conditions dictate otherwise. According to Levy et al. (1983a), the mode of CVC processing favored on RVF/left-hemisphere trials involves phonetic mechanisms that are unavailable to the right hemisphere (see Chapters 2 and 3). In contrast, both hemispheres are able to process the CVC stimuli in a letter-by-letter fashion, which seems to be the mode of processing favored by the isolated right hemisphere (e.g., Reuter-Lorenz and Baynes, 1992). From this perspective, the letter-by-letter mode of processing would be used as long as the viewing conditions either favored the right hemisphere (LVF trials) or encouraged the participation of both hemispheres (redundant BILATERAL trials or CENTER trials). Only when the viewing conditions favored the left hemisphere (RVF trials) would the letter-by-letter mode of processing be supplemented by a more phonetic analysis.

Another possibility comes from the fact that a letter-by-letter mode of processing is usually quite appropriate when the stimuli consist of a vertical arrangement of English letters. In reading English, it is unusual to encounter words (or pronounceable nonwords) with the letters arranged vertically. In fact, a vertical column of letters usually does not spell a word or pronounceable nonword. In this case, there may be some cost associated with using a phonetic mode of processing, making it preferable to use a letter-by-letter mode of processing. On this view, the brain may be biased to favor the mode of processing that would generally be more effective for processing a vertical column of letters, even though the CVC patterns represent exceptions. As a result, a more phonetic mode of processing is limited to the viewing condition that favors the left hemisphere.

The possibility just outlined is consistent with the results of a preliminary experiment just finished in my laboratory. The experiment was identical to Experiment 1 reported by Hellige et al.

(in press), which presented CVCs to LVF, RVF, and CENTER locations. The only difference was that, in the new experiment, the letters making up the CVC were arranged horizontally rather than vertically. The qualitative error patterns on RVF/left-hemisphere and LVF/right-hemisphere trials differed from each other in the same way as shown in Figure 6.5 for vertical stimulus presentations. However, the qualitative error pattern obtained on CENTER trials was different from both of the unilateral patterns. Specifically, the normalized percentage of FEs was similar on LVF/right-hemisphere and CENTER trials whereas the normalized percentage of LEs was similar on RVF/left-hemisphere and CENTER trials. That is, CENTER stimuli seemed to be processed using some combination of the two modes of processing favored on LVF and RVF trials. While this change in the pattern of results on CENTER trials is consistent with the hypothesis outlined earlier, other interpretations are possible. For example, the apparent collaboration of the two hemispheres (at least in terms of the mode of processing) may have less to do with the fact that the horizontal arrangement is more typical of English words than with the fact that when the CVCs were arranged horizontally and presented in the CENTER, the first letter projected directly to only the right hemisphere whereas the last letter projected directly to only the left hemisphere. This raises the interesting possibility that the functioning of metacontrol mechanisms is determined, in part, by whether or not relevant perceptual information is divided between the hemispheres.

Determinants of Metacontrol

The experiments reviewed earlier suggest that mechanisms of metacontrol sometimes influence interhemispheric interaction in the intact brain, at least insofar as the mode of processing favored by one hemisphere is sometimes chosen over the mode of processing favored by the other hemisphere. In this section, I speculate about some of the determinants of metacontrol. It is useful to distinguish between two issues about the determinants of metacontrol. The first concerns the factors that determine when metacontrol mechanisms are likely to determine hemispheric interaction and when, instead, the hemispheres interact in a more collaborative fashion. The second issue concerns the factors that determine the direction of hemispheric dominance for those situations that produce a pattern of results suggestive of metacontrol mechanisms.

The fact that the qualitative mode of processing on BILATERAL and CENTER trials sometimes matches the qualitative pattern obtained on one of the unilateral trials must not be taken as an indication that such a pattern will always occur. There are certain to be cases in which the mode of processing that is used when viewing conditions are not biased in favor of one hemisphere or the other is either a mixture of the left- and right-hemisphere modes of processing or some completely different mode of processing. In fact, such a pattern was obtained in the experiment described earlier that used CVC stimuli with the letters arranged horizontally. As already noted, the results of that experiment (and the contrast to the experiments with the CVCs oriented vertically) suggests that hemispheric collaboration rather than hemispheric dominance may be particularly likely when relevant perceptual information is divided between the two hemispheres.

Another important factor is likely to be whether or not the two modes of processing preferred on LVF/right-hemisphere and RVF/left-hemisphere trials require the brain to engage in mutually inconsistent processes. To some extent , the experiments reviewed earlier were biased purposely in this direction by using tasks that are performed in qualitatively different ways as a result of the stimuli being projected to one hemisphere or the other. For example, Levy and Kueck (1986) argue that the phonetic mode of CVC processing favored by the left hemisphere leads to a very rapid or global distribution of attention across the three letters of a CVC stimulus, but that the letter-by-letter mode of CVC processing favored by the right hemisphere leads to a slower, letter-by-letter, sequential distribution of attention. It may be that the difficulty of simultaneously distributing attention in these two different ways makes it unlikely that both modes of processing will be used simultaneously. Perhaps when the modes of processing are not mutually inconsistent, the two hemispheres are better able to process the same information in parallel, with each hemisphere using its preferred mode.

Some evidence for this possibility in split-brain patients comes from recent studies reported by Sergent (in press). She examined the processing of different types of spatial relationships in three split-brain patients. Of particular relevance to the present discussion was the inclusion of trials that presented the same information to both visual fields/hemispheres and required patients to produce a single response (e.g., is a dot to the right or left of the center of a circle). One important finding was that performance on these redundant BILATERAL trials was superior to performance on LVF/right-hemisphere or RVF/left-hemisphere trials. As Sergent argues, this result by itself suggests that both hemispheres of these split-brain patients contributed to overall performance. Further evidence that both hemispheres contributed to the final response came from the observation that certain qualitative aspects of the results were different on BILATERAL trials than on either of the two types of unilateral trials. For example, in the left/right task a spatial-compatibility effect was present on unilateral trials (patients responded faster with the hand ipsilateral to the field stimulated), but the response-hand effect on BILATERAL trials was different from that of either type of unilateral trial.

Sergent (in press) contrasts her results with those obtained in earlier studies of split-brain patients in which a typical outcome was that only one hemisphere responded after simultaneous presentation of different information to the two hemispheres (e.g., Levy et al., 1972). Sergent suggests that the earlier results occurred because the two hemispheres have difficulty producing two different responses simultaneously, probably because both hemispheres must converge on the same neural structures underlying the preparation and organization of a motor response. By presenting the same information to both hemispheres on BILATERAL trials and requiring only a single response, Sergent's studies removed this response competition problem and found evidence that both hemispheres contributed to the final response. The possibility suggested here is that the response-production level is just one place that can lead to a blocking of performance by one hemisphere or at least to a single mode of processing that can be used by both. For example, to rephrase a possibility considered earlier, the blocking of one hemisphere's preferred mode of processing may occur if both hemispheres must converge on the same neural structures that underly the distribution of attention across space.

With respect to the second issue about determinants of metacontrol, the fact that some of the experiments reviewed in this section found a qualitative similarity between redundant BILATERAL trials and RVF/left-hemisphere trials whereas others found a qualitative similarity between BILATERAL and LVF/right-hemisphere trials indicates that the direction of hemispheric dominance produced by any metacontrol mechanisms also depends on aspects of the task that is being performed. While more research is needed to clarify exactly what aspects of a task are relevant, it is useful to note the following similarities among those tasks that found similar qualitative patterns on BILATERAL and RVF/left-hemisphere trials. None of these tasks produced an overall advantage for one hemisphere or the other on unilateral trials. That is, neither hemisphere's preferred mode of processing was uniformly superior. In addition, there is no evidence that either of the two modes of processing used in these tasks was completely unavailable to the hemisphere that was not biased in its favor. As a result, the overall level of performance on BILATERAL trials was not likely to depend on which mode of processing was used. In contrast, for the CVC experiments that found similar qualitative patterns on BILATERAL and LVF/right-hemisphere trials, there was a large difference in hemispheric ability on unilateral trials, and there is good reason to believe that one mode of processing was available only to the left hemisphere whereas the other mode of processing was available to both. It would be useful for additional studies to investigate whether these factors are relevant.

Summary and Conclusions

When thinking about hemispheric asymmetry, we must be mindful of the unity of the brain (theme 2 in Chapter 1). This is so because both cerebral hemispheres seem to be involved in virtually everything we do, creating the potential for conflicts in perception, cognition, emotion, and action. Despite this potential, conflicts seem rare because the two hemispheres interact with each other in a variety of ways. Together with the corpus callosum, other neocortical commissures, and various subcortical structures, the left and right hemispheres form a single, integrated information-processing system. The unity of this system is preserved by several varieties of interhemispheric interaction, with such things as the information-processing demands of a task determining how the hemispheres will interact while performing that task.

For many tasks, the two hemispheres are dominant for different task-relevant processing components. Under these conditions the two hemispheres seem to coordinate their activities so that each can take the lead for those components of processing that it handles best. This coordination involves the transfer of relevant information from one hemisphere to the other and, at the same time, the insulation of certain hemisphere-specific processes from each other so that those processes can proceed efficiently in parallel.

Investigations of the biological mechanisms of interhemispheric cooperation have focused on the possible roles of the corpus callosum and various subcortical structures. An important issue that has emerged concerns the extent to which the corpus callosum is best conceptualized as excitatory or inhibitory, especially at a functional level. One interesting hypothesis is that the corpus callosum produces homotopic inhibition, which tends to produce mirror-image patterns of activation and inhibition in the two cerebral hemispheres. On this view, the corpus callosum not only permits information to be transferred from one hemisphere to the other but also leads the hemispheres to become dominant for complementary functions. The corpus callosum has also been proposed to regulate the state of asymmetric activation or arousal between the two hemispheres, thereby helping to maintain the hemispheres in a state of mutual inhibitory balance. Related to this is the hypothesis that the corpus callosum serves as an inhibitory barrier between the hemispheres, preventing maladaptive cross-talk between the processes for which each hemisphere is dominant. Whether it is by inhibition or excitation at the neural level, the corpus callosum certainly plays an important role in the transfer of at least certain types of *information* from one hemisphere to the other, including information about the identity of stimuli. At the same time, it may also serve to reduce maladaptive cross-talk between mutually inconsistent *processes* by functioning as something of a barrier between the two hemispheres.

Although the corpus callosum may play several roles related to interhemispheric cooperation, it is also clear that some interhemispheric interaction can take place subcortically. Although information about the name or identity of a stimulus cannot be transferred from one hemisphere to the other subcortically, various other types of information can: for example, connotative and contextual information about an object, information about the categories to which an object belongs, and information about the location of objects in space. In addition, subcortical structures can play an essential role in coordinating the activity of the two hemispheres, permitting a single unified response to be produced by the dual-brain system.

Although the two hemispheres are capable of sharing many types of information in the intact brain, cooperation at all levels does not always take place. For example, when a stimulus is presented directly to only one hemisphere, it is sometimes the case that the hemisphere receiving the stimulus information carries out virtually all of the necessary processing—perhaps sharing only its final decision with the other hemisphere. Consequently, it is important to understand the factors that determine when it is more efficient for the two hemispheres to operate in this independent manner and when it is more efficient for them to operate in a more collaborative manner. With this issue in mind, researchers have undertaken a variety of studies that have examined the benefits and costs of interhemispheric cooperation. The strategy used in these studies has been to compare performance of the same task under conditions that demand interhemispheric cooperation (the between-hemisphere condition) and conditions that permit a single hemisphere to handle all of the processing (the within-hemisphere condition). One important conclusion to emerge from these studies is that distributing processing across both hemispheres becomes beneficial as the processing requirements of a task become more demanding. When the processing requirements are sufficiently simple, the costs associated with the need to transfer information across the hemispheres and the need to coordinate the activities of the two hemispheres may be sufficiently detrimental to offset the benefits of cooperation. Other factors that may be important are the amount of practice on a task and the extent to which the two hemispheres can restrict their activities to mutually exclusive processes.

In addition to cooperating to perform a task, the two hemispheres sometimes interact in a way that allows one of them to dominate processing in a more general way, even when stimuli are experienced in a way that does not favor one hemisphere or the other. Of particular interest is the finding that the hemisphere that is dominant in this sense is not always the hemisphere with greater ability to perform the task. This has been demonstrated in split-brain patients under conditions that present different stimuli to each hemisphere and allow either hemisphere to make the final response. The neural mechanisms that determine the extent to which each hemisphere tends to assume control of processing and behavior have been referred to as metacontrol mechanisms.

Metacontrol mechanisms sometimes operate in the intact brain in the performance of tasks for which both hemispheres have competence but for which there is evidence that the two hemispheres are biased to go about processing in qualitatively different ways. In the intact brain it is not so much that metacontrol mechanisms prevent either hemisphere from contributing to performance when stimuli are presented in a way that permits each hemisphere to receive the relevant information (bihemispheric stimulation). Instead, under these conditions of bihemispheric stimulation, metacontrol mechanisms sometimes lead the brain to favor one hemisphere's preferred mode of processing over the mode of processing preferred by the other hemisphere. Of particular importance is the fact that the mode of processing used on bihemispheric trials is not always the mode of processing associated with the hemisphere that has greater ability to perform the task.

There are likely to be several determinants of metacontrol. A particularly interesting possibility is that the mode of processing favored by either one hemisphere or the other will dominate on bihemispheric trials to the extent that the two modes of processing require the brain to engage in mutually inconsistent processes at the same time (e.g., simultaneously distributing attention across space in both a global fashion and in an item-by-item fashion or commanding the same neural structures to program and execute two conflicting responses at the same time). When the modes of processing used by the two hemispheres do not require mutually inconsistent processes, various forms of interhemispheric cooperation may be more likely.

It should be apparent from the recency of the research reviewed in this chapter that the study of interhemispheric interaction in all its forms and complexity is beginning to receive far more attention than it has in the past. To some extent, the emergence during the last thirty years of so much research dealing with the *differences* between the left and right cerebral hemispheres has resulted in a particular view of the brain: the brain has been "taken apart," and there has been a tendency to analyze and conceptualize the pieces as separate processing systems. Indeed, the same thing could be said about the highly fine-grained modularity that has come to characterize much of cognitive neuroscience. The time has come to put the brain back together again. In fact, one of the most important challenges facing cognitive neuroscience is to account for the emergence of unified processing from a brain consisting of a variety of processing subsystems. The left and right hemispheres can be characterized as two very general processing subsystems with different processing biases and propensities. Coming to understand the varieties of interhemispheric interaction, the determinants of those types of interaction, and the biological mechanisms that support them will provide important clues about the unity of perception, cognition, emotion, and action.

7

Individual Differences

Previous chapters in this book have discussed hemispheric asymmetry for what might be characterized as a prototypical human brain, as if we were all identical with respect to the anatomy and function of the two cerebral hemispheres. In fact, there is sufficient homogeneity to make this characterization of the prototypical brain a worthwhile enterprise. Individuals sometimes differ a great deal in hemispheric asymmetry, however. For example, most of the research reviewed in the first six chapters of this book was based on studies in which participation was restricted to right-handed individuals because it is generally acknowledged that the pattern of hemispheric asymmetry is related to handedness. The purpose of the present chapter is to consider several dimensions of individual variation in hemispheric asymmetry and to consider whether such variation is related to differences in cognitive abilities and propensities.

The existence of individual variation in the hemispheric asymmetry of humans should come as no surprise in view of recent studies of asymmetry in nonhuman species (see Chapter 5). For example, individual rats differ in the preferred direction of circling and in the magnitude of circling bias, with the bias determined by asymmetries in the dopamine content of the left and right sides of the nigrostriatal pathway of the rat brain. The existence of such individual differences in other species makes it reasonable to suppose that similar individual differences exist in humans. Furthermore, to the extent that individual differences in brain asymmetry have behavioral consequences in other species (e.g., the direction of preferred circling), it is likely that individual differences in brain asymmetry in humans also have behavioral manifestations.

It is also interesting to note that anatomical studies of the human brain show considerable individual variation in the direction and magnitude of certain asymmetries (see Chapter 4). For example, the sylvian fissure is typically longer in the left hemisphere than in the right hemisphere (65 percent of the time in right-handers, as reported by Geschwind and Levitsky, 1968); but even among right-handers there is considerable individual variation—in this case, the direction of the asymmetry is reversed for 11 percent of the individual brains studied by Geschwind and Levitsky. Although the precise relationship between these anatomical features and behavioral asymmetries remains to be established, the fact that they differ so much across individuals raises the possibility that there are equally detectable individual variations in functional aspects of hemispheric asymmetry.

Studies of individual differences in other species and of anatomical asymmetries in human brains provide a backdrop against which it is instructive to consider individual differences in functional aspects of hemispheric asymmetry. In order to do so, we must acknowledge that there are several dimensions of individual variation related to hemispheric asymmetry and interhemispheric interaction. Accordingly, the first section of this chapter considers several relevant dimensions along which individuals have been shown to differ. This is followed by a discussion of the possible relationship between individual variation in these factors related to hemispheric asymmetry and a number of between-subject factors. These between-subject factors include handedness, sex, intellectual abilities (including intellectual giftedness and dyslexia), and psychopathology. The chapter ends with a critical examination of the notion that a person can be "right-brained" or "left-brained," a concept that has come to be known as "hemisphericity."

Dimensions of Individual Variation

Individual differences in hemispheric asymmetry are usually discussed as if there is only one important dimension along which individuals can differ. In fact, there are several logically independent dimensions of what could generally be termed hemispheric asymmetry, and individuals have been shown to differ along a number of them. The present section considers five such dimensions that seem particularly relevant: (1) direction of hemispheric asymmetry, (2) magnitude of hemispheric asymmetry, (3) asymmetric arousal of the two hemispheres, (4) complementarity of asymmetries, and (5) interhemispheric communication.

Direction of Hemispheric Asymmetry

One dimension along which individuals differ is the *direction* of hemispheric asymmetry. That is, some individuals show a pattern of hemispheric asymmetry that is opposite what would be considered the prototypical direction. For example, in Chapter 1 we saw that results from patients with unilateral brain injury and from people injected with barbituate anesthetic into one hemisphere or the other both indicated that the left hemisphere is dominant for speech production in 90–95 percent of right-handed adults; the right hemisphere is dominant for speech production in the remainder of right-handed adults. As discussed in more detail later, the percentage of individuals with patterns of hemispheric asymmetry that are opposite the prototypical pattern is even larger for certain subgroups of the human population, for example, non-right-handers.

Studies of perceptual and response asymmetries in neurologically intact individuals also indicate asymmetries in opposite directions for different individuals. For example, in the study of CVC nonsense-syllable recognition discussed in Chapter 1 (see Figure 1.3) it was found that 85 percent of the right-handers tested recognized more stimuli from the RVF/left-hemisphere trials than from the LVF/right-hemisphere trials, whereas 12 percent showed an asymmetry in the opposite direction. Likewise, in a dichotic-listening experiment discussed in Chapter 1, 80 percent of the right-handers tested recognized more CV stimuli from the right ear (RE) than from the left ear (LE), whereas 19 percent showed an asymmetry in the opposite direction. Additional studies indicate that a good deal of this type of variation in performance asymmetries is reliable, even within the right-handed population. For example, it is not unusual to find reliability estimates for perceptual asymmetries in the range from .50 to .90 (e.g., Boles, 1989; Hellige and Wong, 1983; Nestor and Safer, 1990).

It is also interesting to note that individual differences in the direction of at least some perceptual asymmetries are predictive of individual differences in the direction of asymmetry for other tasks. For example, Hellige, Bloch, and Taylor (1988) found a relationship between the direction of ear dominance on a dichotic-listening task and the direction of hemisphere-specific interference in a dual-task paradigm using interference with motor activity. The dichotic-listening task required 120 right-handed individuals to identify CV syllables, and the right-minus-left ear difference score was used to classify individuals into three groups: those with a left-ear advantage (N = 10), those with the largest right-ear advantage (N = 10), and those with a right-ear advantage within one standard deviation of the group mean (N = 100). The dual-task paradigm examined the effects of concurrent repetition of CV syllables and concurrent solving of anagrams on the rate of finger tapping by the index fingers of the left and right hands. Recall from the discussion of dual-task experiments in Chapter 3 that concurrent verbal activity interferes with right-hand tapping more than with left-hand tapping for most right-handed individuals, and that this asymmetry is consistent with the hypothesis that verbal processing requires more capacity from the left hemisphere than from the right hemisphere. Such an asymmetry was, in fact, present in the data reported in this study.

Of particular importance for present purposes is the fact that the hand difference in interference during the dual-task paradigm was reversed for the group of individuals that showed a left-ear advantage on the dichotic-listening task. This is illustrated in Figure 7.1, which shows the percentage reduction in finger tapping for each hand and for each of the two concurrent activities (CV syllables and anagram solution). The results are shown for each of the three groups defined by performance on the dichotic-listening task. Note that for the largest group of individuals (the midrange group in Figure 7.1), both concurrent activities interfered more with the right than with the left hand. This asymmetry was reversed in the left-ear-advantage group and enhanced in the large right-ear-advantage group when the concurrent task was CV recognition. The pattern was similar when the concurrent task was solving anagrams, except that the hand difference for the left-ear-advantage group was eliminated rather than reversed.

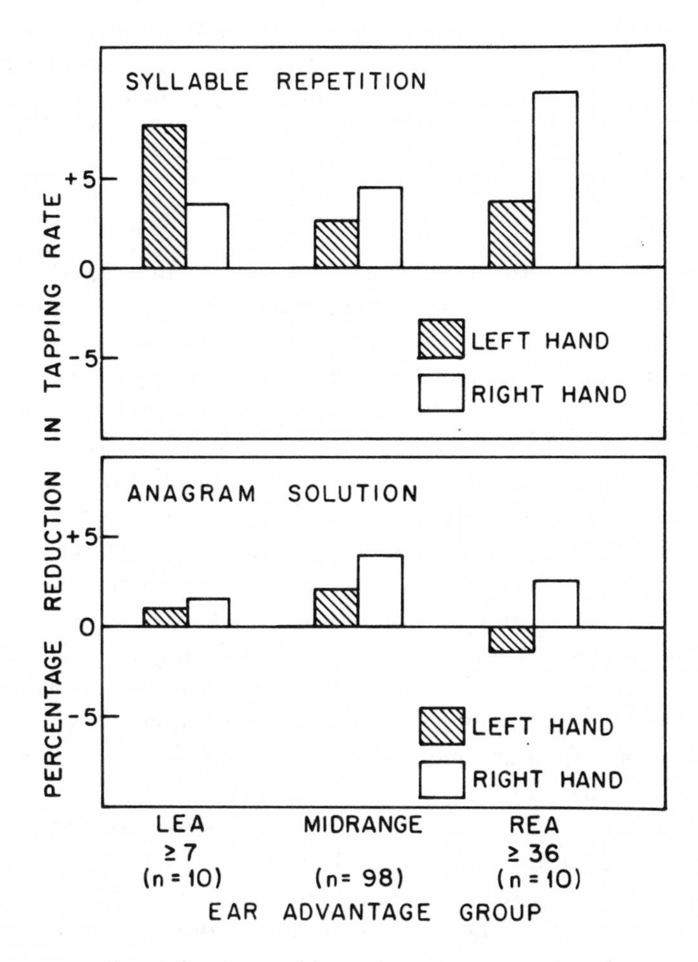

Figure 7.1. Percentage reduction in finger-tapping rate for each hand while subject performs a syllable-recognition task *(upper panel)* and solves anagrams *(lower panel)*. The results are shown for each of three groups defined by their ear asymmetries on a verbal dichotic-listening task: those who showed a left-ear advantage of at least 7 items (LEA ⩾ 7), those who showed a right-ear advantage of at least 36 items (REA ⩾ 36), and those whose ear differences were between these values (the Midrange group). [Reprinted from J. B. Hellige, M. I. Bloch and T. L. Eng, "Multitask Investigation of Individual Differences in Hemispheric Asymmetry," *Journal of Experimental Psychology: Human Perception and Performance,* 14 (1988):176–187. Copyright 1988 by the American Psychological Association. Reprinted by permission.]

The interpretation favored by Hellige et al. (1988) is that both dichotic listening and the dual-task paradigm are sensitive to individual differences in the direction (and perhaps in the magnitude) of cerebral dominance for certain common aspects of linguistic processing. Although this interpretation remains plausible, we shall see later that other dimensions of individual variation may also contribute to these effects.

Magnitude of Hemispheric Asymmetry

For most tasks, both hemispheres have some ability, and we use the term *hemispheric asymmetry* to refer to the different ability levels of the two hemispheres rather than to a situation in which one hemisphere has absolutely no ability at all. This being the case, it is possible to make a distinction between the direction of asymmetry (i.e., which hemisphere has greater ability) and the magnitude of asymmetry (i.e., how great is the difference in ability between the two hemispheres). In the previous section we saw that individuals can differ in the direction of hemispheric asymmetry. Now we acknowledge that a second dimension along which individuals can differ is the magnitude of hemispheric asymmetry. Although it is often difficult in practice to distinguish differences in magnitude from differences in direction, the two are logically independent. That is, two individuals can be equally asymmetric, regardless of whether their asymmetry is in the same direction or in opposite directions.

Before we move on to a discussion of some indications that humans differ in the magnitude of hemispheric asymmetry, it will be useful to recall that there is evidence for the independence of magnitude and direction of asymmetry from studies of other species. For example, in certain species of mice it has been possible to breed selectively for the magnitude of paw preference, creating strongly lateralized versus weakly lateralized strains. However, in these same species it has not been possible to breed selectively for the direction of paw preference. In addition, in several species the magnitude of paw preference and of other behavioral asymmetries is related to the ability of the individual to learn tasks that require discrimination of one side from the other—and this is largely independent of the direction of the behavioral asymmetry (see Chapter 5). These results and others argue that in other species there are important differences between magnitude and direction of brain asymmetry.

It is virtually impossible to use data from brain-injured patients to study individual differences in the magnitude of hemispheric asymmetry. This is true for several reasons, not the least of which is the fact that each individual has a unique injury in terms of size, precise location, and so forth. In studies with intact individuals, there is no doubt that the magnitude of performance asymmetries differs from person to person (e.g., see Figures 1.3 and 1.4) and, as noted earlier, the reliability of some asymmetry scores is sufficiently high to prevent these differences from being attributed completely to error variance. However, there has been little motivation to classify individuals according to the magnitude of their performance asymmetries (regardless of direction) and look for relationships of the resulting magnitudes to cognitive abilities and to other dimensions of individual variation. Despite the lack of studies that take care to separate the direction of asymmetry from the magnitude of asymmetry, we shall see that conclusions have been reached that one subpopulation or other of humans is only "weakly lateralized" compared to some prototype. Clearly, the implication is that the magnitude of asymmetry is unusually small but the direction may be prototypical. We shall examine some of these statements in subsequent sections of the present chapter and attempt to determine whether they are based on a clear separation of direction from magnitude of asymmetry.

Asymmetric Arousal of the Hemispheres

It has been suggested that the two cerebral hemispheres can be at different levels of activation or arousal, and recent studies suggest that individuals differ with respect to their habitual or characteristic pattern of arousal asymmetry (e.g., Levine, Banich, and Kim, 1987; Levy et al., 1983a,b; Kim, Levine, and Kertesz, 1990). These differences in arousal-asymmetry have been hypothesized to influence performance on a variety of perceptual laterality tasks and to be independent of individual differences in the direction and magnitude of hemispheric asymmetry as discussed earlier.

Predictions from a strong version of the arousal asymmetry hypothesis are illustrated in Figure 7.2 (from Kim et al., 1990). This figure shows hypothetical results for three different visual-half-field tasks that show different average performance asymmetries for a group of right-handed individuals: a word-recognition task that produces a RVF/left-hemisphere advantage, a chair-recognition task that produces no visual-field difference, and a face-recognition task that produces a LVF/right-hemisphere advantage. Figure 7.2 shows the hypothetical frequency of individuals who show different magnitudes of performance asymmetry: the amount of RVF/left-hemisphere advantage decreases from left to right across the horizontal axis until the 0 point (no advantage) is reached and, thereafter, the amount of LVF/right-hemisphere advantage increases. Note that each task produces a different mean asymmetry and that for each task there is vari-

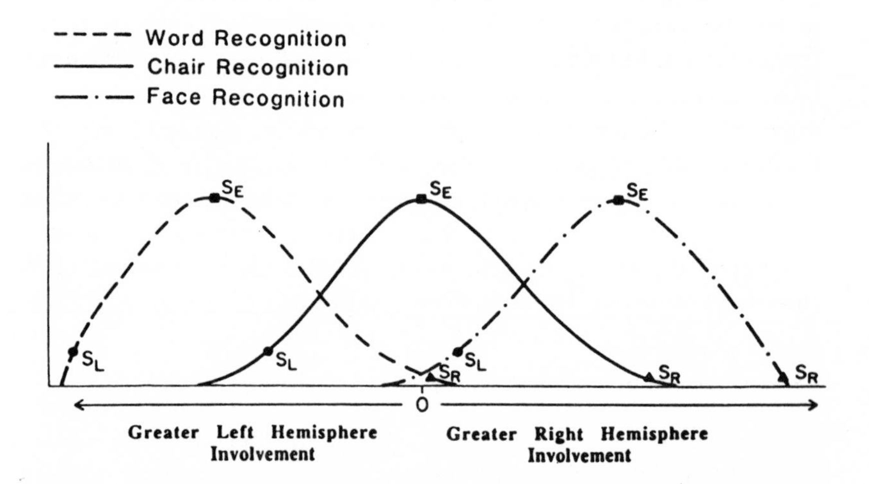

Figure 7.2. Hypothetical asymmetry scores predicted by a strong version of the arousal asymmetry hypothesis. Predictions are given for a subject with greater right- hemisphere arousal (S_R), a subject whose hemispheres are equally aroused (S_E), and a subject with greater left-hemisphere arousal (S_L). Hypothetical difference scores are shown for a word-recognition task, a chair-recognition task, and a face-recognition task. [Reprinted from H. Kim, S. C. Levine, and S. Kertesz, "Are Variations among Subjects in Lateral Asymmetry Real Individual Differences or Random Error in Measurement? Putting Variability in Its Place," *Brain and Cognition,* 14 (1990):220–242. Copyright 1990 by Academic Press, Inc. Reprinted by permission.]

ability of the individual asymmetry scores around the mean (much as in Figures 1.3 and 1.4). Of particular interest are scores for three hypothetical individuals: one with a more aroused right hemisphere (S_R), one with a more aroused left hemisphere (S_L) and one with equally aroused hemispheres (S_E).

All three of these individuals are hypothesized to have the same pattern of hemispheric asymmetry for the three tasks. Consequently, for all three it is the case that left-hemisphere involvement decreases and right-hemisphere involvement increases from the word-recognition task to the chair-recognition task to the face-recognition task. In this sense, the pattern of interaction between visual field or hemisphere and task is identical for all three individuals. However, for any one task the specific asymmetry score differs for the three individuals in a particular way. That is, for each task S_E scores near the mean, S_L's score is displaced toward a value indicating greater left-hemisphere involvement, and S_R's score is displaced toward a value indicating greater right-hemisphere involvement. It is this consistency of individual position across tasks that is predicted by the hypothesis that the individuals differ in the pattern of characteristic arousal asymmetry.

Evidence for the existence of individual variation in characteristic arousal asymmetry has come from a variety of studies (e.g., Levine et al., 1987; Levy et al., 1983a,b; Kim et al., 1990). Consider, for example, a study reported by Kim et al. (1990), who required right-handed and left-handed individuals to complete a battery of four visual-half-field tasks and one free-vision task. Each of the visual-half-field tasks used the display of two different stimuli on each trial, one to each visual field, and individuals attempted to identify both stimulus items. Different tasks used different types of stimuli that produced different average perceptual asymmetries: words (RVF/left-hemisphere advantage), photographs of chairs (no perceptual asymmetry), line drawings of common objects (LVF/right-hemisphere advantage), and photographs of faces (LVF/right-hemisphere advantage). The free-vision task used a 36-item set of chimeric face stimuli prepared by Levy et al. (1983b). One item from this set is pictured in Figure 7.3. As illustrated in the figure, each item consisted of a chimeric face and its exact mirror image, with the two versions arranged one under the other. Half of each chimeric face came from a

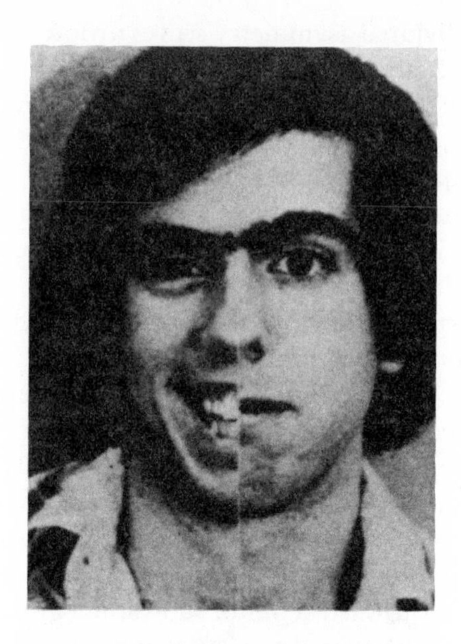

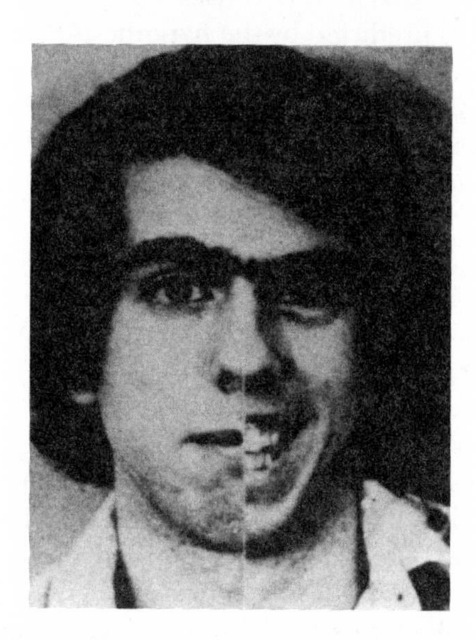

Figure 7.3. Item 1 from the free-vision face task developed by Levy et al. (1983b). [Reprinted from J. Levy, W. Heller, M. T. Banich, and L. A. Burton, "Asymmetry of Perception in Free Viewing of Faces," *Brain and Cognition,* 2 (1983):404–419. Copyright 1983 by Academic Press, Inc. Reprinted by permission.]

photograph of a male posing with a neutral expression and the other half came from a photograph of the same poser with a happy expression. The task of the observer is to indicate for each pair of faces whether the top or bottom version looks happier. When right-handed individuals are given this task, they show a significant bias to choose the face with the happy expression on the viewer's left as looking happier than its mirror image. Levy et al. (1983a,b) argue that this occurs because the requirement to attend to facial emotions selectively arouses the right hemisphere, thereby biasing attention toward the left side of space. Furthermore, they argue that individual differences in the extent of bias reflects individual variability in characteristic arousal asymmetry (see also Luh, Rueckert, and Levy, 1991).

In order to test the predictions illustrated in Figure 7.2, Kim et al. (1990) computed perceptual asymmetry scores for each of these five tasks and entered the scores into a principal component analysis, which is a statistical technique used to extract common components that explain the maximum amount of variance of the scores in the sample. In accordance with the predictions outlined earlier, this analysis indicated that about half of the variance in asymmetry scores was accounted for by a single component that was a reflection of individual differences in characteristic perceptual asymmetry. That is, an individual's asymmetry score for any single task was partly a function of factors unique to that task but also a function of the individual's characteristic bias toward one side of space. It is also interesting to note that the amount of variance accounted for by this characteristic perceptual-asymmetry component was the same for left-handers as for right-handers. Such results are consistent with the hypothesis that, regardless of handedness, individuals differ in their patterns of characteristic hemispheric arousal; they also indicate that it is possible to isolate individual variation along this dimension from variation along other dimensions related to hemispheric asymmetry. With respect to the latter point, it is interesting to note that Kim et al. refer to other studies indicating that characteristic arousal asymmetry contributes less to performance when only a single item is presented to one visual field or the other on each trial than when two different items are presented, one to each visual field (see also Kim and Levine, 1991a). This interesting finding suggests that characteristic arousal asymmetry may modulate the direction in which multiple items are processed.

Complementarity of Asymmetries

Yet another dimension of individual variation involves what might be called a "complementarity of function." Throughout this book we have seen that at the population level each cerebral hemisphere is dominant for certain components of processing. For example, at least in right-handed adults the left hemisphere is dominant for speech production and for the perception of phonemic information whereas the right hemisphere is dominant for identifying emotion on the basis of intonation or facial expression. To the extent that the percentage of individuals who show the prototypical dominance pattern is high, it will be the case that most individuals have opposite hemispheric dominance for such things as speech production and identifying emotion. This will be true even if hemispheric asymmetry for these different functions are statistically independent (see Chapter 2). However, there are almost certain to be individual differences in the extent to which various processing components that show hemispheric asymmetry at the population level are in fact lateralized to opposite hemispheres within an individual brain.

It has been suggested that there are disadvantages to having functions that are typically lateralized to opposite hemispheres uncharacteristically "crowded" into the same hemisphere (e.g., Geschwind and Galaburda, 1987; Levy, 1969; O'Boyle and Hellige, 1989). Earlier we have considered the hypothesis that one advantage of functional hemispheric asymmetry is that it reduces the amount of maladaptive interaction between processing components for which opposite hemispheres are dominant. On this view, it would be disadvantageous for an individual to have a neural organization that does not permit the component processes involved in a task to be spread over both hemispheres. Despite the potential importance of individual variation in complementarity of function, heretofore very little attention has been given to this dimension of individual variation. However, it would seem that the investigation of such differences could contribute significantly to the study of individual differences related to hemispheric asymmetry and the relation of such differences to cognitive ability.

Interhemispheric Communication

The final dimension of individual variation considered here concerns variation in the ability of the two cerebral hemispheres to communicate with each other and to coordinate their activities. In Chapter 6 we examined several issues related to interhemispheric interaction. There is likely to be reliable individual variation in several aspects of such interaction. As illustration, consider the following example.

In Chapter 6 we saw that the benefits and costs associated with interhemispheric cooperation depend on task difficulty. This idea was examined in experiments that required observers to compare various types of stimuli, with the critical stimulus information sometimes presented to one visual field (the within-hemisphere condition) and sometimes split between the two visual fields (the between-hemisphere condition). When the task is sufficiently easy, performance is better under the within-hemisphere condition than under the between-hemisphere condition. When the task is sufficiently difficult, however, the pattern of results is reversed: performance is better under the between-hemisphere condition than under the within-hemisphere condition. For both easy and difficult tasks, a measure of the efficiency of interhemispheric communication and coordination is given by the magnitude of the difference in performance between the within- and between-hemisphere conditions. An individual whose brain is efficient in these respects should demonstrate fewer costs associated with the need to transfer information from one hemisphere to the other than would an individual whose brain is inefficient in these respects. Consequently, the between-hemisphere disadvantage for easy tasks should be smaller for the "efficient-transfer" individual than for the "inefficient-transfer" individual. At the same time, the between-hemisphere advantage for difficult tasks should be larger for the "efficient-transfer" individual than for the "inefficient-transfer" individual because the advantage of spreading processing across both hemispheres is not counteracted so much by the costs of interhemispheric communication. An extreme case of inefficient transfer occurs in split-brain patients, who are completely unable to match stimuli across the hemispheres regardless of whether the task is difficult or easy.

Despite the plausibility of measuring the efficiency of interhemispheric communication in the ways just outlined, the reliability of such things as between-hemisphere minus within-hemisphere difference scores remains to be determined. Nevertheless, later sections of the present chapter discuss several hypotheses about how various subpopulations of humans may differ in the efficiency of relay of information across the corpus callosum. To the extent that data are in accord with these hypotheses, the measures used must have at least a level of reliability that is sufficient for research purposes.

In Chapter 6 we also saw evidence for the concept of metacontrol in the intact brain. Consider a task that the two hemispheres are predisposed to process in qualitatively different ways. *Metacontrol* refers to the ability of neural mechanisms to determine the extent to which each hemisphere tends to exert control over performance when stimulus conditions do not create a bias in favor of either hemisphere (e.g., presenting a stimulus to the center of the visual field). Chapter 6 reviewed data indicating that on such unbiased trials it is sometimes the case that the qualitative mode of processing favored by one hemisphere is used at the expense of the other hemisphere's preferred mode of processing. Furthermore, the favored mode of processing on unbiased trials is not always the mode associated with the hemisphere with greater ability to perform the task. This being the case, it is worthwhile to consider the possibility of reliable individual differences in whether it is the left or right hemisphere's preferred mode of processing that is used on perceptually unbiased trials, with individual variation on this dimension being independent of many of the other dimensions considered here. Some evidence of this possibility will be considered in a later section of the present chapter.

The existence of these logically independent dimensions of individual variation and the existence of ways to separate them does not mean that past investigations of individual differences have done so. In fact, it will become obvious in the subsequent sections of this chapter that most studies to date do not allow unambiguous interpretation in terms of these dimensions, even when reliable individual differences are found. This is the case because such things as performance asymmetries for virtually any single task can be influenced by a number of these dimensions. Nevertheless, as I discuss what is known about how individual differences in hemispheric asymmetry relate to between-subject factors such as handedness and sex, I will try to consider the foregoing dimensions as much as the existing data permit. In future studies designed with these dimensions in mind, it would be worthwhile to have the same individuals perform a variety of tasks chosen so that together they provide an interlocking web of converging evidence about the dimensions of interest.

Handedness

The study of handedness is interesting because individual differences in hand preference are a behavioral manifestation of individual differences in hemispheric asymmetry for certain types of manual activity and also because individual differences in hand preference may have some relationship to hemispheric asymmetry for other functions of a more cognitive nature. We will begin by considering hand preference in its own right as an example of hemispheric asymmetry for manual activity and review briefly what is known about the nature of handedness and about the genetic and environmental determinants of individual differences in handedness. This will be followed by discussion of the relationship between handedness and hemispheric asymmetry for other activities and processes.

As noted in Chapter 2, the most striking behavioral asymmetry in humans is the fact that approximately 90 percent of the population is right-handed. Consequently, deviation from the prototypical pattern of right-handedness is the most striking example of individual variation in those components of hemispheric asymmetry that are responsible for manual dominance. This being the case, it comes as no surprise that there are thousands of published studies comparing handedness groups in various ways. It is beyond the scope of this book to discuss all of the issues that have been addressed in this literature, but it is possible to note some of the conclusions that have been suggested most often.

It is clear that the consistency of hand dominance across behavioral tasks is typically greater for a group of self-described right-handers than for a group of self-described left-handers. In literate societies, most of us regard our dominant hand as the one we use for writing, and it is not unusual to find people who are left-handed for writing doing many other things with their right hand. It is more difficult to find people who are right-handed for writing preferring their left hand for other things. On this basis it has been argued that left-handers typically differ from right-handers in both the direction and magnitude of manual dominance—so much so that many investigators now prefer to talk about right-handers and non-right-handers, as there are very few strongly left-handed individuals. However, both right- and left-handers (by self-report) have different magnitudes of hand dominance for different dimensions of hand preference. For example, Steenhuis and Bryden (1989) found that both groups have a very strong preference for what they term "skilled" activities, including such things as writing, drawing, using a toothbrush, or holding a needle to sew. In contrast, the preferences for both groups were weaker for "less-skilled" activities, such as picking up a variety of small objects. It is interesting to note that there is some similarity of these two major dimensions of handedness in humans to the distinction made in other primates between manipulation and reaching (e.g., MacNeilage et al., 1987, and other studies discussed in Chapter 5).

Given that handedness is a behavioral manifestation of a particular type of hemispheric asymmetry, it is useful to consider what factors determine whether an individual becomes right-handed or left-handed. There is consensus that handedness is determined in complex ways by both genetic and environmental factors, but neither the genetic nor the environmental mechanisms are well understood. Furthermore, it should be kept in mind that the mechanisms of determination may well be different for strength or magnitude of hand dominance than for direction of hand dominance (see Bryden, 1987; Bryden and Steenhuis, 1991; and Chapter 5).

With respect to genetic determinants of the magnitude of handedness, Bryden and Steenhuis (1991; see also Bryden, 1987) suggest that what has been called the degree of handedness is more heritable than is the direction of handedness. In fact, they suggest that what is inherited is a predisposition to be either rigid or plastic in handedness. Individuals who are plastic may alter their hand preference readily in response to environmental situations, whereas individuals who are rigid do not.

With respect to genetic determinants of the direction of handedness, the strongest conclusion to emerge is that direction is not determined in any straightforward way by a single gene with a dominant allele for right-handedness and a recessive allele for left-handedness. This being the case, several more elaborate genetic models have been proposed, none of which are entirely supported by the available data. By way of illustration, consider the following "right-shift" model proposed by Annett (1985), as it has been one of the more promising.

Annett (1985) proposes that handedness is determined by one gene with two alleles, with offspring receiving one allelle from each parent. The dominant allele (*RS*+) produces a bias toward left-hemisphere dominance for language and toward right-handedness, whereas the recessive allelle (*RS*−) leads to the absence of this "right-shift" bias. With random mating, 75 percent of the population is predicted to be subject to the right-shift bias (25 percent of the population will have a genotype of *RS*++ and 50 percent of the population will have a genotype of *RS*+−). The remaining individuals (25 percent of the population will have the genotype *RS*−−) are unbiased genetically, and their handedness will be determined by environmental factors, so that half become right-handed and half become left-handed. This model makes several interesting predictions, one of which is that the observed proportion of left-handers should be approximately 12.5 percent—a value close to that obtained by several investigators. Not all predictions from this model fare as well. For example, by linking hemispheric asymmetry for language so closely with handedness, this model predicts that equal proportions of left-handed individuals should have left- versus right-hemisphere dominance for language. As we have seen earlier, it is estimated that only 20 percent or so of left-handers are right-hemisphere dominant for speech production, not the 50 percent predicted by the model. For additional discussion of this model as well as other genetic models, see Bradshaw (1989), Bryden (1982, 1987), and Bryden and Steenhuis (1991).

One type of environmental factor that is thought to contribute to handedness is pre- or perinatal trauma. In fact, Bakan and colleagues (e.g., Bakan, 1977) have argued for the strong position that all left-handedness is the result of such trauma. A weaker position has been taken by Satz and colleagues (e.g., Orsini and Satz, 1986), who argue that most left-handedness occurs naturally (e.g., determined by genetic factors) and only the remaining left-handedness occurs as a result of early trauma. For various reasons, the strong version of this trauma hypothesis is likely to be incorrect. However, the weaker version is plausible in view of the fact that the incidence of left-handedness is positively correlated with such things as birth stress.

A second type of environmental factor concerns the intrauterine environment during certain critical phases of fetal development. One of the most interesting theories of this type has been proposed by Geschwind and Galaburda and their colleagues (e.g., Geschwind and Galaburda, 1987) and is based on differences in the level of hormones, especially fetal testosterone. According to this theory, higher levels of fetal testosterone promote the intrauterine development of the right hemisphere relative to the left, either by slowing down development of portions of the left hemisphere or speeding up development of the right hemisphere. Given that the distal movements of each hand are controlled by the contralateral cerebral hemisphere, this theory argues that a higher level of fetal testosterone produces a greater chance of left-handedness.

There is at least some circumstantial evidence for this theory. Studies of asymmetry in other species indicate that fetal levels of hormones such as testosterone influence brain development in many ways, including the development of anatomical asymmetries and the behavioral asymmetries they produce (see Chapter 5 and Geschwind and Galaburda, 1987). In addition, in humans the incidence of left-handedness is higher in males than in females, who tend to have lower levels of fetal testosterone than males. Furthermore, de Lacoste, Horvath, and Woodward (1991) have reported sex differences in the developing fetal brain that are consistent with this hypothesis. For example, there was an overall volumetric asymmetry in favor of the right hemisphere for males, but the two hemispheres were equal in size or the left hemisphere was slightly larger in the fetal brains of females. There is also evidence that elevated levels of fetal testosterone predispose an individual toward diseases of the immune system, perhaps by influencing development of the thymus gland. This leads to the interesting prediction that the incidence of certain allergies and autoimmune disorders should be higher for left-handers than for right-handers, a finding that has been confirmed in a number of recent studies. For example, Geschwind and Galaburda summarize evidence showing that the incidence of various autoimmune diseases as well as certain atopic diseases (e.g., allergies, athsma) is over twice as great in left-handers than in right-handers. This theory has also been used to help explain why left-handers tend to die at a younger age than right-handers (e.g., Coren and Halpern, 1991) and to account for certain forms of intellectual precocity, as discussed by O'Boyle and Benbow (1990a,b) and in a later section of the present chapter.

Another intriguing theory of the prenatal origins of hemispheric asymmetry in humans has been proposed by Previc (1991) and will be discussed more fully in Chapter 8. For the moment, it is sufficient to note that according to this theory handedness is related to the asymmetric positioning of the fetus *in utero* during the final trimester of pregnancy. In most cases, the position of the fetus favors the development of the left otolith (part of the vestibular organ used for balance and so forth) and its neural pathways. This, in turn, creates a bias to use the left side of the body for postural control and the right side of the body for voluntary motor behavior. On this theory, left-handedness is the result of a deviation from the typical positioning of the fetus.

In addition to attempting to understand handedness for its own sake, researchers have focused on handedness as a possible marker for other aspects of hemispheric asymmetry. A common theme emerges across a wide variety of paradigms and measures. On the average, hemispheric asymmetry for a group of left-handers is typically in the same direction as for a group of right-handers, but the magnitude is significantly smaller. In part, this is attributable to the fact that a greater proportion of left-handers show an asymmetry opposite the direction considered prototypical of right-handers. This pattern of results has led to the conclusion that left-handers are more variable than right-handers in both the magnitude and direction of hemispheric asymmetry.

We have already seen that the distribution of certain biological asymmetries is related to handedness. For example, Hochberg and Lemay (1975) found that in right-handers the posterior end of the sylvian fissure was higher in the right hemisphere in 67 percent of the brains studied, equal in the two hemispheres in 26 percent of the brains studied, and higher in the left hemisphere in 7 percent of the brains studied (see also Chapter 4). For left-handers the corresponding percentages were 22 percent, 71 percent, and 7 percent, respectively. If we regard the right-handed pattern as prototypical, then the left-handed pattern contains fewer instances of prototypical asymmetry and more instances with no measurable asymmetry.

A similar pattern has been obtained from studies of behavioral deficits in brain-injured patients. For example, in Chapter 1 we saw that such studies lead to estimates that approximately 95 percent of right-handers are left-hemisphere dominant for speech production whereas approximately 5 percent are right-hemisphere dominant, with there being very few if any right-handers with speech represented equally in both hemispheres. Similar studies of left-handers lead to estimates that approximately 60 percent are left-hemisphere dominant for speech, 20 percent are right-hemisphere dominant, and 20 percent have speech represented equally in both hemispheres (e.g., Segalowitz and Bryden, 1983). Bryden and Steenhuis (1991) have used studies of brain-injured patients to provide similar estimates of hemispheric asymmetry for certain visuospatial tasks and report similar differences in the distribution for right- and left-handers. For right-handers, they estimate that 32 percent are left-hemisphere dominant, 68 percent are right-hemisphere dominant, and virtually none have equal ability in both hemispheres. For left-handers, they estimate that 30 percent are left-hemisphere dominant, 38 percent are right-hemisphere dominant, and 32 percent have equal representation in both hemispheres. Notice that in both cases fewer left-handers (compared with right-handers) are estimated to have the prototypical direction of asymmetry and more left-handers (compared with right-handers) are estimated to have no asymmetry.

The same general pattern of results has been obtained in studies using a variety of paradigms that investigate hemispheric asymmetry in neurologically intact individuals. By way of illustration, consider the distributions shown in Figure 7.4 for visual-field differences in CVC recognition. The data for right-handers are reproduced from Figure 1.3 and the data for left-handers were obtained in the same laboratory. For each handedness group, Figure 7.4 shows the percentage of the sample falling into categories defined by visual-field difference scores of different magnitudes, with positive values indicating an RVF/left-hemisphere advantage and negative values indicating an LVF/right-hemisphere advantage. As illustrated by the distributions in Figure 7.4, on the average there is an RVF/left-hemisphere advantage for both groups, but the advantage is larger for the right-handers. This comes about because, relative to the right-handers, there are more left-handers who have better performance on LVF/right-hemisphere trials and who have equal scores for the two visual fields. A similar outcome is shown in Figure 7.5 for a dichotic-listening task requiring the identification of CV syllables.

A very promising approach taken by Kim and Levine (1991b)

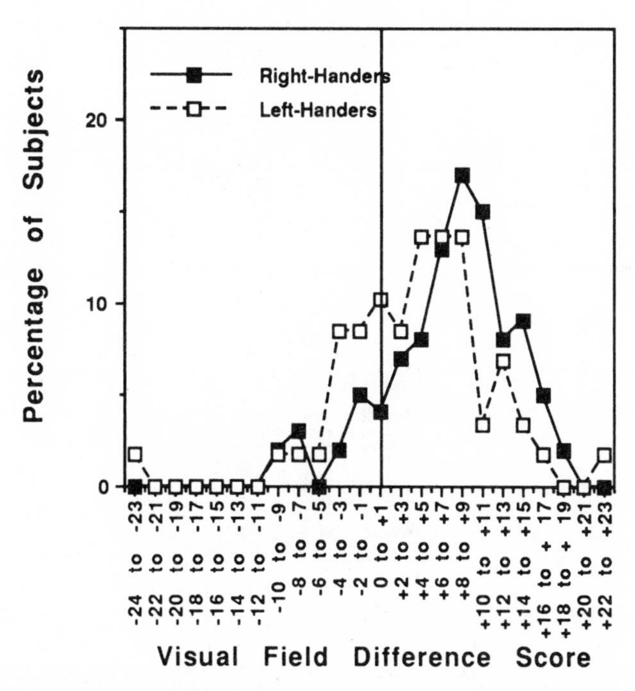

Figure 7.4. Visual-field difference scores for groups of right-handed and left-handed observers. The figure graphs the percentage of subjects who showed RVF − LVF scores of different magnitudes (with a maximum possible score of 36). The vertical line divides positive from negative scores.

has used what they refer to as a "two-task criterion" to infer patterns of hemispheric asymmetry for right-handed and left-handed individuals. Each individual performed two visual-half-field tasks, one involving word recognition and the other involving face recognition. For each task, the number of correct responses was recorded for each visual field and the RVF-minus-LVF difference value was computed as an asymmetry score. Individuals whose asymmetry scores were larger for the word task than for the face task were classified as "typical" in relative asymmetry for the two tasks, and individuals whose asymmetry scores were larger for the face task than for the word task were classified as "atypical." Of particular interest is the fact that approximately 90 percent of right-handed individuals fell into the typical group whereas only approximately 66 percent of left-handers fell into

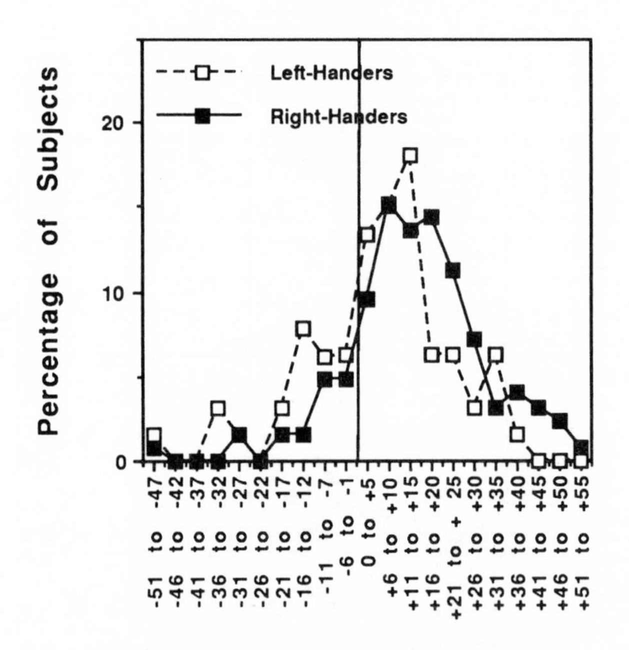

Figure 7.5. Ear difference scores for groups of right-handed and left-handed listeners. The figure graphs the percentage of subjects who showed RE − LE scores of different magnitudes (with a maximum possible score of 120). The vertical line divides positive from negative scores.

that group. Note that these values are very close to the estimates discussed earlier of the proportion of right- and left-handers with left-hemisphere dominance for speech.

All of these results suggest that there is a relationship between handedness and hemispheric asymmetry for other processes, but the relationship is far from perfect. On the average, hemispheric asymmetry is generally in the same direction for left-handers as it is for right-handers, but the magnitude is smaller. In fact, this general pattern is so well accepted that some have argued that it can be used to examine the validity of new tasks designed to measure hemispheric asymmetry (e.g., Bryden, 1982; Bryden and Steenhuis, 1991; Levy et al., 1983b). That is, if a perceptual or performance asymmetry in right-handers is attributable to hemispheric asymmetry, one would expect a group of left-handers to show an average asymmetry in the same direction but smaller in magnitude.

The fact that left-handers are more variable than right-handers in the direction of hemispheric asymmetry for other processes has led to the search for additional variables that might indicate which left-handers have a pattern of hemispheric asymmetry characteristic of right-handers and which do not. By and large, the search has been unsuccessful. One variable that seemed very promising for a time is the hand posture used for writing (for reviews see Levy, 1982, and Weber and Bradshaw, 1981). Levy and Reid (1978) hypothesized that use of an inverted handwriting posture (with the hand positioned above the line of writing) indicates control of linguistic functions by the hemisphere ipsilateral to the writing hand, whereas use of the normal, noninverted posture (with the hand positioned below the line of writing) indicates control of linguistic functions by the hemisphere contralateral to the writing hand. Although initial tests of this hypothesis in visual-half-field experiments were very promising, the critical effects have been difficult to extend to other stimulus modalities and even to a wide range of visual tasks. Consequently, handwriting posture is not a sufficiently reliable index of hemispheric asymmetry of cognitive function in either left- or right-handers.

Another variable that has shown some promise is the presence of left-handedness in the close relatives of an individual (that is, familial sinistrality). Unfortunately, findings with respect to the importance of familial sinistrality are equivocal. There is some evidence that hemispheric asymmetry for certain aspects of language is reduced in both left- and right-handed individuals who have a left-handed parent (e.g., Bradshaw, 1989; Kee, Hellige, and Bathurst, 1983; McKeever, 1991; O'Boyle and Benbow, 1990a). However, evidence of this is not uniformly found, and any effects related to familial sinistrality may be moderated in complex ways by other factors, such as sex.

Most of the investigations of handedness and hemispheric asymmetry have focused on the direction and magnitude of asymmetries. Consequently, it is difficult to know whether handedness is also related to the other dimensions of individual variation discussed earlier. Nevertheless, more recent studies have begun to shed light on the possible relationship of handedness to asymmetric arousal, to complementarity of asymmetries, and to interhemispheric communication.

What little evidence there is at the present time suggests that the characteristic pattern of asymmetric arousal of the two cerebral hemispheres is unrelated to handedness. For example, in the study discussed earlier by Kim et al. (1990), the distribution of individuals with a leftward or rightward perceptual bias was not significantly different for right- versus left-handers. Furthermore, their principal-component analyses indicated that characteristic perceptual bias accounted for approximately the same amount of variance within each handedness group. It should be noted, however, that the mean visual-field difference was smaller for left-handers than for right-handers for only one task (face recognition), contrary to the expectations described earlier and to the patterns for other tasks shown in Figures 7.4 and 7.5. Thus, additional investigations of the relationship between handedness and characteristic perceptual asymmetry are needed before any definitive conclusion is drawn.

Research to date does not seem to have addressed explicitly the question of whether complementarity of asymmetries differs with handedness, but there are sufficient data to suggest the following as a working hypothesis. There is reasonable evidence that complementarity of function arises as a by-product of different processes becoming asymmetric independently of each other—rather than because of a direct causal link (see Chapters 2 and 5). Thus, if two functions are strongly asymmetric in opposite directions, then a large proportion of the population will have opposite hemispheric asymmetry for the two tasks as a statistical by-product of their independence. This distribution is illustrated in the upper portion of Table 7.1, which shows the hypothetical incidence of hemispheric asymmetries for language and visuospatial processes for 100 individuals. When the two functions are only weakly asymmetric, however, the same statistical independence leads to a much smaller proportion of individuals who have opposite hemispheric asymmetry for the two tasks. This distribution is illustrated in the lower portion of Table 7.1. In fact, the marginal totals used in the upper portion of Table 7.1 are based on estimates of the propor-

Table 7.1. Hypothetical outcomes of patients tested on language tasks and on visuospatial tasks: predicted number of subjects showing each possible outcome

Right-Handed Patients

	Hemispheric advantage for language			
	Left	Neither	Right	Total
Hemispheric advantage for visuospatial processing				
Left	30.8	0	1.2	32
Neither	0	0	0	0
Right	65.2	0	2.8	68
Total	96	0	4	

Left-Handed Patients

	Hemispheric advantage for language			
	Left	Neither	Right	Total
Hemispheric advantage for visuospatial processing				
Left	19.8	5.1	5.1	30
Neither	21.1	5.4	5.4	32
Right	25.1	6.5	6.5	38
Total	66	17	17	

tion of right-handers who have left-hemisphere dominance, right-hemisphere dominance, or no asymmetry for certain aspects of language and for certain aspects of visuospatial processing (e.g., Bryden and Steenhuis, 1991). The corresponding marginal totals in the lower portion of Table 7.1 are estimates given by Bryden and Steenhuis for left-handers. What this illustration suggests is that complementary asymmetry for certain processes is likely to occur more often in right-handers than in left-handers, a hypothesis that merits explicit investigation.

In an explicit investigation of interhemispheric communication in right- and left-handers, Banich, Goering, Stolar, and Belger (1990) compared within-hemisphere and between-hemisphere performance for matching tasks similar to those used by Banich and Belger (1990) and discussed earlier. In each of two experiments, the within- versus between-hemisphere performance differences were identical for both handedness groups. Furthermore, the pattern of results did not depend on writing-hand posture, familial sinistrality, or sex. This suggests that, at least in the visual modality, interhemispheric communication and coordination are unrelated to handedness. Similar results have been reported by Clarke (1990) for a visual matching task performed under either within- or between-hemisphere conditions and also for a tactile "roughness" matching task performed under either same- or different-hand conditions. However, in a study that examined the extent to which vibrotactile information was integrated across the two hands, Verrillo (1983) reported equal integration in strongly right- and strongly left-handed subjects and no integration in ambidextrous subjects, despite the fact that performance under same-hand control conditions was equivalent for all three groups. In addition, Potter and Graves (1988) have reported better performance by non-right-handers than by right-handers on tasks that require the integration of either tactile information or motor control across the two hands. Unfortunately, it is difficult to interpret their results in terms of interhemispheric integration because there were no same-hand control tasks.

Sex

In humans there are relationships between biological sex and cognitive abilities. For example, females tend to score higher than males on tests of verbal fluency and manual skill, whereas males tend to score higher than females on tests of visual perception and spatial ability (see Bradshaw, 1989; Halpern, 1986; O'Boyle and Benbow, 1990a; O'Boyle and Hellige, 1989). Of course, there is substantial overlap in the performance of the two sexes for all of these tasks, and the differences in the means are typically small relative to the variance. The determination of these sex differences in cognitive ability is likely to involve complex interplay between biological and environmental factors, and it is beyond the scope of the present book to present a systematic review of the many hypotheses that have been advanced. Instead, the cognitive differences are noted here because some have suggested that they come about because various aspects of hemispheric asymmetry tend to be different for males and females, and it is this possibility that will be explored.

The possibility that hemispheric asymmetry in humans is somewhat different for males and females is plausible in view of the many effects of fetal hormones on brain development in other species and in view of the many behavioral effects of circulating sex-hormone levels in other species (see Chapter 5). Although there are clearly effects of sex-related hormones in other species, it should be noted that the pattern of effects is not at all simple —at least with respect to biological and functional brain asymmetries. On this basis, one might expect that whatever effects are found in humans will be similarly complex. Some indication of the complexity at the biological level comes from studies of the size of the corpus callosum reviewed in Chapter 4. For example, to the extent that there are sex differences in the size of the isthmus of the corpus callosum, the differences are moderated by handedness. Of course, it remains to be determined how the size of different regions of the corpus callosum relates to functional hemispheric asymmetry, but there is some indication that the nature of the relationship may be different for males and females (e.g., Clarke, 1990).

At the behavioral or functional level, probably the most well-established sex difference is that the incidence of left-handedness is slightly higher in males than in females. As noted earlier, this is consistent with the hypothesis that higher levels of fetal testosterone tend to promote development of the right hemisphere relative to the left. On this basis, one might expect that measures of asymmetry for other behaviors would also be shifted toward an advantage for the right hemisphere in males relative to females. To the extent that this shift happens at all, it would seem to be an artifact of the slightly higher incidence of left-handedness in males. For example, in studies restricted to right-handers there is no evidence whatsoever of such a uniform shift in asymmetry related to sex. In fact, the hypothesis most often advanced regarding sex differences in hemispheric asymmetry (holding handedness constant) is that males are more asymmetric than females. However, this hypothesis is controversial and the evidence in its favor is equivocal—so much so that a few investigators have even claimed that females are more asymmetric than males!

One way to examine sex differences in hemispheric asymmetry is to examine the incidence of certain disorders (e.g., aphasia) after unilateral brain injury and determine whether different distributions are obtained for males and females. Probably the most encouraging data for the hypothesis that there are sex differences in hemispheric asymmetry come from such studies, although there is still controversy surrounding their interpretation. For example, on the basis of studies by Hecaen (cited in Bryden, 1982) and by Hecaen, DeAgostini, and Monzon-Montes (1981), Bryden (1982) estimates that the left hemisphere is dominant for speech in approximately 95 percent of right-handed males and in approximately 79 percent of right-handed females, with the right hemisphere being dominant for speech in the remaining right-handed males and females (see also McGlone, 1980, 1986). Bryden also estimates that the incidence of right-hemisphere dominance for spatial processes is higher for right-handed males than for right-handed females. As he notes, such results may indicate that fewer females than males show the prototypical pattern of hemispheric asymmetry, but they should not be taken as evidence that females have a smaller magnitude of hemispheric asymmetry than males. In fact, on the basis of other data he argues that females have asymmetries just as large in *magnitude* as those for males, but there are more females with an atypical *direction* of asymmetry (see also Bradshaw, 1989). As Bryden also notes, the hypothesis that more females than males have an atypical direction of hemispheric asymmetry for such things as speech and spatial processes has implications for sex differences in complementarity of asymmetry.

Using the same logic applied earlier to handedness, we would expect fewer females than males to have complementary hemispheric dominance for speech and spatial processes. This hypothesis merits explicit investigation in future studies.

It should be noted that there are alternative explanations for such things as the lower rate of aphasia after left-hemisphere injury in females than in males (a major factor that contributes to the hypothesis that fewer females are left-hemisphere dominant for speech). One of the most interesting alternatives argues that this effect occurs because of differences between males and females in the organization of language *within* the left hemisphere (Kimura, 1987). On this view, speech and manual activities are represented more focally in the female left hemisphere, with anterior regions of the left hemisphere being especially important. For males, speech is proposed to be represented more diffusely in the left hemisphere, with the critical regions extending somewhat more posteriorly than is the case for females. Most studies of brain-injured patients use individuals who have had vascular accidents, and there are indications that when such accidents cause restricted injury they affect posterior more than anterior regions. Thus, the reason that the rate of aphasia after left-hemisphere injury is lower in females is that the type of injury usually studied is more likely to spare the anterior regions of the left hemisphere. This hypothesis must be subjected to a great deal more empirical testing before it can be evaluated (see Bradshaw, 1989). Nevertheless, it provides a useful illustration of the difficulties that arise in trying to interpret sex differences in brain function.

If aspects of hemispheric asymmetry tend to be different for males and females, it should be possible to see evidence of these sex differences in studies of perceptual asymmetry in neurologically intact individuals. There is certainly no shortage of relevant studies in the literature. Unfortunately, almost every conceivable pattern of results has been reported so that no clear pattern emerges. When attempts are made to draw systematic conclusions, the most common claim is again that, on the average, females show smaller perceptual asymmetry scores than males—with the results more likely to reflect sex-related differences in the direction of asymmetry rather than in the magnitude of asymmetry. However, even those who make such attempts based on their own reviews of the literature are quick to point out the existence of at least some contradictory data (e.g., Bradshaw, 1989; Bradshaw and Nettleton, 1983; Bryden, 1982; Springer and Deutsch, 1989).

The lack of robust sex differences in perceptual asymmetries is illustrated by interesting reviews of sex differences in dichotic-listening studies, visual-half-field studies and studies of tactile asymmetry provided by Hiscock and colleagues (e.g., Hiscock, Hiscock. and Kalil, 1990; Hiscock, Hiscock, and Inch, 1991, 1992). In each case, the entire contents of several neuropsychology journals were surveyed for relevant studies that included a statistical analysis of sex differences in perceptual asymmetry (e.g., hemisphere-by-sex interaction). In their survey of approximately 350 dichotic-listening studies, Hiscock et al. (1990) found that 40 percent of studies provided information about sex differences. Of those, only 8 studies met the authors' criteria for finding sex differences in ear asymmetry (e.g., a statistically significant ear-by-sex interaction in analysis of variance). Of those 8 studies, 6 were consistent with the hypothesis of greater hemispheric asymmetry in males. Similar results were obtained in a survey of approximately 500 visual-half-field experiments. Of those, Hiscock et al. (1991) found that 42 percent provided information about sex differences and, of those, 24 studies met the authors' criteria for finding sex differences in visual-field asymmetry. Of those 24 studies with sex differences, 17 were consistent with the hypothesis of greater hemispheric asymmetry in males. In a survey of 59 investigations of tactile asymmetry, 30 provided information about sex differences. Of those, only 5 met the authors' criteria for finding sex differences in tactile asymmetry. Two of those investigations produced results that were consistent with the hypothesis of greater asymmetry in males, one showed the opposite, and two were ambiguous. As Hiscock and colleagues point out, the incidence of sex differences is so small in each of these cases that it is difficult to rule out the possibility that those sex differences that were obtained are chance findings. Of course, in fairness to those who argue for at least weak evidence in favor of the hypothesis that hemispheric asymmetry is greater in males than in females, it should be noted that when sex differences were found, results were in a direction consistent with the hypothesis approximately 68 percent of the time (in 25 out of 37 cases, $p < .05$ by binomial sign test).

With respect to the dimensions of individual variation discussed at the beginning of the present chapter, it is clear that most investigations of sex differences have focused on the direction and magnitude of hemispheric asymmetry and have often not made a clear distinction between the two. In addition, we have seen that the existence of any sex differences in the tendency to deviate from the prototypical pattern of asymmetry would be expected to have consequences for the complementarity of asymmetries. To date there have been too few studies of possible sex differences in characteristic arousal asymmetry to reach strong conclusions, but it is the case that there were no sex differences in the study discussed earlier by Kim et al. (1990).

With respect to possible sex differences in interhemispheric communication, the few relevant findings are contradictory. Recall that in visual experiments neither Banich and Belger (1989) nor Banich et al. (1990) found sex effects in comparisons of within-hemisphere and between-hemisphere conditions. In addition, in a study discussed earlier, Clarke (1990) found no sex differences in the ability to transfer either visual or tactile information from one hemisphere to the other. By way of contrast, Potter and Graves (1988) found that females outperformed males on tasks that required interhemispheric integration of either visual or tactile information but, as noted earlier, they did not include appropriate within-hemisphere control conditions. From experiments that did include appropriate control conditions, there is some indication males outperform females when both hemispheres are required to work together in making haptic discriminations (e.g., Cohen and Levy, 1986) or auditory discriminations (e.g., McRoberts and Sanders, 1992), but more work is clearly needed before any definitive conclusion can be drawn.

Even when sex differences are found in performance that is thought to reflect hemispheric asymmetry, it does not necessarily mean that males and females differ in relatively permanent aspects of hemispheric asymmetry. For example, it may be the case that males and females use different strategies to approach a variety of tasks. It is now well known that the pattern of hemispheric asymmetry observed in an experiment is determined more by the processing strategies used than by the nature of the stimuli per se (see Chapter 3). Consequently, whenever such things as perceptual asymmetries are different for males and females, explanations in terms of preferred strategy must be considered in addition to explanations in terms of brain structure per se. Viewed another way, uncontrolled variation in the mixture of strategies from study to study may account for some of the equivocal findings in the literature.

Another variable that may be important in determining whether or not sex differences are obtained in a particular study concerns the point at which women are tested during their menstrual cycles. For example, Hampson and Kimura (1988) examined the performance of women on a variety of tasks during the luteal and menstrual phases of their cycle. Among other things, they found better performance on the spatially demanding rod-and-frame task during the menstrual phase than during the luteal phase, and just the opposite for a variety of manual-dexterity tasks. They suggest that high levels of female hormones (e.g., estrogen) during the luteal phase enhance performance on tasks that females generally perform better than males (e.g., manual sequencing) and impair performance on tasks that males generally perform better than females (e.g., the rod-and-frame task). Because impaired manual sequencing is associated with left-hemisphere injury (see Chapter 2), Hampson and Kimura suggest that the left hemisphere may be particularly sensitive to levels of sex hormones in adult humans.

Chiarello, McMahon, and Schaefer (1989) provided a preliminary investigation of this hypothesis by examining changes in visual-half-field asymmetry across the menstrual cycle. One of their tasks required individuals to indicate whether or not a string of letters spelled a word. For this lexical-decision task there was an RVF/left-hemisphere advantage both for a measure of discriminability and for reaction time of correct responses, and in neither case did the asymmetry change over the menstrual cycle. In contrast, a measure of the response criterion did change over the menstrual cycle on RVF/left-hemisphere trials but not on LVF/right-hemisphere trials. In this sense, there is at least some preliminary support for the Hampson and Kimura (1987) hypothesis. For a task that required the discrimination of line orientation, there were no significant visual-field differences in discriminability at any point in the menstrual cycle. Analysis of the measure of the response criterion indicated a more lax criterion on LVF/right-hemisphere trials than on RVF/left-hemisphere trials, but there was no indication that this measure changed over the menstrual cycle for either visual field. The contrast to performance on the lexical-decision task suggests that effects of sex hormones on measures of hemispheric asymmetry are specific to certain processing components. In addition to being interesting for what they tell us about the effects of sex hormones on functional hemispheric asymmetry, studies such as these illustrate how sex differences may well depend on when during their menstrual cycles females are tested.

Intellectual Abilities

The present section considers the possibility that individual differences in hemispheric asymmetry are related to individual differences in intellectual ability and performance. It should be noted at the outset that, to the extent such relationships exist, they are subtle and complex and do not seem likely to account for a very large proportion of the normal variation in ability. However, there may be a stronger relationship between patterns of hemispheric asymmetry and cognitive performance outside of the "normal" range. This possibility will be illustrated by considering various hypotheses about intellectual precocity and dyslexia.

Over the years, a number of hypotheses about the relationship between handedness and intellectual functioning have been proposed. Given that there is at least some relationship between handedness and hemispheric asymmetry for a variety of cognitive factors, relationships between handedness and intellectual functioning provide at least indirect support for a relationship between intellectual functioning and hemispheric asymmetry. The most common hypotheses advanced typically propose small deficits for left-handers compared with right-handers, especially for certain spatial functions (for reviews see Hardyck and Petrinovich, 1977; O'Boyle and Benbow, 1990a; O'Boyle and Hellige, 1989). On the basis of her own empirical findings, Levy (1969) suggested that the deficits occur because areas of the right hemisphere that are normally involved in spatial functions have been encroached upon by bilateral representation of language in left-handers. Despite a certain appeal to this hypothesis, there are many studies that fail to find spatial deficits in left-handers, and recent reviews generally call into question the entire notion of intellectual deficits in left-handers.

Interesting work by Harshman and colleagues (summarized by Harshman and Hampson, 1987) clarifies some of the conflicting reports in the literature and suggests a complex relationship between handedness and intellectual ability. In several studies, they divided samples of male and female right- and left-handers into "high-reasoning" and "low-reasoning" groups on the basis of tests of reasoning ability. For the high-reasoning group, left-handed males scored lower than right-handed males on each of 15 tests of spatial ability, a pattern that is consistent with Levy's (1969) hypothesis. For high-reasoning females, however, left-handers performed better than right-handers on 12 of the 15 tests of spatial ability. In the low-reasoning group, the patterns for both males and females tended to be opposite what was found in the high-reasoning group. That is, with respect to performance on tests of spatial ability, there was a three-way interaction involving handedness, sex, and reasoning ability (for converging evidence of such an interaction, see Gordon and Kravetz, 1991). It was also found that in groups with high reasoning ability larger right-ear/left-hemisphere advantages on a verbal dichotic-listening test were associated with enhanced performance on tests of verbal ability, whereas larger left-ear/right-hemisphere advantages on a nonverbal dichotic-listening test were associated with enhanced performance on tests of spatial ability. In groups with low reasoning ability these relationships were absent or reversed.

Given the complexity of the results just noted, it is not surprising that there is little support for any simple relationship between aspects of hemispheric asymmetry and intellectual ability within the right-handed population (see O'Boyle and Hellige, 1989). However, in a notable series of experiments, Bever and colleagues (e.g., Bever, Carrithers, Cowart, and Townsend, 1989) have reported that in right-handers the relative dependence on different aspects of language is influenced by familial sinistrality. Specifically, right-handers with no left-handed family members depend more on grammatical relations between words than on conceptual information provided by individual words. The relative dependence on these aspects of language is reversed for right-handers with one or more left-handed family members. Bever et al. suggest that these effects of familial sinistrality may be related to individual differences in how extensively lexical knowledge is represented throughout both hemispheres, a hypothesis worthy of additional investigation. In addition, there are indications that extreme intellectual precocity and dyslexia are related to unusual aspects of hemispheric asymmetry, and it is to these more extreme levels of intellectual performance that we now turn.

Intellectual Precocity

Guided by the theoretical work of Geschwind and Behan (1982; see also Geschwind and Galaburda, 1987), Benbow and colleagues have found a link between extreme intellectual precocity, left-handedness, sex, and behaviorally measured aspects of hemispheric asymmetry (e.g., Benbow, 1986; O'Boyle and Benbow, 1990a,b). For example, the incidence of left-handedness is higher than expected for boys and girls who score at least 630 on the verbal portion of the Scholastic Aptitude Test and at least 700 on the math portion of the same test before age 13. Approximately 1 in 10,000 children score in this range, so we may assume that such a sample is extremely gifted. It has been proposed that this atypical distribution of handedness is related to enhanced right-hemisphere functioning in the gifted, possibly as a result of early exposure to fetal testosterone. The notion is that the sort of reasoning that is necessary to reach this extremely gifted range depends on superior spatial abilities of the sort for which the right hemisphere is dominant.

The fetal-testosterone hypothesis receives some support from the fact that females are greatly underrepresented in the extremely gifted group, primarily because of their inability to achieve sufficiently high scores on the math portion of the test (the ratio of boys to girls may be as high as 13:1). Support also comes from the fact that certain allergies and immune disorders are more prevalent in the extremely gifted population than in appropriate control populations (e.g., Benbow, 1986). Recall that it was noted earlier that higher levels of fetal testosterone are correlated with an increase in the incidence of such disorders.

Even within the right-handed population, there is some evidence of enhanced right-hemisphere involvement in this extremely gifted population. For example, O'Boyle and Benbow (1990b) tested right-handers from the extremely gifted population and from a control population of age-matched children with average intellectual ability on a dichotic-listening task with CV syllables and on the free-vision face task developed by Levy et al. (1983a). On the dichotic-listening task, the control group showed a significant right-ear/left-hemisphere advantage. However, the smaller advantage in the same direction for the gifted children was not statistically significant. This was primarily caused by the fact that left-ear performance was better for the gifted group than for the control group, whereas right-ear performance was more nearly equal for the two groups. Consequently, overall performance was significantly better for the gifted group than for the control group. On the free-vision face task, both groups showed the expected leftward bias discussed earlier in this chapter, but the bias was significantly greater for the gifted group than for the control group. Recall that Levy et al. (1983a,b; see also Kim et al., 1990) argue that individual variation in bias on the face task is related to variation in characteristic arousal asymmetry of the two hemispheres. Thus, the results obtained by O'Boyle and Benbow are consistent with the hypothesis of greater right-hemisphere involvement in gifted than in control children.

A converging test of this possibility comes from a preliminary EEG study comparing gifted and control children (O'Boyle, Alexander, and Benbow, 1991). Interestingly, when children were looking at a blank slide (the baseline condition), the left hemisphere was found to be more active than the right hemisphere for gifted children but not for control children. Relative to baseline, the gifted children showed increased activation over the right hemisphere (especially over the right temporal lobe) when they were performing the free-vision face task, whereas the control children did not. No group difference of this sort was obtained for a task that required children to judge whether a word was a noun or a verb. Given the group differences in the baseline condition, the results of this study are difficult to interpret. On one hand, the baseline results seem inconsistent with the hypothesis of greater right-hemisphere activation in gifted children. On the other hand, relative to baseline, the results obtained during the free-vision face task offer some support for the hypothesis—at least when children are performing a task thought to reflect asymmetric hemispheric arousal. In any event, these preliminary results are provocative and sufficiently interesting to provide motivation for additional studies of the relationship between various dimensions of hemispheric asymmetry and extreme intellectual ability.

Dyslexia

A specific reading disorder known as dyslexia is characterized by extreme difficulty in learning to read, despite above-average intelligence, the absence of social or educational deprivation, and the absence of any noticeable brain trauma. Over 50 years ago, Orton (1937) hypothesized that dyslexia was accompanied by incomplete or reversed hemispheric dominance for language, which was consistent with his observation that left-handedness was higher among dyslexics than among normal readers. Many of Orton's specific ideas about the neurological correlates of dyslexia have proven wrong, and some of his observations about asymmetry have not been confirmed in more recent studies. For example, despite popular views to the contrary, neither the incidence of left-handedness nor the incidence of non-right-handedness is consistently higher among dyslexics than among normal readers (e.g., Gilger, Pennington, Green, S. M. Smith, and S. D. Smith, 1992; Hugdahl, Synnevag, and Satz, 1990). Nevertheless, the search for neurological correlates has continued, some having to do with hemispheric asymmetry. For example, various investigators have hypothesized that dyslexics have reduced hemispheric asymmetry relative to normal readers, specific left-hemisphere problems, specific right-hemisphere problems, deficits of interhemispheric communication, and so forth. The literature on dyslexia has grown so large and varied that it has become difficult to evaluate these possibilities, a situation that is compounded by the uneven quality of the empirical studies and by the fact that dyslexia is not a single disorder with one single cause or set of neurological correlates (see O'Boyle and Hellige, 1989). Despite problems with some of the research, the remainder of this section reviews two hypotheses that are sufficiently promising to warrant additional empirical investigation.

One promising hypothesis is that at least a sizable subgroup of dyslexics has left-hemisphere dominance for aspects of language (like normal readers) but has deficiencies in the left-hemisphere language mechanisms. A number of subtle abnormalities in the brains of dyslexics are consistent with this possibility. For example, in Chapter 4 we saw that an area of the left hemisphere known as the planum temporale is typically larger in the left hemisphere than in the right hemisphere in the brains of adults and children who learn to read normally. We also saw that this asymmetry is reduced, absent, or even reversed in dyslexic adolescents—largely because of a reduction in the size of the left planum relative to normal readers. In addition, Galaburda and colleagues (see Geschwind and Galaburda, 1987) have reported abnormal cytoarchitecture (the specific arrangement of neurons) in the temporal language areas of the left hemisphere in dyslexics. Furthermore, Landwehrmeyer, Gerling, and Wallesch (1990) have examined task-related brain potentials in dyslexics and in normal readers. They found that when normal readers were performing a number of language tasks (e.g., reading, orthographic error detection, finding antonyms) there was greater cortical activity over the left hemisphere than over the right hemisphere. This pattern was reversed for dyslexics.

In view of these findings, it is interesting that a number of behavioral studies suggest that left-hemisphere dominance for certain aspects of language is just as characteristic of dyslexics as it is of normal readers but that the language mechanisms within the left hemisphere are deficient. For example, Moscovitch (1987) provides a critical review of visual-half-field studies that compare dyslexics and normal readers and concludes that dyslexics show an RVF/left-hemisphere advantage for the identification of words and pseudowords that is just as large as the advantage obtained with normal readers. However, the overall performance of dyslexics is far below that of normals, which is consistent with the hypothesis that the language capacities of the left hemisphere are below normal. Similar results have been obtained with dichotic-listening tests (e.g., Bloch, 1989; Witelson, 1977), although there is some indication that the pattern of results varies with such things as dyslexic subtype, type of stimuli, and instructions to attend selectively to one ear (e.g., Morton and Siegel, 1991). In addition, Gordon (1980) used a battery of tasks designed on the basis of deficits in brain-injured patients to tap left- versus right-hemisphere functions. He found that dyslexics and their immediate family members were impaired on the left-hemisphere tasks but not on the right-hemisphere tasks. Thus, the hypothesis that dyslexics suffer from problems of the left-hemisphere language areas has received sufficient report to merit additional investigation.

A second promising hypothesis is that at least a subgroup of dyslexics has normal hemispheric asymmetry of cognitive functions but has unusual difficulty with interhemispheric communication. This hypothesis has a great deal of intuitive appeal. Dyslexia is a disorder that is largely restricted to reading, which is a task that demands a great deal of interplay between visuospatial and linguistic information of all sorts. This being the case, it is reasonable to suppose that reading requires efficient communication between the hemispheres and that efficient interhemispheric communication is particularly important during the *acquisition* of reading skill, when relatively novel and arbitrary visual patterns have to be associated with phonetic information (e.g., Gladstone and Best, 1985).

Gladstone and Best (1985) make a plausible case for this hypothesis by reviewing suggestive evidence, but there is clearly a need for more direct empirical tests. In one of the few studies available, Badian and Wolff (1977) required dyslexics and normal readers to tap along with a metronome and to continue tapping at the same tempo when either the metronome was turned off or switched to a different tempo. Sometimes individuals tapped with one hand and sometimes they alternated hands (a condition that requires interhemispheric communication and coordination). Relative to the normal readers, the dyslexics were particularly impaired in the alternating-hand condition. This finding led the authors to suggest that the primary motor deficiency that is sometimes reported in dyslexics is caused by a disturbance in interhemispheric cooperation. Work summarized by Kinsbourne (1988; cited in O'Boyle and Hellige, 1989) also suggests that dyslexics have difficulty with interhemispheric communication and cooperation, though a recent study by Bloch (1989) finds no evidence of such a difficulty in the tactile modality. Given the highly visual nature of reading, it would seem particularly worthwhile to examine interhemispheric communication in the visual modality, perhaps using tasks like the one used by Banich and Belger (1990) and illustrated in Figure 6.2.

It would also seem worthwhile in future studies to determine whether dyslexics differ from normal readers on the other dimensions of individual variation discussed earlier in the present chapter. For example, there may be at least a subgroup of dyslexics whose problem arises from the fact that in their brains the same hemisphere is dominant for certain critical processes for which opposite hemispheres are usually dominant (i.e., whose brains show an unusual pattern of complementarity). In addition, some dyslexics may have such a highly aroused right hemisphere that they have difficulty applying the appropriate left-hemisphere processes in learning to read, and this arousal asymmetry may stem from subtle biological abnormalities of the left hemisphere. At the present time, there are virtually no data directly relevant to these possibilities, which means that any evaluation of their merit must be postponed.

Psychopathology

In Chapter 2 we saw that the emotional behavior of patients with unilateral brain injury depends on which hemisphere is injured. Left-hemisphere injury leads to more catastrophic reactions, whereas right-hemisphere injury leads to more reactions of indifference. These and other findings have led to the hypothesis that the right hemisphere is more involved than the left in producing "negative" emotions, whereas the left hemisphere is more involved than the right in producing "positive" emotions. The idea of hemispheric asymmetry for the production of emotions suggests the possibility that emotional disorders are associated with unusual aspects of hemispheric asymmetry. Consider as an example recent research that investigates the neurological correlates of schizophrenia.

Several hypotheses have been advanced about the relationship between hemispheric asymmetry and schizophrenia. At the present time support is growing for the general hypothesis that schizophrenia is accompanied by subtle abnormalities of the left hemisphere. Evidence has come from studies using a variety of techniques described earlier in this book, including various measures of regional cortical activity and measures of perceptual and response asymmetry. Furthermore, the specific pattern of hemispheric asymmetry seems to depend on such things as the nature of schizophrenic symptoms (for reviews see Flor-Henry, 1987; Walker and McGuire, 1982; Wexler, Giller, and Southwick, 1991).

For example, Wexler et al. (1991) tested healthy control subjects, schizophrenic patients, bipolar manic patients, schizo-affective patients, and depressed patients on two dichotic-listening tasks that required subjects to make a single response on each trial. One test presented two very similar nonsense stimuli on each trial (e.g., *aka, aba*), one to each ear, and required subjects to make a single response. The other test was similar except that the stimuli were two similar words (e.g., *coat, goat*). In order to examine hemispheric asymmetry for these two tasks, Wexler et al. computed an asymmetry score by subtracting the number of correct left-ear responses from the number of correct right-ear responses and dividing by the total number of correct responses. Using this asymmetry score, they found that the right-ear advantage was significantly smaller for schizophrenic patients than for healthy control subjects, depressed patients, and schizo-affective patients. This was equally true for both the nonsense task and the word task and is consistent with the hypothesis of subtle left-hemisphere deficits in schizophrenia. Unfortunately, Wexler et al. do not present the accuracy scores for the left and right ears, so it is impossible to determine whether the schizophrenic asymmetry pattern was produced by changes in only one ear or in both. It should be noted that earlier dichotic-listening studies using somewhat different tasks were more equivocal in their results when accuracy was the dependent variable (for review see Walker and McGuire, 1982, as well as Wexler et al.). In an extensive review of earlier studies of perceptual asymmetry, however, Walker and McGuire conclude that when reaction time was the dependent variable in studies of perceptual asymmetry, there was a nearly uniform elevation in schizophrenics' response time to right-ear and RVF stimulation, relative to appropriate control subjects.

Other aspects of the study by Wexler et al. (1991) illustrate the importance of specific task demands and of the nature of schizophrenic symptoms. For example, a subset of schizophrenic patients was retested after symptom remission and showed a recovery-related increase in right-ear advantage for the nonsense task but a recovery-related decrease in right-ear advantage for the word test, a finding that has been replicated. In addition, at the time of initial testing, those schizophrenic patients with a smaller right-ear advantage for the nonsense task than for the word task had predominantly "negative" symptoms (e.g., emotional withdrawal, blunted affect). By way of contrast, those schizophrenic patients with the opposite pattern of ear asymmetry for the two tasks had predominantly "positive" symptoms (e.g., grandiosity, hostility). Although the interpretation of these complex results is not completely clear, they indicate the importance of subdividing schizophrenics along such dimensions as symptom type (see also Flor-Henry, 1987).

As noted earlier in the present chapter, perceptual asymmetries are known to be influenced by such things as attentional asymmetries in favor of one hemisphere or the other as well as the direction and magnitude of hemispheric asymmetry for the task being performed (see also Chapter 2). This may be especially true when subjects are required to generate a single response to a stimulus based on two fused stimuli, as in the tasks used by Wexler et al. (1991). Thus, the reduced right-ear advantage for schizophrenics could reflect difficulty in attending to the right side of space rather than a deficit with left-hemisphere language processing per se.

The foregoing possibility is consistent with results from a visual task reported by Posner, Early, Reiman, Pardo, and Dhawan (1988). Posner et al. required schizophrenics and control subjects to respond as quickly as possible to the onset of a "star" pattern presented to either the LVF or RVF on each trial. On different trials, subjects were presented with either no cue to indicate which location would be stimulated or with a cue to attend to one side or the other. On cued trials, the cue was valid 80 percent of the time (that is, the star appeared in the cued location) and invalid 20 percent of the time (that is, the star appeared in the location opposite the one cued). For the control subjects, reaction time was faster with a valid cue than with an invalid cue and there was never any hint of a visual-field/hemispheric difference. For the schizophrenics, reaction time was faster for LVF than for RVF stimulation when attention was not already directed to the correct location (with no cue or with an invalid cue). Note that this result is consistent with earlier reports of elevated reaction time to RVF stimulation in schizophrenics. However, this visual-field/hemispheric difference in reaction time was eliminated when a valid cue was presented 100 or 800 milliseconds prior to the target, allowing subjects time to shift attention toward the correct location before the target was presented. Posner et al. note that the pattern of effects shown by schizophrenics is similar in this respect to results obtained in earlier studies with patients who had unilateral lesions of the left hemisphere. A second experiment suggested that the pattern of effects shown by schizophrenics can be mimicked to some extent by control subjects who perform a verbal shadowing task at the same time as the star-detection task. In a final experiment, schizophrenic patients were found to be slow in processing visual word cues about direction (e.g., "right") in comparison to symbolic directional cues (e.g., an arrow pointing to the right). Similar results have been found in patients with left-hemisphere injury. On the basis of this entire pattern of results, Posner et al. speculate that schizophrenics might have a deficit in an anterior region of the left hemisphere that is important for coordinating attention to language. The same area is thought to be involved in coordinating attention to visual location by interacting with the posterior/parietal visuospatial system. For discussion of the neural circuits thought to be involved in different aspects of attention, see Posner (1992).

As noted by Posner et al. (1988), the notion that schizophrenics have a difficult time attending toward the right side of space is consistent with the observation of Bracha (1987) that 10 unmedicated schizophrenic patients tended to circle toward the left side (neglecting the right?), whereas 85 normal control subjects showed nearly equal right and left circling (see Chapter 5 for a discussion of this paradigm). In Chapter 5 we saw that preferred direction of circling in rats is related to asymmetry in dopamine content and in sensitivity to dopamine, and in earlier research Bracha has found evidence that humans prefer to circle in a direction opposite the side of the brain with greater dopaminergic activity. Consequently, Bracha considers the possibility of a dopaminergic asymmetry in unmedicated schizophrenics. Specifically, he suggests that in schizophrenics the amount of dopaminergic activity is relatively higher in right anterior cortical or subcortical structures than in left anterior structures. Regardless of whether Bracha's speculations are entirely correct, it is interesting to note that anti-schizophrenic medications are also anti-dopaminergic (e.g., Flor-Henry, 1987).

There is also some indication that the corpus callosum is dysfunctional in schizophrenics, although the evidence is mixed. For example, Craft, Willerman, and Bigler (1987) tested schizophrenics, schizo-affective patients, and normal controls on four tasks shown to be influenced by callosal agenesis (a disease in which the corpus callosum does not develop). Relative to normal controls, both patient groups were impaired on all measures but one, but it must be noted that other tasks that are known to be unaffected by callosal agenesis were not included and without this comparison the schizophrenic deficits must be interpreted with caution. By way of contrast, Raine, Harrison, Reynolds, Sheard, Cooper, and Medley (1990) failed to find any behavioral evidence of impaired interhemispheric transfer in schizophrenics performing tasks that required transfer of either auditory or tactile information (see Chapter 4). Using MRI data, however, Raine et al. did find that the corpus callosum is thicker in female schizophrenics than in male schizophrenics, which is a pattern that tends to be reversed in normal controls. A nonschizophrenic psychiatric control group showed the same pattern as schizophrenics, suggesting that any sex-dependent callosal abnormalities are not specific to schizophrenia.

Although there has been far more work with schizophrenics than with patients with other forms of psychopathology, it should be noted that it has been hypothesized that other psychopathologies are also related to unusual aspects of hemispheric asymmetry. For example, there have been suggestions of right-hemisphere disturbance in patients with major depressions (for review see Flor-Henry, 1987). In addition, Wexler et al. (1991) found that bipolar manic patients with smaller right-ear advantages on nonsense and word tasks had more symptoms of thought disorder than of mood disturbance, whereas patients with larger right-ear advantages showed the reverse pattern of symptoms. Using a dichotic-listening task that required identification of CV nonsense syllables, Raine, O'Brien, Smiley, Scerbo, and Chan (1990) found a reduced right-ear advantage in adolescent psychopaths compared with nonpsychopaths. This occurred because psychopaths had a reduced right-ear score and an elevated left-ear score relative to nonpsychopaths.

Even this brief review illustrates the experimental approaches and some of the theoretical alternatives that have been considered in studies of the relationship of hemispheric asymmetry to psychopathology. Several of the hypotheses have enough support to warrant additional investigation, but none of them has enough unequivocal support for the issue to have been decided. Future behavioral studies would benefit from a self-conscious attempt to separate the various dimensions of individual variation discussed earlier in the present chapter. Given that a number of hypotheses have to do with the overactivation or overuse of one hemisphere or the other, it would seem particularly worthwhile to determine whether the characteristic pattern of arousal asymmetry or the pattern of metacontrol biases is related to certain forms of psychopathology, independently of any differences in the direction and magnitude of hemispheric dominance for specific functions.

Hemisphericity

No discussion of individual variation in hemispheric asymmetry would be complete without consideration of a popularized concept that has come to be known as "hemisphericity." In its most controversial form, *hemisphericity* refers to the idea that different individuals have different preferred modes of cognitive processing that they use almost exclusively, with the mode of processing determined by which cerebral hemisphere is more active. An especially popular version maintains that "left-brained" people are rational and analytic in their approach to cognitive problems whereas "right-brained" people are intuitive and creative. This notion of hemisphericity has led to various questionnaire measures of cognitive style, and the dichotomous classifications that often result are claimed to be firmly rooted in contemporary neuroscience. In addition to classifying individuals as left-brained or right-brained, another result of the acceptance of this notion of hemisphericity is the creation of programs that are advertised to train one side of the brain or the other or to train "whole-brain thinking." (For thoughtful reviews of this particular concept of hemisphericity and of some of these programs, see Beaumont, Young, and McManus, 1984; Druckman and Swets, 1988; Hines, 1987; Kinsbourne, 1982; O'Boyle, 1986; Thompson, 1984).

It should be clear from earlier chapters that there can be very little support for this particular notion of hemisphericity. Hemispheric asymmetries are rarely of the all-or-none type that are assumed by the concept. Furthermore, when asymmetries *are* found they have more to do with dominance for specific processing components than with such large, ill-defined constructs as "rational thought" and "creativity." As we have seen time after time, many areas of both hemispheres as well as many subcortical areas are activated by almost every cognitive or behavioral task we perform. Consequently, a strong notion of hemisphericity or "thinking with one side of the brain" has been rejected by most neuroscientists, although it will undoubtedly live on for some time in the popular press.

Although the view of hemisphericity just described has little to recommend it, certain weaker versions may, in fact, have merit. For example, the Cognitive Laterality Battery developed by Gordon (1980, 1986) consists of a variety of verbal, sequential, and visuospatial neuropsychological tests that were selected because they were especially impaired by injury to one hemisphere or the other. Consequently, the battery has at least some validity as a measure of cognitive functions that are more associated with one hemisphere or the other. Furthermore, Gordon has reported that factor analyses confirm the existence of two main factors, with visuospatial tasks loading on one factor and with verbal and sequencing tasks loading on the other. While he is careful to avoid presenting this battery as a test of the relative activation of the two hemispheres, Gordon argues that the pattern of results obtained from an individual does provide a cognitive profile of the relative performance of functions that are known to be associated with left- versus right-hemisphere injury. As such, it has provided some useful information, including the finding noted earlier that children with dyslexia and members of their families are impaired on the verbal and sequential tasks associated with left-hemisphere injury but not on the visuospatial tasks associated with right-hemisphere injury.

The term *hemisphericity* has also been used to refer to individual differences in the pattern of characteristic arousal asymmetry discussed earlier. Results such as those reported by Kim et al. (1990) leave little doubt that individuals differ in such things as characteristic perceptual asymmetry, although there are also other intertask correlations that are not readily accounted for by this weaker form of hemisphericity (e.g., Boles, 1989, 1991). However, it remains to be determined whether individual variation on this dimension is related to individual differences as measured, for example, by the Cognitive Laterality Battery. As noted earlier, Kim et al. found no relationship between a measure of characteristic perceptual asymmetry and performance on various tests of cognitive ability.

Regardless of whether characteristic arousal asymmetry is related to cognitive *ability*, it is possible that arousal asymmetry is related to an individual's propensity to *use* a particular mode of processing when given a choice—although on a much less grandiose scale than is claimed by the strong version of hemisphericity. Preliminary data obtained by Hellige, Bloch, and Eng (1992) are consistent with this possibility and illustrate the type of relationship that might exist.

Hellige et al. required 60 right-handed individuals to identify dichotically presented CV nonsense syllables (e.g., *ba, da, ga*) and examined the right-ear-minus-left-ear difference score as a measure of asymmetry. These same individuals also performed the visual CVC recognition task used by Hellige et al. (1989) in their studies of interhemispheric interaction and metacontrol (see Chapter 6). Of particular importance in this task is the qualitative pattern of errors on LVF/right-hemisphere trials, RVF/left-hemisphere trials, and redundant BILATERAL trials that present the same CVC simultaneously to both LVF and RVF locations. Recall that Hellige et al. (1989) found the pattern of errors to be qualitatively different on LVF/right-hemisphere and RVF/left-hemisphere trials, indicating that a qualitatively different mode of CVC processing is associated with each hemisphere. The qualitative error pattern on redundant BILATERAL trials provides some indication of the mode of processing utilized when stimulus presentation does not create a bias in favor of one hemisphere or the other.

In order to determine whether ear differences in dichotic listening are related to the mode of processing utilized during the CVC recognition task, Hellige et al. (1991) compared the CVC performance of those individuals who showed a left-ear advantage ($n = 8$) and a like number of individuals who showed the largest right-ear advantage. Recall that ear differences on this task are influenced by the direction and magnitude of hemispheric dominance for phonetic processing and also by individual variation in characteristic hemispheric arousal. Figure 7.6 shows the normalized percentage of three different error types for LVF/right-hemisphere trials, for RVF/left-hemisphere trials, and for redundant BILATERAL trials. The results for the left-ear-advantage group are shown in the upper panel, and the results for the right-ear-advantage group (those with the largest advantage) are shown in the lower panel.

For both groups, the results on LVF/right-hemisphere and RVF/left-hemisphere trials differ in exactly the way that they have differed in earlier studies (see Figure 6.5). That is, the difference between the normalized percentage of first-letter and last-letter errors is larger on LVF/right-hemisphere trials than on RVF/left-hemisphere trials. However, the relationship between the results on BILATERAL trials and the results on each of the two types of unilateral trials differs for the two groups. Specifically, for individuals who show a left-ear advantage, the BILATERAL pattern is more similar to the LVF/right-hemisphere pattern than to the RVF/left-hemisphere pattern, suggesting a bias toward the mode of processing associated with the right hemisphere. Although the distribution is not pictured in Figure 7.6, the same is true for most of the individuals with a small to moderate right-ear advantage—a result consistent with those pictured in Figure 6.5. However, for those individuals whose right-ear advantage was unusually large (the top 8 of 60 right-handers tested), the BILATERAL pattern is virtually identical to the RVF/left-hemisphere pattern, suggesting a bias toward the mode of processing associated with the left hemisphere. This relationship suggests, at

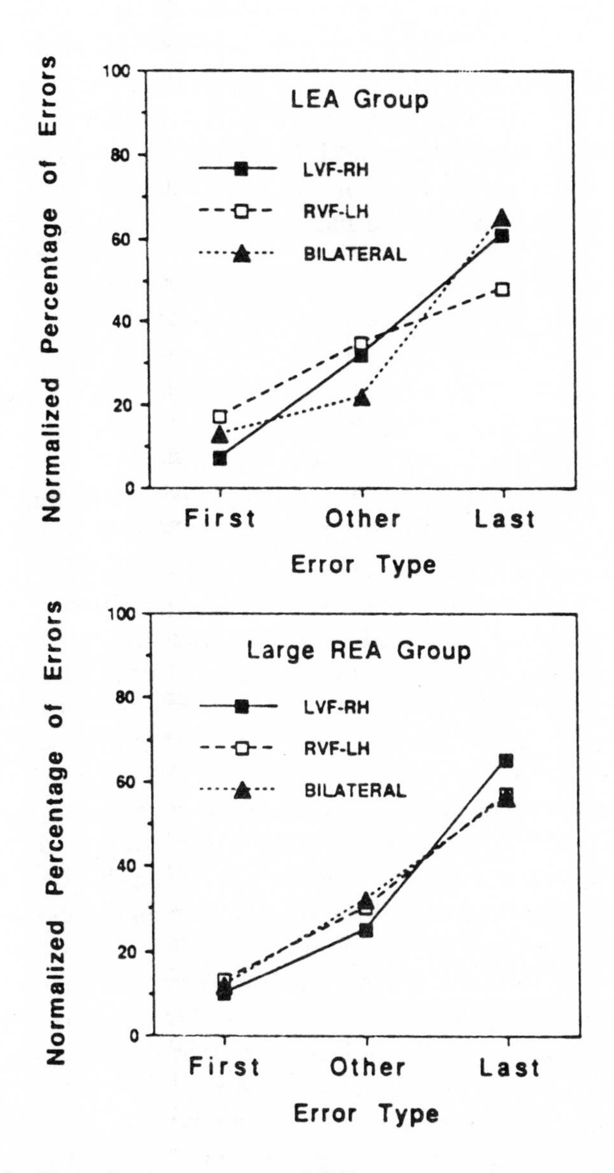

Figure 7.6. Normalized percentage of different types of errors (First, Other, Last) for LVF/right-hemisphere (LVF/RH), RVF/left-hemisphere (RVF/LH), and BILATERAL trials. The results in the upper panel are from the Left-Ear-Advantage (LEA) Group and the results in the lower panel are from the Large Right-Ear-Advantage (REA) Group identified by Hellige et al. (1992).

the very least, that there are reliable individual differences in a bias toward the mode of processing associated with one hemisphere or the other. Beyond that, the results indicate that such a bias is related to either hemispheric asymmetry for phonetic processing or characteristic arousal asymmetry. Although these results are far less dramatic than the claims made for strong hemisphericity, they point to the value of additional investigation of individual differences in aspects of metacontrol.

Summary and Conclusions

Individuals differ in several aspects of hemispheric asymmetry (theme 4 in Chapter 1). There are differences in anatomical asymmetry and differences in several dimensions of functional asymmetry. Among the functional dimensions along which individuals have been shown to differ are (1) direction of hemispheric asymmetry, (2) magnitude of hemispheric asymmetry, (3) asymmetric arousal of the two hemispheres, (4) complementarity of asymmetry for certain tasks and processes, and (5) interhemispheric communication. The recent development of behavioral tasks that measure individual variation on each of these dimensions allows the study of individual differences in "hemispheric asymmetry" to become far more precise than has heretofore been the case. Furthermore, if individuals are required to perform more than one task, it will be possible in future studies to provide an interlocking web of converging evidence about the dimensions of interest and about whether variation on those dimensions is related to cognitive abilities and propensities.

Individual variation on one or more of these dimensions has been hypothesized to be related to a number of between-subject factors, although the research to date does not always provide a clear separation of the dimensions. Among the between-subject factors that have been studied are handedness, sex, intellectual abilities (including intellectual giftedness and dyslexia), and psychopathology. This is not an exhaustive list, but it serves to illustrate the range of empirical approaches taken to study individual variation. In addition, potential relationships between hemispheric asymmetry and these particular between-subject variables have been among the most widely studied and the hypotheses to emerge have been among the most promising.

Handedness is noteworthy because it is a behavioral manifestation of hemispheric asymmetry for certain manual activities and also because it has at least some relationship to hemispheric asymmetry for other functions. An individual's hand dominance is determined by a variety of genetic and prenatal and perinatal environmental influences. Furthermore, the direction and magnitude of hand preference may be determined by different factors, and there is some indication that magnitude is more heritable than direction. Among the environmental influences that may contribute to an individual's handedness are pre- and perinatal trauma, levels of testosterone and other hormones during certain critical stages of fetal development, and asymmetric positioning of the fetus *in utero* during the final trimester of pregnancy.

Handedness is also something of a marker for other aspects of hemispheric asymmetry. For example, on the average, hemispheric asymmetry for a group of left-handers is typically in the same direction as for a group of right-handers, but the magnitude of the difference is smaller. In part, this is attributable to the fact that a greater proportion of left-handers show an asymmetry opposite to the direction considered prototypical. As a result, complementary asymmetry for specific linguistic and visuospatial processes is likely to occur more often in right-handers than in left-handers. At the present time, there is little indication that handedness is related to asymmetric arousal of the hemispheres or to the ability of the hemispheres to communicate with each other.

In view of the many effects of sex-related hormones on brain development in other species, it would not be surprising to find that hemispheric asymmetry in humans is somewhat different for males and females. In fact, the incidence of left-handedness is slightly higher in males than in females. In addition, it has been suggested that the brains of males are more asymmetric than the brains of females, although the evidence in favor of this hypothesis is equivocal. One possibility is that males and females have asymmetries of equal magnitude but there are more females than males with an atypical direction of asymmetry. If this were the case, it would be likely to produce sex differences in complementarity of asymmetries, in much the same way that is true for handedness differences. At the present time, there is little to indicate that males and females differ in either asymmetric arousal of the two hemispheres or in interhemispheric communication, but there have been too few studies to permit strong conclusions about these dimensions of individual variation. Even when sex differences are found, it is important to consider explanations in terms of the tendency for males and females to prefer different cognitive strategies that may depend on a different balance of processing between the two hemispheres.

There are some indications that individual variation in hemispheric asymmetry is related to intellectual abilities. For example, relative performance on tests of verbal and visuospatial abilities is related to handedness, although the precise nature of the relationship is moderated by sex and by reasoning ability. In addition, there is some indication that extreme intellectual precocity is related to advanced development of and reliance on the right hemisphere, perhaps as a result of relatively high levels of testosterone during critical stages of fetal development. At the same time, it has been hypothesized that some forms of dyslexia (impaired reading ability) are related to subtle dysfunctions of language areas of the left hemisphere and, perhaps, to difficulty in interhemispheric communication. Future studies would do well to test these hypotheses and also to consider whether dyslexia is related to other dimensions of individual variation, especially to an unusual pattern of complementary asymmetry for those processes necessary for learning to read.

The existence of hemispheric asymmetries related to emotion has led to consideration of possible relationships between asymmetries and psychopathology. One promising hypothesis is that schizophrenia is related to dysfunction of an anterior region of the left hemisphere that is important for controlling the posterior/parietal visuospatial system and also for attending to language. This hypothesis remains somewhat controversial, however, and may only apply to a subset of schizophrenics. It has also been suggested that the corpus callosum is dysfunctional in schizophrenics, but the evidence is mixed at best. Other forms of psychopathology have also been linked to unusual patterns of hemispheric asymmetry (e.g., there is some indication of right-hemisphere disturbance in depressives). Future studies would benefit from an attempt to separate the various dimensions of individual variation noted earlier. In view of the fact that several of the working hypotheses have to do with the overactivation of one hemisphere or the other, it would seem particularly worthwhile to examine patterns of asymmetric arousal of the two hemispheres.

The term *hemisphericity* has been used to refer to the idea that different individuals have different preferred modes of cognitive processing that they use almost exclusively, with the mode of processing determined by which cerebral hemisphere they tend to rely on. Despite a great deal of popular appeal, there is virtually no evidence for this strong distinction between "left-brained" and "right-brained" people. However, there is some indication that individuals do, in fact, differ on their relative performance on tasks that are known to be associated with left- versus right-hemisphere injury. There is also evidence of individual variation in characteristic arousal asymmetry and in an individual's propensity to use a mode of processing associated with one hemisphere or the other when given a choice. While these results are far less dramatic than the claims made for strong hemisphericity, they argue for continued investigation of individual differences on all of the dimensions of hemispheric asymmetry.

8

Hemispheric Asymmetry across the Life Span

The patterns of hemispheric asymmetry discussed in this book are a product of the complex interplay of a number of genetic and environmental factors. The purpose of the present chapter is to consider how some of those factors influence the unfolding of hemispheric asymmetry across the life span, from conception to death or, if you will, from womb to tomb. The view taken here is that the seeds of functional hemispheric asymmetries are sown long before an individual's birth and are likely to date back to the ontogenetic formation of the first neural structures, to asymmetries of the ovum or even to various molecular asymmetries. From these early origins, functional hemispheric asymmetries are shaped by the interaction of many biological and environmental factors, beginning with the fetus as it develops *in utero* and continuing into old age. Here I will review the factors that seem most relevant, beginning with a discussion of prenatal asymmetries as precursors of later functional hemispheric asymmetry, moving to a discussion of hemispheric asymmetry from birth through young adulthood, and ending with a discussion of hemispheric asymmetry in old age.

Before we turn to those factors that influence the emergence of hemispheric asymmetry across the life span, it is instructive to acknowledge a distinction that has been made by others between what has been termed the *development* of hemispheric asymmetry versus the *unfolding* of hemispheric asymmetry. Although this distinction is not always as clear as I would like, a brief consideration of whether or not hemispheric asymmetry "develops" or "unfolds" will serve to illustrate a number of issues that arise in the study of possible age-related changes in hemispheric asymmetry.

Does Hemispheric Asymmetry Develop?

For a time, the predominant question asked by investigators of age-related changes in hemispheric asymmetry was, "Does hemispheric asymmetry develop?" As it happens, this is something of a loaded question, and the answer depends on exactly what is meant by the word *develop*. At the outset, it is instructive to consider the kind of behavioral result that would have led the investigators who posed this question to an affirmative answer. Suppose that some cognitive or motor function (e.g., the production of speech) appears initially without any hemispheric asymmetry whatsoever. That is, at first both hemispheres are equally involved in all the relevant processes. Then, after a time, one hemisphere becomes dominant over the other for that function. For example, suppose it could be shown that when speech first emerges in human infants, it is controlled equally by both hemispheres, and only later does the left hemisphere become dominant for speech in most people. Given this sort of result, it would be reasonable to say that hemispheric asymmetry (at least for speech) "develops."

The fact is that there is no unequivocal evidence for this type of development of hemispheric asymmetry. On the contrary, the hemispheres are asymmetric for certain computational processes from the time that the organism is sufficiently developed for those processes to occur and be measured. To be sure, until a specific computational component has appeared in the course of general cognitive development, there can be (by definition) no functional hemispheric asymmetry for that specific component. In other words, the appearance of functional hemispheric asymmetry for specific functions or processing components is related to increasing cognitive abilities, but cognitive development is generally regarded as being distinct from the development of hemispheric asymmetry per se (at least as development was defined above). While it is certainly the case that cognitive abilities increase dramatically from birth through young adulthood, there is little evidence for an increase in hemispheric asymmetry except as a by-product of cognitive development. Figure 8.1 illustrates graphically these hypotheses about increasing cognitive ability and unchanging hemispheric asymmetry as they have been applied to language by Witelson (1987).

When one cerebral hemisphere is injured, the opposite (uninjured) hemisphere can sometimes take over the functions for which the now-injured hemisphere is usually dominant. This ability reflects what has been called the *functional plasticity* of the nervous system. There is evidence that the amount of functional plasticity changes with age, and care must be taken to separate developmental changes in functional plasticity from changes in

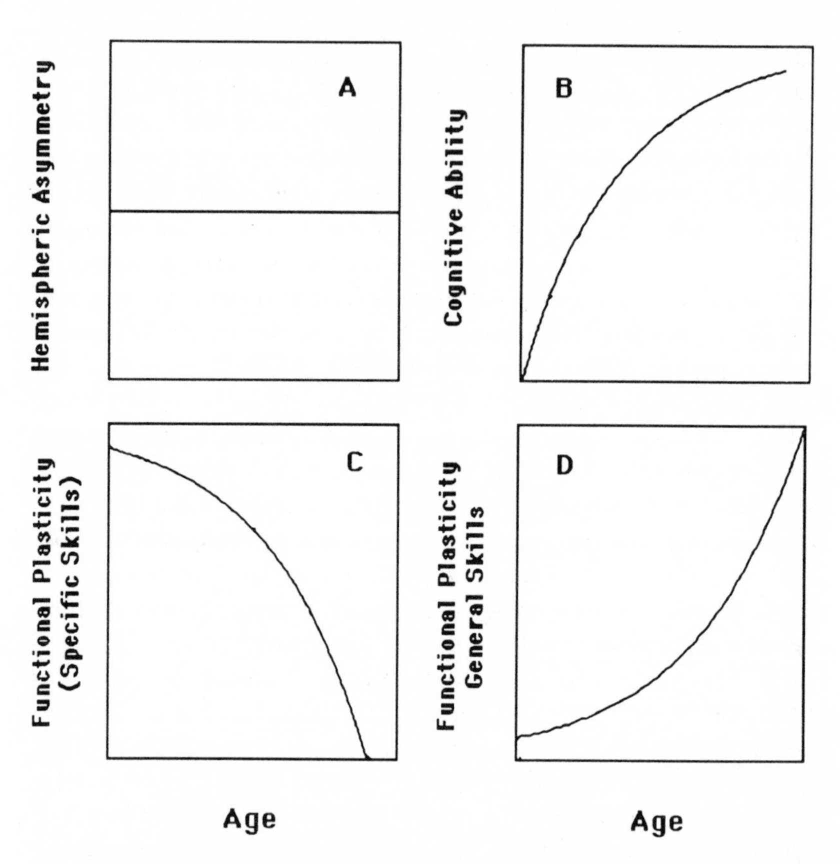

Figure 8.1. Illustration of hypotheses about the manner in which four different characteristics related to behavioral asymmetry change with age. Panel A: Hemispheric asymmetry remains constant from birth through adulthood. Panel B: Cognitive ability increases from birth through adulthood. Panel C: Functional plasticity for specific skills diminishes from birth through adulthood. Panel D: Functional plasticity for general skills increases from birth through adulthood. [Adapted from S. F. Witelson, "Neurobiological Aspects of Language," *Child Development,* 58 (1987):653–688. Copyright 1987 by The Society for Research in Child Development, Inc. Adapted by permission.]

hemispheric asymmetry per se. The difference between the two will be illustrated later in the present chapter in a discussion of hemispheric asymmetry for language from birth through young adulthood. For the moment, it is sufficient to note that functional plasticity for specific skills that are usually mediated by relatively localized areas of one hemisphere has been hypothesized to *decrease* with age (e.g., Witelson, 1987). Thus, the extent to which such a skill can be performed normally after injury to the hemisphere that is dominant for it depends on age at the time of brain injury; the earlier the injury, the better the chances of recovery. This hypothesis is illustrated graphically in Figure 8.1, panel C. By way of contrast, early brain injury appears to be more detrimental to general cognitive ability than is brain injury at an older age. Consequently, it has been hypothesized that functional plasticity for general cognitive capacity *increases* with age (see Figure 8.1, panel D).

In order to illustrate the potential problems of interpretation caused by changes in functional plasticity, consider studies of the effects of early brain injury on specific skills such as language. In adults, recovery of various language functions is far more likely after right-hemisphere injury than after left-hemisphere injury. In fact, this observation has been central to the discovery of left-hemisphere dominance for many aspects of language (see Chapters 1 and 2). By way of contrast, recovery of those same language functions may be excellent after injury to either hemisphere if the injury is sustained at a sufficiently young age (e.g., Witelson, 1987). For the sake of argument, assume that even if the left hemisphere is removed completely (i.e., left hemispherectomy), language will develop normally as long as the hemispherectomy occurs at a sufficiently young age. (As discussed later, this assumption is controversial at best, but for the moment we will assume it is valid.) What would this allow us to conclude about whether hemispheric asymmetry "develops" in the manner described earlier?

At first glance, one might conclude that at an early age there are no hemispheric asymmetries for language—that is, the hemispheres are *equally involved* in language—and that hemispheric asymmetry for language develops with increasing age in the manner described earlier. Despite the initial appeal of this conclusion, the fact is that recovery of function reveals far less about hemispheric asymmetry at the time of injury than about the brain's capacity for functional plasticity at different ages. For example, even if there had been left-hemisphere dominance for language processing at the very early age, it would be hidden completely if the right hemisphere were sufficiently plastic to assume control of language during the period of recovery. Viewed in this way, the hypothetical results that have been described indicate that functional plasticity for language skills is much greater at a very young age than in adulthood. They do not indicate that at that very young age language was processed equally by both hemispheres or that the two hemispheres ever had equal potential for acquiring language skills (for further discussion of this point see Bradshaw, 1989; Witelson, 1987).

Evidence from a variety of experiments—such as studies of perceptual and response asymmetries and the use of techniques, like the EEG, for measuring localized brain activity—suggests that many hemispheric asymmetries are present at birth and possibly before. Some of this evidence will be reviewed in subsequent sections of the present chapter. In addition, when the same experimental paradigm can be applied to subjects of a variety of ages, there is little indication that either the direction or the magnitude of hemispheric asymmetry for a particular skill changes systematically from its earliest appearance through old age. This has led to a predominant view that is expressed nicely by John Bradshaw (1989, pp. 192–193):

> In considering the question of developmental aspects of cerebral asymmetry, perhaps the key point to be made is that cerebral asymmetry almost certainly does not develop; asymmetries, morphological and behavioral, are present *ab initio,* and the most that can be said is that they may unfold.

We turn now to some of the ways that asymmetries begin to unfold before birth and to suggestions of how these prenatal asymmetries provide the seeds of hemispheric asymmetry in later life.

Prenatal Asymmetries

There are now a number of well-documented asymmetries in the fetus as it develops *in utero.* The existence of some of these asymmetries has been known for a long time, and merely documenting their existence is not particularly noteworthy. What are noteworthy, however, are suggestions that at least some of these prenatal asymmetries are at least precursors and perhaps causes of functional hemispheric asymmetry in later life. Although there is disagreement about exactly which prenatal asymmetries are critical and which are not, the idea that later hemispheric asymmetries have their roots in prenatal asymmetries is emerging as a dominant theme in current theories of the ontogenetic unfolding of hemispheric asymmetry. Furthermore, the potential importance of prenatal environmental influences is illustrated by Rogers (1986), whose work with chick embryos shows that later asymmetry for a variety of behaviors depends on light striking one eye and not the other at precisely the right stage of embryonic development (see Chapter 5). Here I will review some of the prenatal asymmetries in humans that seem most promising as precursors of later hemispheric asymmetry. These include asymmetries in the rate of maturation of the two hemispheres *in utero,* asymmetries of cranio-facial development, and asymmetries of fetal position. Then follows a discussion of so-called snowball mechanisms that, from an initial asymmetry that seems quite small, may produce a much larger range of hemispheric asymmetries as neural form and function continue to unfold.

The Rate of Maturation of the Two Hemispheres

There is growing evidence that during the course of fetal development, certain areas of the right hemisphere mature more quickly than homologous areas of the left hemisphere. Evidence comes from a variety of sources and concerns a variety of specific brain regions, and reviews have been provided by Geschwind and Galaburda (1987), Turkewitz (1988), and de Schonen and Mathivet (1989). For example, there is a delay in the appearance of cortical markings surrounding the sylvian fissure on the left compared with the right side, and folds surrounding the sylvian region appear later on the left side than on the right (e.g., Fontes, 1944; Chi, Dooling, and Gilles, 1977). In addition, Chi et al. also found that some of the cortical landmarks in the superior temporal region appeared 1–2 weeks later on the left than on the right side. Higher-order dendritic branches in areas important for language in adults also develop later in the left hemisphere than in the right hemisphere (Scheibel, 1984; see also Simonds and Scheibel, 1989), although the extent of dendritic branching is ultimately greater in the left hemisphere. With respect to the ultimate ontogenetic origins of these maturational asymmetries, Geschwind and Galaburda suggest that they may date back to the formation of the first primitive neural elements, to asymmetries of the ovum itself, or even to various molecular asymmetries (see also Bradshaw, 1989; Corballis, 1991; Previc, 1991).

What is most interesting is that various asymmetries of growth may serve as an important mechanism for the emergence of functional hemispheric asymmetries in later life. In order to illustrate this, I will consider possible scenarios having to do with left-hemisphere dominance for speech, hemispheric differences for the processing of low versus high visual-spatial frequencies (and for global versus local aspects of the visual world), right-hemisphere dominance for face processing, and hemispheric differences for the control of fine motor movements.

With respect to hemispheric asymmetry for language, an interesting scenario has been proposed by Turkewitz (1988, p. 74):

> In short, I will suggest that (1) the fetus is sensitive to its acoustic environment, (2) that there are systematic changes in the nature of this environment, (3) that there are differences in the rate of development of the two hemispheres, and (4) that the changes in the acoustic environment and neurological substrate interact to produce the hemispheric specialization in function that appears to characterize even very early stages of human development.

More specifically, according to this scenario the right hemisphere is more highly developed than the left at a time when various nonlanguage noises are presented to the fetal brain (e.g., fetal heartbeat, mother's heartbeat, sounds of digestion). It is proposed that the more advanced hemisphere has priority in dealing with classes of input for which there is not already an established hemispheric dominance. Thus, the more advanced right hemisphere "learns" more from these noises than does the less advanced left hemisphere. This early experience is the foundation for later right-hemisphere dominance for processing a variety of nonlanguage sounds. Later in fetal development, when the left hemisphere is more developed than it was earlier, the fetal brain is presented not only with the same noises as before but also with the sounds of the mother's voice. Turkewitz argues that this new information is relayed to the fetus as a result of changes in the acoustic transmission properties of the uterus as it expands.

According to Turkewitz, two factors contribute to emergence of left-hemisphere dominance for this new acoustic information. First, the right hemisphere would continue to be more responsive to the noises for which it has become dominant, which means that it is less available than the left hemisphere to respond to maternal speech. Second, the left hemisphere is now more advanced in its development than it was earlier. This early, prenatal left-hemisphere dominance for responding to maternal speech is proposed to be the precursor of left-hemisphere dominance for processing speech and would account for the fact that certain asymmetries are already present at birth and, perhaps, in premature infants. What Turkewitz does not propose (but might have) is that the early right-hemisphere dominance for processing noise might bias the right hemisphere to be particularly responsive to intonation and prosody in maternal speech during this later stage of fetal development. Thus, this scenario might also account for certain aspects of right-hemisphere dominance for intonation and prosody in later life (see Chapters 2 and 3).

Although the fetal brain experiences auditory information prior to birth, the presentation of information about visual patterns begins only at birth. Other interesting scenarios for the emergence of asymmetry are based on the idea that development of areas important for visual cognition are more advanced in the right hemisphere than in the left hemisphere at the moment of birth and for at least a short time thereafter. Following logic similar to that used by Turkewitz, assume that, as a consequence of this hemispheric difference in development, the right hemisphere is initially more responsive than the left hemisphere to the incoming visual information and is modified in response to the information that it receives. The visual sensory system of newborns presents the brain with highly degraded visual input (for reviews, see Banks and Dannemiller, 1987; de Schonen and Mathivet, 1989). Of particular interest is the fact that degradation is not uniform across all visual-spatial frequencies. Specifically, the visual system of neonates is especially limited in its transmission of visual information carried by relatively high spatial frequencies, with the transmission of high spatial frequencies improving considerably over the first 6 months of life. Thus, the more advanced right hemisphere would have priority in dealing with the highly degraded visual input shortly after birth and could, as a consequence, become dominant for processing the "global" information conveyed by relatively low visual-spatial frequencies (see Hellige, 1989; Sergent, 1987a). Later (perhaps only a few weeks or months), when the relevant areas of the left hemisphere are more developed than they were at birth, the more advanced visual sensory system presents the brain with visual information conveyed by a higher range of spatial frequencies in addition to the the low frequencies that were presented earlier. Following the earlier logic of Turkewitz (1988), at least two factors could encourage left-hemisphere dominance for processing the "local" information conveyed by relatively high visual-spatial frequencies. The first is that the right hemisphere would continue to be more responsive than the left hemisphere to information conveyed by relatively low spatial frequencies, which means that the right hemisphere is less available than the left for processing information conveyed by relatively high spatial frequencies. Second, the left hemisphere is now more advanced in development than it was earlier. To these reasons one could add a third factor. Perhaps a certain "critical period" for being modified by visual input occurs earlier for the right hemisphere than for the left hemisphere. Although admittedly speculative, this scenario is consistent with the hemispheric asymmetries found in adults for components of visual processing (see Chapter 3) and with the fact that those asymmetries emerge shortly after birth (see later sections of the present chapter).

De Schonen and Mathivet (1989) use logic similar to this as part of a scenario for the development of right-hemisphere dominance for face processing. However, their scenario includes the idea that the hemispheres contain neural networks that are specialized for the identification of faces, with these specialized networks being embedded in regions of the brain that are involved in visual pattern processing. To the extent that the identification of faces is best accomplished on the basis of information conveyed by relatively low visual spatial frequencies, it might not be necessary to propose neural networks that are specialized for face processing per se (see Hellige, 1989). In either case, asymmetries in the rate of maturation of the two hemispheres may well be an important factor in producing right-hemisphere dominance for identification of faces in infants as young as 4 months of age (e.g., de Schonen and Deruelle, 1991; de Schonen and Mathivet, 1989).

It is instructive to consider whether a scenario similar to those presented for auditory and visual processing could also be proposed for the development of left-hemisphere dominance for aspects of fine motor control and sequencing of movements. Suppose that relevant areas of the right hemisphere are more advanced in development than homologous areas of the left hemisphere at a time when the motor system is incapable of producing precise movements and sensorimotor feedback is likely to be very coarse (either *in utero* or for a short time after birth). Using logic similar to that used earlier, we might conclude that the right hemisphere becomes dominant for gross movements, postural control, and so forth. By developing slightly later, the left hemisphere would be saved to become dominant for more precise movements and for sequencing those movements. This possibility is admittedly quite speculative, but it follows naturally from other scenarios based on the notion of asymmetric rates of maturation. In addition, it is at least consistent with the observation in nonhuman primates that the left hand is favored for reaching whereas the right hand is favored for manipulation (see Chapter 5) and with the fact that hand preference is generally stronger for skilled movements than for unskilled movements (see Chapter 7).

It is also interesting to note that negative emotions have been hypothesized to differentiate out of a more general approach/avoidance spectrum sooner than positive emotions (e.g., Bridges, 1932; see also Sroufe, 1979). To the extent that this is the case, one could wonder whether earlier maturation of the right hemisphere predisposes it to play a more dominant role in the processing of information related to earlier-appearing negative emotions than to later-appearing positive emotions.

Geschwind and Galaburda (1987) also suggest that earlier maturation of the right hemisphere is related to what has been referred to as the greater *conservatism* of right-hemisphere function.

We have seen earlier that even though the left hemisphere is typically dominant for speech and handedness, there are individuals in whom the right hemisphere makes significant contributions or for whom the right hemisphere is dominant (e.g., 10 percent or so of the population is left-handed, and aphasia sometimes follows injury to the right hemisphere). It has been argued that it is less common to find impairments typically associated with injury to right-hemisphere posterior areas (e.g., unilateral neglect; see Chapters 2 and 3) occurring after injury to left-hemisphere posterior areas. Furthermore, even when injury to the right hemisphere leads to aphasia, the aphasia is likely to be accompanied by spatial impairments and emotional reactions typical after right-hemisphere injury. It is this greater uniformity of right-hemisphere dominance for certain functions that is referred to as the greater conservatism of right-hemisphere function. Geschwind and Galaburda suggest that, because the right hemisphere develops earlier, the normal course of its development is subject to influence for a shorter period of time than is the development of the left hemisphere. As a result, the development of the right hemisphere is less likely to be impaired, and so there is greater uniformity across individuals in the functions for which the right hemisphere is dominant.

Despite the appeal of certain scenarios involving earlier maturation of the right hemisphere, it must be noted that some have argued that it is the *left hemisphere* that develops earlier and that such things as left-hemisphere dominance for linguistic processing arise as a by-product (see Corballis and Morgan, 1978; Corballis, 1991). As noted by Corballis (1991), such proposals are not necessarily at odds with the assertion that *prenatal* development favors the right hemisphere. In fact, we have seen that in the scenario proposed by Turkewitz (1988) it is entirely possible that left-hemisphere language areas are more highly developed than homologous areas of the right hemisphere by the time of birth or shortly thereafter. It is also possible that which hemisphere develops more quickly differs from area to area within the hemispheres and that asymmetric growth spurts continue to occur during critical periods after birth, with some spurts favoring the right hemisphere and some favoring the left hemisphere. For example, in a large-scale cross-sectional study of 577 children,

Thatcher, Walder, and Giudice (1987) used electrophysiological methods to estimate developmental changes in the cerebral hemispheres from 2 months of age to young adulthood. Although overall developmental trajectories appeared to be continuous, there were many examples of what the investigators refer to as growth spurts, and some of them were asymmetric. For example, in both frontal and occipital areas, the left hemisphere developed more rapidly than the right hemisphere during the period from 3 to 6 years. It is interesting that this is something of a critical period in the acquisition of language, especially in the acquisition of grammar. As noted by Corballis (1991), it is possible that this growth spurt in the left hemisphere that begins between 2 and 3 years of age is attributable to release from a factor that inhibits earlier growth of the left hemisphere, a mechanism suggested by Geschwind and Galaburda (1987) to account for earlier, prenatal growth gradients in favor of the right hemisphere. At the very least, the existence of asymmetric rates of maturation both before and after birth provides a potential mechanism by which functional hemispheric asymmetries continue to unfold throughout the life span.

Cranio-Facial Development

In developing a general theory of the prenatal origins of hemispheric asymmetry in humans, Previc (1991) reviews evidence of several asymmetries in the skull and face. For example, in approximately two-thirds of humans most cranial bones are larger on the right side than on the left side, whereas the area of the cheek and jaw is somewhat larger on the left side than on the right side. These and other cranio-facial asymmetries begin to emerge during the first trimester of pregnancy, and Previc argues that they constitute the prenatal origins of left-hemisphere dominance for perceiving and producing speech. Briefly, the argument goes like this.

Previc (1991) suggests that the slight enlargement of the left side of the face tends to restrict motion of the lower jaw (mandible) on that side of the face, and he reviews evidence that mandibular dysfunction is linked to hearing loss as a result of poorer middle-ear conduction. As a result, monaural sensitivity is slightly less in the left ear than in the right ear, especially in the 1,000–4,000 Hz range. Furthermore, male faces tend to have larger diameters than female faces, magnifying the left-sided enlargement and consequent right-ear advantage in monaural sensitivity. As Previc notes, there is evidence for such a right-ear advantage in monaural sensitivity for frequencies greater than 1,000 Hz and this advantage is, in fact, larger for males than for females (see also Chapters 2 and 3). According to Previc's scenario, the foundation of this aural asymmetry is laid down very early in fetal development, but the functional consequences are not felt until birth (or shortly thereafter). At that time, there is a right-ear advantage for speech processing because of the monaural right-ear advantage for the range of auditory frequencies that convey critical speech features, such as second and third formant transitions. By way of contrast, there is no right-ear advantage for processing such things as environmental sounds, prosody, and music because they are based on frequencies below 1,000 Hz. The eventual emergence of left-hemisphere dominance for speech *production* is proposed to be a consequence of the earlier left-hemisphere dominance for speech *perception* that comes about as a result of these monaural asymmetries.

In contrasting his scenario with that proposed by Turkewitz (1988) and discussed earlier in the present chapter, Previc (1991) argues that there is no objective evidence that "noise" is perceived prior to speech during the last trimester of pregnancy. In addition, he argues that the scenario proposed by Turkewitz cannot explain the monaural right-ear advantage (which Previc assumes would be uninfluenced by different rates of maturation for the two hemispheres) and that such a scenario is inconsistent with the fact that the left-ear advantage for environmental sounds is less robust than the right-ear advantage for speech sounds. Of course, as Previc himself points out, his own cranio-facial scenario has some difficulty explaining why there is any left-ear advantage at all for such things as musical sounds because there is no monaural left-ear advantage in the low frequency range. As a result, he considers other factors, such as involvement of the right hemisphere in emotion and the subsequent development of right-hemisphere dominance for low-frequency sounds, as a consequence of earlier left-hemisphere dominance for speech perception.

As noted earlier, Previc (1991) argues that the particular craniofacial asymmetry discussed above is characteristic of approximately two-thirds of humans. On that basis, one would predict that approximately two-thirds of the population would show a right-ear advantage for speech recognition. Although estimates vary, that value is a reasonable approximation to the results reported from many studies (e.g., see Figures 1.4 and 7.5). One would also predict left-hemisphere dominance for speech production to occur in approximately two-thirds of the population. However, the current estimates are much greater than this. For example, it is estimated that the left hemisphere is dominant for speech production in approximately 95 percent of right-handers and in approximately 66 percent of left-handers (see Chapter 7). In order to account for this discrepancy, Previc argues that for the one-third or so of individuals whose aural asymmetry is indeterminate, cultural factors influence hemispheric dominance for speech, at least indirectly. In particular, he argues that in Western cultures the left-to-right bias in reading and writing creates a predisposition toward left-hemisphere dominance for speech production in those individuals with little or no aural asymmetry. While Previc reviews some suggestive evidence that hemispheric dominance for speech production differs for literate and nonliterate cultures, there is no unequivocal evidence that either ear asymmetries or hemispheric asymmetry for speech production is systematically different for native readers of languages like Hebrew (which are read from right to left) than for native readers of languages like English (which are read from left to right). Thus, more research is clearly needed to test the specific predictions of Previc's scenario.

Fetal Position

With respect to prenatal origins of hemispheric asymmetry, one of the most-discussed asymmetries of fetal position concerns the position during much of the last trimester of pregnancy—especially the last month. As illustrated in Figure 8.2, in approximately two-thirds of the cases the fetus is positioned with the head down and the back to the mother's left side. As a result, the left arm is positioned against the mother's pelvis and backbone and the right arm is positioned against the front of the uterus. It has been suggested that this asymmetric positioning of the fetus is responsible for the predominance of right-handedness in the human population. Although the evidence for a relationship between this particular fetal asymmetry and later handedness is indirect, the evidence is sufficiently compelling to warrant more direct investigations. For example, about two-thirds of infants emerge from the uterus head first with their backs to the mother's left side, and it has been argued that fetal position is related rather strongly to asymmetries of birth presentation (for review, see Previc, 1991). It is the case that those infants who emerge in this position show a preference after birth for lying with their heads turned to the

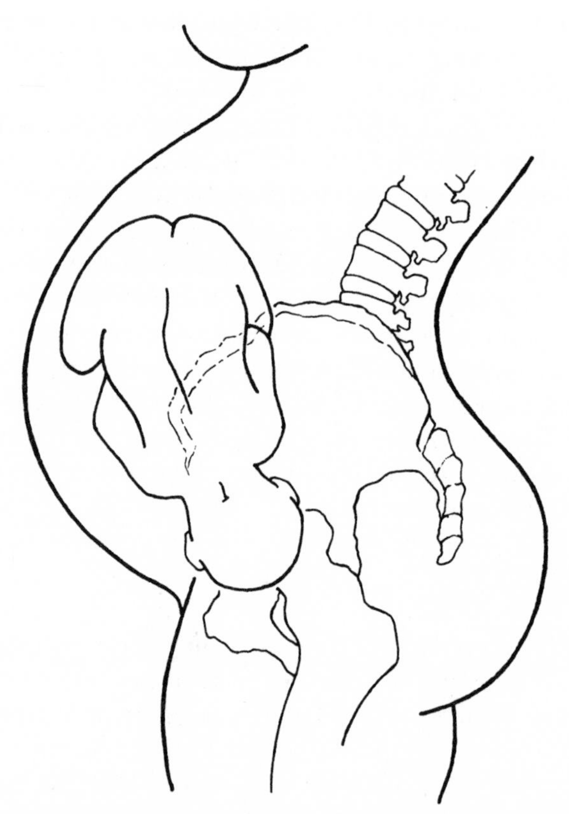

Figure 8.2. The typical position of the fetus during much of the last trimester of pregnancy, especially the last month.

right rather than to the left whereas, infants who emerge from the uterus head first with their backs to the mother's right side show no systematic head-turn preference in either direction. This very early head-turn preference is, in turn, related to hand preference in infancy. Specifically, infants tend to prefer the hand ipsilateral to the direction of head-turn preference (for review and discussion of the evidence for these assertions, see Corballis, 1991, and Previc, 1991).

Why might this particular asymmetry of fetal position be related to handedness? According to one view, a critical factor is which hand and arm is freer to move late in pregnancy (for discussion, see Corballis, 1991, and Previc, 1991). Note that the fetal position illustrated in Figure 8.2 would allow the right hand and arm to move more freely than the left, because the left is to some extent pinned against the rigid barrier provided by the mother's pelvis and backbone. Another extremely interesting view is that asymmetry of fetal position contributes to various vestibular asymmetries and, in turn, these vestibular asymmetries create asymmetric postural tendencies that lead to handedness and footedness (Previc, 1991). The gist of this scenario is as follows.

When the fetus is positioned in the uterus in the manner illustrated in Figure 8.2, the acceleration forces caused by the mother's walking create an asymmetric shearing that favors development of the left otolith (the organ of the inner ear responsible for maintaining balance) and its neural pathways. The exact manner in which this occurs is discussed in detail by Previc (1991). This preferential development leads to a left-otolithic advantage in most humans, which is presumed to underlie the fact that most humans rely more on the left side of the body for postural control. Reliance on the left side for postural control frees the right side of the body (and the left hemisphere) to become dominant for voluntary motor behavior.

In addition, Previc (1991) argues that the left-otolithic advantage also underlies right-hemisphere dominance for visuospatial processing. According to this scenario, such right-hemisphere dominance emerges in the following way. One major projection from the otolith organ on each side is to an area on the contralateral side of the brain known as the primary vestibular cortex, which is located just posterior to the primary auditory cortex of each hemisphere. From here, there is a presumed projection stream into the parietal lobe. Consequently, asymmetric development of the otolith organs may lead to asymmetries of parietal-lobe function. Thus, the dominance of the right parietal lobe for maintaining spatial orientation is proposed to be related to the greater involvement of the right (compared with the left) parietal lobe in vestibular function. This is an extremely interesting possibility for which Previc presents a good deal of circumstantial evidence.

Despite possible effects of fetal position, there is some evidence for handedness in the human fetus well before the last trimester of pregnancy and independent of fetal position. For example, Hepper, Shahidullah, and White (1991) have made ultrasound observations of fetuses from 15 weeks of gestational age to term and reported a strong bias for thumb sucking of the right hand. Specifically, of 274 observations of fetal thumb sucking, 252 (92 percent) involved the right hand. Furthermore, this proportion was constant from 15 weeks to term. In addition, for 50 fetuses observed at 32–38 weeks of gestational age, the hand preference for thumb sucking was unrelated to fetal position *in utero*. However, hand preference for fetal thumb sucking was correlated with head turning when infants were tested shortly after birth, with infants showing a preference to turn their heads to same side preferred for fetal thumb sucking. Given these results, it seems unlikely that asymmetries of fetal position are the sole determinants of later handedness. In fact, Hepper et al. even speculate that the early right-side bias for thumb sucking may be caused by advanced right-side neuromuscular development and that this early manual bias may, itself, differentially stimulate the left and right hemispheres and contribute to other aspects of hemispheric specialization.

"Snowball" Mechanisms

The preceding review indicates how several prenatal asymmetries have been proposed as precursors of functional hemispheric asymmetry in later life. There are important differences among the various scenarios in the prenatal asymmetries that are seen as important and in the mechanisms that are proposed for their importance. Perhaps the most radical suggestions come from Previc (1991), who argues that hemispheric asymmetry in humans is derived either completely or in large part from asymmetries in the prenatal development of the ear and vestibular organ. At the present time, none of these scenarios has accumulated so much evidence in its favor that the others can be discarded. In fact, it is not clear that the different scenarios are necessarily mutually exclusive. There is at least the possibility that a number of these prenatal asymmetries interact (perhaps each acting in the ways that are proposed) to determine the pattern of hemispheric asymmetry. What is most needed now is additional research that is designed to test these ideas explicitly.

Despite differences in the scenarios that have been proposed, they all agree that asymmetries that are very small when they emerge during the course of prenatal development can eventually have profound consequences for functional hemispheric asymmetry. That is, the effect of small asymmetries at one moment in ontological development can "snowball" into larger and more complex asymmetries at later moments. Although how this happens is not yet known in detail, it is instructive to consider in a general way the notion of a "snowball" mechanism that is at least implicit in the various scenarios that have been discussed.

As the fetus (and, later, the infant) is presented with new and richer types of sensorimotor information, the extent to which neural networks in the two hemispheres are responsive is likely to depend on how those networks have already been shaped by input that has come before. Even if previous input has led to only slightly different modifications of the neural networks in the two hemispheres, this difference may make one hemisphere's networks slightly more hospitable as a neural substrate for the new and richer information being presented to the brain. Even the slightest difference in hospitality may well be sufficient for the neural networks in the two hemispheres to be modified differently by the the new information that is presented to them. When this process is repeated over and over, the functional differences between the hemispheres grow and grow, even though the entire process was set in motion by a few small seeds. Consequently, as cognitive development proceeds (see Figure 8.1), new abilities may depend more on the activities of one hemisphere than the other because one hemisphere has come to provide a more appropriate neural substrate. We have already seen how such hemispheric asymmetries may begin to unfold before birth, and we now consider how that process continues throughout the life span.

Hemispheric Asymmetry from Birth through Young Adulthood

The present section traces the unfolding of hemispheric asymmetry from birth through young adulthood. For convenience of exposition, the discussion of behavioral asymmetries is divided into the same four general categories used to discuss adult asymmetries in Chapter 2: (a) handedness and the control of motor activities, (b) language, (c) visuospatial processing, and (d) emotion. There follows a brief discussion of the unfolding of biological differences between the two cerebral hemispheres.

Handedness and the Control of Motor Activities

One of the most dramatic asymmetries in adults is left-hemisphere dominance for the control of certain motor activities: approximately 90 percent of the population is usually classified as right-handed (see Chapter 2). To a large extent, handedness in children is established by 3 years of age, with the ratio of right- to left-handers being similar to that of adults (for review, see Annett, 1985; Bradshaw, 1989). Before that time, differences in the abilities of the two hands and preferences for one hand versus another are not so uniform. However, it is not possible to assess handedness per se in very young infants, and the tests for handedness that are used at different ages often consist of different, age-appropriate items. Consequently, the examination of developmental changes in handedness and in asymmetric dominance for motor activities is difficult to determine in a completely straightforward manner until approximately 3 years of age. Nevertheless, there are a variety of motor asymmetries that can be observed from birth and at least some of them are related to later handedness. On this basis it is reasonable to hypothesize that the seeds for handedness and other motor asymmetries are sown very early (perhaps before birth) and that the functional asymmetries unfold as the organism matures.

When placed on their stomachs, newborns typically show a preference to lie with their head turned to one side rather than the other. After reviewing several large-scale studies of newborns, Previc (1991) estimates that approximately 70 percent show a rightward preference and perhaps as many as 20 percent show no preference toward one side or the other. Infants who show a rightward bias for head turning are more likely than others to show right-hand preferences when tested later, although the relationship between head-turning preference and hand preference is only moderately strong. One problem is that motor asymmetries in newborns and very young children are far less stable than in older children and adults. Nevertheless, by 4 months of age most infants show a right-hand preference for directed, target-related activity (e.g., reaching for a visually presented object), whereas the left hand is preferred for passive holding and for activity that is not directed toward any particular object (e.g., Young, Segalowitz, Misek, Alp, and Boulet, 1983).

Results such as those just noted indicate that various motor asymmetries are present from birth onward. Of course, the specific manifestations of such things as handedness change dramatically as the sensorimotor system develops, but this does not necessarily mean that hemispheric asymmetry for motor control "develops" in the manner discussed earlier. In fact, there is some suggestion that from 3 years of age onward the magnitude of certain performance asymmetries remains constant. For example, Annett (1985) examined hand differences on a task that involved moving pegs from one row of holes to another and found that children as young as 3 years of age and adults were faster with the right hand than with the left hand. Furthermore, the magnitude of the right-hand dominance was approximately the same from age 3 years to age 50 years. In addition, the proportion of individuals who show left-hand dominance for this task ranged between 10 percent and 20 percent for all ages tested, with there being no systematic effect of age. It appears that once hand dominance is established, its direction and magnitude remain constant throughout the life span.

Language

In an influential book published in 1967, Erich Lenneberg proposed that hemispheric asymmetry for linguistic functions is nonexistent during the first 2 years of life, after which it *develops* until achieving the typical adult pattern at the time of puberty. He based this hypothesis primarily on early studies of recovery of linguistic function in children who sustained brain injury at different ages. In particular, he argued that when injury is sustained prior to the age of 2 years, language is acquired normally (or very nearly so) after injury to either hemisphere. When injury is sustained after the age of 2 years, the extent of recovery is related to which hemisphere is injured. Specifically, recovery tends to be better after right- than after left-hemisphere injury, with the extent of this asymmetry in recovery of function becoming progressively greater until puberty, when the typical adult pattern of asymmetric recovery is obtained. As a result, Lenneberg hypothesized that the two hemispheres are equipotential for language until approximately 2 years of age, at which time asymmetries begin to develop gradually.

As discussed earlier in the present chapter, well-done studies of recovery of function can indicate that functional plasticity for linguistic functions decreases with age, but they cannot provide evidence about developmental changes in hemispheric asymmetry. Consequently, evidence about potential changes in hemispheric asymmetry must be sought elsewhere. Furthermore, the hypothesis of early hemispheric equipotentiality for linguistic functions is even contradicted by more recent studies of recovery of function that examine specific components of linguistic processing with greater precision. For example, greater deficits in syntactic processing have been reported following left- compared with right-hemisphere injury, even when injury is sustained well before the age of 2 years (for reviews, see Hahn, 1987; Witelson, 1987). Such results indicate that there are important limits of functional plasticity, even when injury is sustained at a very young age.

Given the logical problems involved in trying to infer hemispheric asymmetry from studies of functional recovery, in recent years more emphasis has been placed on clinical studies that assess language skills at the time of injury (rather than the extent of eventual recovery). In relatively early studies of brain-injured children, the incidence of speech loss was reported to be more frequent after left-hemisphere than after right-hemisphere injury, even for children as young as 2 years of age (for review, see Witelson, 1987). This suggests that left-hemisphere dominance for "language" is present by at least 2 years of age. However, the results of these studies also indicated that the incidence of speech loss after right-hemisphere injury was higher in children (as much as 30 percent or so) than in adults (5–10 percent). On this basis it could be argued that left-hemisphere dominance for speech is less complete in children than in adults—offering some support for a developmental view. However, at least some of these early studies were subject to a number of serious methodological problems, including the possibility that a number of children actually had bilateral injury or bilateral involvement as the result of secondary infections (see Woods and Teuber, 1978; Witelson, 1987). In their own study of children (aged 2–14 years) with verifiable unilateral brain injury, Woods and Teuber found the incidence of speech loss after right-hemisphere injury to be only 7 percent for right-handers—a result virtually identical to that found for right-handed adults. In addition, their own review of studies conducted after the introduction of antibiotics and mass immunization (which tend to reduce the possibility of widespread bilateral effects) leads to similar values. Thus, what seem to be the best clinical estimates of the incidence of left-hemisphere dominance for speech production in young children argue against the developmental hypothesis advanced by Lenneberg (1967). Instead, the estimates are consistent with the hypothesis that hemispheric asymmetry for linguistic functions is present very early (probably from birth) and that the magnitude of that asymmetry does not change from birth through young adulthood.

Additional evidence for this hypothesis comes from a variety of studies using neurologically intact infants and children. For example, electrophysiological measures have been used to examine electrical activity over a variety of areas of the two hemispheres in response to a variety of stimuli, with such measures coming from adults, children, and infants as young as a few hours old (for discussion of some of these techniques, see Chapter 1). Of particular importance is the finding of greater electrical activity over areas of the left hemisphere than over the right hemisphere in response to a variety of speech sounds, with such effects extending to full-term infants less than 24 hours old and to preterm infants averaging 35 weeks of conceptional age at the time of testing (e.g., Molfese and Molfese, 1979,1980; for review, see also Hahn, 1987). It is important to note that such asymmetries are not usually reported in response to nonspeech sounds such as noise bursts, tones, and musical chords. In fact, for at least some of these nonspeech stimuli, there is more electrical activity over the right hemisphere than over the left hemisphere (for review, see Hahn, 1987). Thus, the asymmetry in favor of the left hemisphere in response to speech sounds is not simply an asymmetry in responding to auditory stimuli in general. Instead, the asymmetry appears to be restricted to human speech or to sounds with certain acoustic properties that are characteristic of human speech. It is also interesting to recall that certain components of the auditory evoked response recorded over the right hemisphere (but not over the left hemisphere) have been been found to discriminate between speech sounds in a categorical manner in human adults and children and in 15-week-old border collies (see Chapter 5 and Adams et al., 1987). This particular asymmetric response has also been reported in human infants as young as 2 months of age (e.g., Molfese and Molfese, 1979).

Dichotic-listening studies requiring the identification of a variety of speech sounds (e.g., syllables, digits, words) have found a right-ear advantage for children as young as 3 years of age, and there is no clear indication that the magnitude of the advantage changes in any systematic way from age 3 years on (for reviews, see Hahn, 1987; Witelson, 1987). It is not possible to use standard stimulus-identification procedures with children younger than about 3 years of age. However, Entus (1977) demonstrated a right-ear advantage for the discrimination of CVC syllables in a group of infants ranging in age from 3 weeks to 3 months. She did this using the following nonnutritive high-amplitude sucking paradigm. Specifically, infants are trained to suck a nipple (which provides no nutrients) in order to hear two speech sounds, one in each ear (e.g., *ba* in the left ear and *da* in the right ear). The rate of an infant's sucking gradually diminishes, and when it falls below a predetermined level the sound in one ear is changed (e.g., *da* in the right ear might be changed to *ga*). Typically, this change in sound increases the infant's interest in maintaining the stimulus and the rate of sucking increases. Entus found that the increase in sucking rate following a stimulus change was greater when the speech syllable was changed in the right ear than when it was changed in the left ear. That is, there was a right-ear advantage. By way of contrast, when the stimuli were musical notes, there was a left-ear advantage, indicating that the effect obtained with speech sounds does not reflect a general left-hemisphere dominance for processing all auditory input.

In a study patterned after the work of Entus (1977), Vargha-Khadem and Corballis (1979) were unable to replicate her results with speech sounds. One possible difference between the studies that may have been important is that most of the infants in Entus's study were over 2 months of age whereas many of the infants in the Vargha-Khadem and Corballis study were under 2 months of age. The potential importance of this seemingly small difference in age is noted by Best, Hoffman, and Glanville (1982), who used a heartbeat-deceleration paradigm to examine ear differences in dichotic listening in infants. Although they report a left-ear advantage for musical sounds in infants ranging in age from 2 to 4 months, a right-ear advantage for speech was found in 3- and 4-month-old infants but not in 2-month-old infants.

Additional indications that left-hemisphere dominance for linguistic processing is established very early come from dual-task studies that use children ranging in age from 3 years through adolescence. In Chapter 1 we saw that concurrent verbal activity of various sorts interferes with manual activity (e.g., rapid finger tapping) of the right more than the left hand and that this asymmetrical interference is often attributed to left-hemisphere dominance for linguistic processing. This pattern of asymmetric interference has been reported for children as young as 3 years of age (the youngest age that could be tested) and the magnitude of asymmetry has been found not to change from age 3 through young adulthood (e.g., Hiscock and Kinsbourne, 1978, 1980).

At approximately the time that children first begin to read, it becomes possible to examine visual-half-field asymmetries for the processing of linguistic material. As a recent review (Hahn, 1987) indicates, an RVF/left-hemisphere advantage for processing words has been found in children as young as 7 or 8 years of age, but there are several discrepancies. Part of the problem involves the need for proper eye fixation to be maintained and the difficulty that young children sometimes have with this. Other problems arise because some studies have used the identification of single letters as their "linguistic" task, and the identification of single letters does not show a consistent RVF/left-hemisphere advantage in adults. In addition, such stimulus variables as duration, luminance, and clarity and such task variables as processing strategy are known to influence the magnitude and even the direction of visual-half-field asymmetries (see Chapter 3 and also Hellige and Sergent, 1986; Sergent and Hellige, 1986). Nevertheless, there is nothing in these developmental studies of visual-half-field asymmetry that offers any clear support for the hypothesis that hemispheric dominance for linguistic processing develops in the manner proposed by Lenneberg (1967).

So, what can be concluded about hemispheric asymmetry for language from birth through adolescence? The empirical findings are most consistent with the hypothesis that hemispheric asymmetry for linguistic processing is present from birth (and, perhaps, before). As proposed by Hahn (1987) and by Witelson (1987) and illustrated in Figure 8.1, the magnitude of this hemispheric asymmetry does not appear to change over the course of a lifetime. Of course, it is difficult to measure the magnitude of behavioral asymmetries across this entire age range because different tasks must be used with infants than with adults, and even when the same task is used across a large age range there are frequently age-related changes in overall difficulty that make interpretation difficult. Despite these problems, many of the studies that have been done could have shown systematic changes in hemispheric asymmetry, but they did not. Of course, as cognitive abilities increase with age (see Figure 8.1), so does the number and complexity of linguistic skills that are available to be processed differently by the cerebral hemispheres. Furthermore, changes in cognitive ability can lead to age-related changes in processing strategy. However, there is little to indicate that the two hemispheres are ever equally involved in linguistic processing or that they ever have equal potential for language acquisition.

Visuospatial Processing

Earlier in this book I reviewed evidence of hemispheric asymmetry for certain aspects of visuospatial processing in adults (see Chapters 2 and 3). Although the right hemisphere is frequently cited as being dominant over the left for visuospatial processing, more recent studies have indicated that the direction of hemispheric dominance in adults differs for different components of visuospatial processing. For example, the right hemisphere is dominant for processing global aspects of visual stimuli, whereas the left hemisphere is dominant for processing local detail; these differences may be related to hemispheric asymmetry for processing information carried by different ranges of visual-spatial frequency. There is growing evidence that at least some of these visuospatial asymmetries appear as early as the first few months of life.

There is some indication that flashes of light produce hemispheric asymmetries in the average evoked response, with the right hemisphere generally being more responsive. Such results have been obtained with newborns and with infants as young as 2 weeks, in addition to older children and adults (for review and discussion, see Hahn, 1987). Of course, light flashes can hardly be said to carry much "visuospatial" information, but the results may reflect a greater readiness of the right hemisphere to respond to visual information. Evidence for hemispheric asymmetry in processing certain types of patterned stimuli has been obtained with visual-half-field techniques adapted for use with infants. By way of illustration, consider the following experiment reported by de Schonen and Mathivet (1990).

De Schonen and Mathivet (1990) tested infants aged 18–42 weeks using an operant conditioning technique that required the infants to discriminate between a photograph of their mother's face and a photograph of the face of a stranger. In order to eliminate superficial cues, women wore a bathing cap and black scarf when photographed, no eyeglasses were worn, the two women whose photographs were paired had similar complexions, the photographs had similar luminances, and so forth. During an initial training phase, the photograph on each trial was presented consistently to only the LVF/right hemisphere or to only the RVF/left hemisphere, and the infants were assigned randomly to one of the two visual-field conditions. A schematic diagram of the stimulus display is shown in Figure 8.3. A trial was initiated only when the infant (seated on its mother's lap facing the screen) had its eyes fixated on a blinking fixation point (with eye position monitored from a screen showing the output of an infrared camera that was photographing the reflection of reference points on the infant's cornea). Depending on whether the face shown on a trial was that of the infant's mother or of a stranger, the infant could cause a mechanical toy to become activated by looking at it within 5 seconds. For one face, the infant had to respond by looking at a toy located above the screen (the upper toy in Figure 8.3) and for the other face the infant had to respond by looking at a toy located below the screen (the lower toy in Figure 8.3). Various measures of learning performance were obtained (e.g., trials to meet a criterion of at least 7 correct responses out of 8 consecutive trials). Of particular importance was the investigation of whether learning depended on which visual field/hemisphere was stimulated.

Various measures of learning indicated better performance when the faces were presented to the LVF/right hemisphere than when the faces were presented to the RVF/left hemisphere; this was the case even for the youngest infants that were tested. For example, of the 40 infants in the LVF/right-hemisphere group 21 (52.5 percent) reached the learning criterion within 28 trials, whereas of the 40 infants in the RVF/left-hemisphere group only 5 infants (12.5 percent) did so. It is important to note that, for at least two reasons, this result cannot be attributed to hemispheric asymmetry for the operant conditioning procedure used in this study. First, de Schonen and Bry (1987) used this same operant-conditioning paradigm in an experiment that required infants to discriminate between schematic drawings of normal faces and drawings with the features of the faces scrambled. For this easier and qualitatively different visual discrimination, there was no hemispheric asymmetry. Second, de Schonen, Gil de Diaz, and Mathivet (1986) used half-field displays similar to those shown in Figure 8.3, but instead of using an operant-conditioning procedure they merely recorded the latency of the infant's saccadic eye

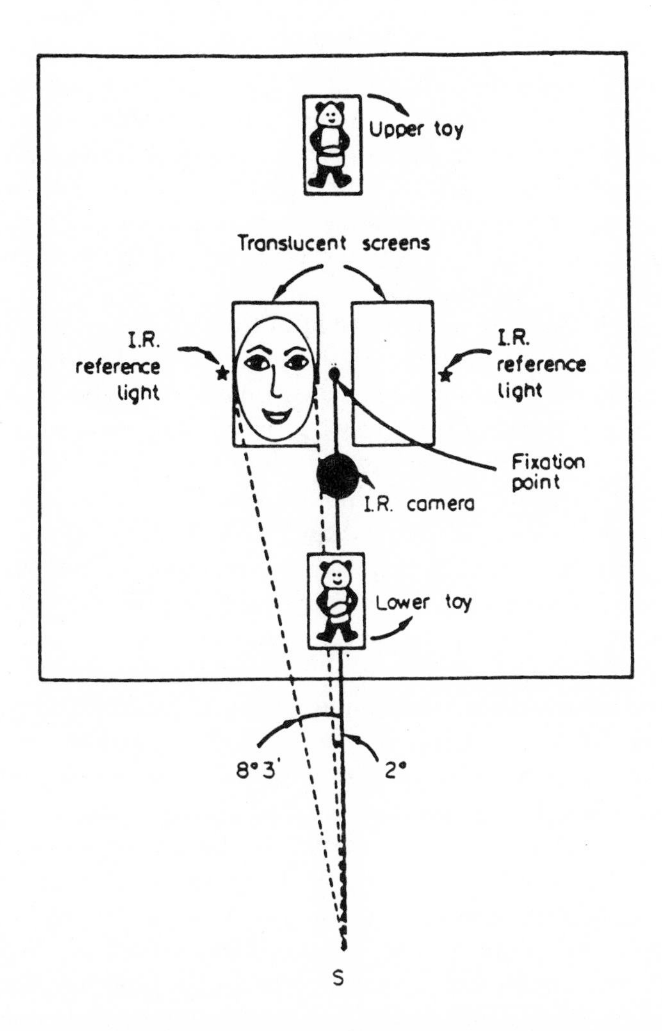

Figure 8.3. The presentation board used by de Schonen and Mathivet (1990). A slide projector and infrared (I.R.) camera were located behind the board. The infant *(s)* and its mother sat in front of the board. [Reprinted from S. de Schonen and E. Mathivet, "Hemispheric Asymmetry in a Face Discrimination Task in Infants," *Child Development,* 61 (1990):1192–1205. Copyright 1990 by The Society for Research in Child Development, Inc. Reprinted by permission.]

movements to the faces when they were presented. Infants as young as 4 months of age (the youngest age tested) reacted more quickly to their mother's face than to the face of a stranger, and this effect was significantly larger on LVF/right-hemisphere trials than on RVF/left-hemisphere trials. Similar results were found when, instead of the mother's face, a completely unfamiliar face was contrasted with a face with which the infant had become familiar during an earlier phase of the experiment. However, no hemispheric asymmetry was found when the stimuli were familiar and unfamiliar simple geometric patterns.

It is also interesting to note that after the initial operant-training phase of their experiment, de Schonen and Mathivet (1990) switched the stimuli to the opposite visual field to determine the extent of interhemispheric transfer of training in their infants. Consistent with earlier reports, they found no evidence of interhemispheric transfer in their infants aged 18–42 weeks.

The results reported by de Schonen and Mathivet and their colleagues suggest that hemispheric asymmetry for whatever processes are involved in discriminating faces is present by about 4 months of age (the earliest age tested). The existence of right-hemisphere dominance in infants is consistent with many reports of LVF/right-hemisphere advantages for face processing in adults. However, these results do not necessarily imply that the right-hemisphere dominance is specific to faces. As noted by de Schonen and Deruelle (1991), another possibility is that the two hemispheres of infants (as well as adults) are biased toward different modes of processing visual information, with the mode preferred by the right hemisphere better adapted for discriminating between two very similar faces. In fact, Deruelle and de Schonen (1991) have obtained data from infants aged 4–10 months indicating that, for visual patterns that have a coherent global form, the right hemisphere discriminates between patterns on the basis of global form whereas the left hemisphere tends to discriminate between patterns on the basis of local components. Note that there is a great deal of evidence that something similar occurs in adults (see Chapter 3).

The foregoing studies indicate that hemispheric asymmetry for certain aspects of visuospatial processing emerges at a very young age. Visual-half-field studies with older children (ages 6 years through young adulthood) present a mixed picture about the extent to which asymmetries in using nonlinguistic stimuli are constant or change with age. At least one reviewer (Hahn, 1987) finds no unequivocal evidence for developmental changes, a conclusion that is supported by more recent studies. Even when there has been some indication of developmental changes, they may be more related to changing cognitive strategies than to changes in hemispheric asymmetry per se. For example, Levine (1985) suggests that children below the age of 10 years or so often recognize unfamiliar faces on the basis of single features or unusual characteristics (e.g., eyeglasses or moustache), and for that reason they may not show an LVF/right-hemisphere advantage for recognizing unfamiliar faces. She also suggests that, starting at about age 10 years, children process unfamiliar faces using multiple features and the relationships among those features. This emphasis on the global or configural properties of faces then leads to an LVF/right-hemisphere advantage. Note that this hypothesis is consistent with that illustrated in Figure 8.1, where hemispheric asymmetry per se emerges early and does not change, yet the behavioral manifestations of that asymmetry change with the emergence of new cognitive abilities and strategies.

Emotion

There is clear evidence of hemispheric asymmetry for the production and perception of emotion in adults, although there is some controversy over which of two competing views provides a better characterization of that asymmetry (see Chapter 2). According to one view, the right hemisphere is dominant over the left for the production of both positive and negative emotions. According to the other view, the right hemisphere is more involved in producing and perceiving negative emotions whereas the left hemisphere is more involved in producing and perceiving positive emotions. Although there has been no final reconciliation of these two views (but see Chapter 2), evidence more consistent with the second view has been obtained in 10-month-old infants and indirect evidence has been obtained from infants as young as 2 days old (for reviews, see Davidson, 1992, and Fox, 1991). Such findings indicate that hemispheric asymmetry related to emotion is probably present at least from the time of birth.

By way of illustrating the kind of evidence obtained with infants, consider the following study reported by Davidson and Fox (1982). Ten-month-old infants sat on their mothers' laps and watched videotaped segments of an actress spontaneously generating a happy or sad facial expression. Electrodes placed over the frontal and parietal scalp regions of the two hemispheres recorded brain activity (EEG measures) while the infants watched the video segments. In response to happy segments, the EEG measures showed more activity over the left frontal region than over the right frontal region, consistent with the hypothesis that the left hemisphere is more involved than the right in perceiving (and perhaps producing) positive emotions. This asymmetrical brain activation in response to the happy facial expressions did not extend to the parietal areas, consistent with the hypothesis that frontal areas are more important than parietal areas for perceiving and producing emotion (see Chapter 2). There was no significant asymmetry over either the frontal or parietal area when infants viewed the sad facial expressions, which is consistent with the view that hemispheric asymmetry related to emotion depends on whether the emotion is positive or negative.

In a variety of follow-up experiments, Fox and Davidson and their colleagues (see Fox, 1991) have shown that infants display greater relative activation of the left frontal region when producing spontaneous facial expressions judged to be positive (e.g., joy) and greater relative activation of the right frontal region when producing spontaneous facial expressions judged to be negative (e.g., disgust). Furthermore, Davidson and Fox (1989) have even reported a relationship in 10-month-old infants between *resting* EEG asymmetry and temperament. In their study, EEG was recorded during a baseline period and measures were taken over the frontal and parietal areas of both hemispheres. Later, the infants were classified into those who cried (Criers) versus those who did not (Non-Criers) during a period of maternal separation. Left/right EEG asymmetries recorded over frontal (but not parietal) areas were related to the infants' behavior. For the Criers (who displayed "negative" emotion), activation during the resting phase was relatively greater in the right hemisphere than in the left hemisphere. For the Non-Criers (who displayed more "positive" emotion), the left/right difference during the resting phase was reversed. Such results suggest that infants may differ in the characteristic pattern of hemispheric arousal asymmetry (see Chapter 7 and Levy et al., 1983a,b) and that the pattern of arousal asymmetry may be predictive of temperament.

Fox (1991) suggests that the development and differentiation of emotions during the first year of life occur through the addition of new motor programs associated with fundamental tendencies toward approach or withdrawal. It may thus be possible that hemispheric asymmetry for producing and perceiving emotions of different valence stems from an early hemispheric asymmetry for primary states of approach or withdrawal. Some evidence for this possibility is provided by Fox and Davidson (1986), who recorded EEG responses from the frontal and parietal areas of the left and right sides of the scalp in 2-day-old infants who were presented with solutions of distilled water, sugar water, and citric acid. An analysis of videotapes of the infants' expressions indicated a clear "approach" reaction to the sugar water and a clear "withdrawal" reaction to the citric acid. Of particular interest is the fact that Fox and Davidson reported relatively greater activation of the left hemisphere after infants tasted sugar water than after they tasted citric acid, with the hemispheric asymmetry occurring over both frontal and parietal areas. Thus, hemispheric asymmetry related to approach versus withdrawal is present for infants shortly after birth and provides at least indirect evidence that the seeds of hemispheric asymmetry for producing and perceiving emotions are sown very early.

Biological Asymmetry

In Chapter 4, various biological asymmetries were seen to characterize the brains of adult humans. Many of those asymmetries begin to appear in the brains of children and infants and at least some begin to appear prenatally. For example, in adults the sylvian fissure in the right hemisphere curls upward more than the sylvian fissure in the left hemisphere. This particular asymmetry begins to appear by about the sixteenth week of gestation (e.g., LeMay and Culebras, 1972). As noted in Chapter 4, the left planum temporale is typically larger than the right planum temporale, and the percentage of asymmetric brains is similar for adults and children (e.g., Hynd et al., 1990; Larsen et al., 1990). Indeed, this asymmetry is also present in newborns (e.g., Witelson, 1987) and has been observed as early as the thirty-first week of gestation (e.g., Chi, Dooling, and Gilles, 1977). Despite the early appearance of certain biological asymmetries, those asymmetries continue to unfold from birth through young adulthood and, perhaps, throughout the entire life span. For example, earlier in the present chapter it was noted that developmental changes in the two cerebral hemispheres from age 2 months to young adulthood are characterized by growth spurts, with at least some of those spurts being asymmetric (e.g., Thatcher et al., 1987). Consequently, at different moments of ontogenetic development, homologous areas of the two hemispheres may be at different levels of maturity and, for that reason, may respond differently to incoming information and to sensorimotor feedback. However, exactly how these asymmetric growth spurts might influence the final adult pattern of biological asymmetry is unknown.

The unfolding of functional hemispheric asymmetry depends not only on the emergence of biological differences between the two hemispheres but also on developmental changes in the corpus callosum and other channels of interhemispheric communication. The corpus callosum matures relatively slowly; for example, it is one of the last areas of the brain to undergo myelination, a process which is not complete until at least the time of puberty (e.g., Lecours, 1975). As a result, the brains of infants and young children are similar to those of split-brain patients in that certain types of interhemispheric transfer is either impossible or greatly impaired. For example, de Schonen and Mathivet (1990) found no evidence of interhemispheric transfer for a face-discrimination task in infants ranging in age from 18 to 42 weeks. In addition, relative to adults, children have more difficulty matching tactile designs between hands (and hemispheres) than within hands (and hemispheres), presumably because of the immaturity of the corpus callosum (e.g., Galin, Johnstone, Nakoff, and Herron, 1979). As noted in Chapter 6, there is debate about whether the corpus callosum should be thought of as a neural structure that permits transfer of information from one hemisphere to the other or as a neural structure that permits activity in one hemisphere to inhibit activity in the other hemisphere. Regardless of which view is more correct (or if both are partially correct), the biological status of the corpus callosum is likely to interact with the biological differences between the two hemispheres to influence the manner in which functional hemispheric asymmetries unfold from birth through young adulthood.

Hemispheric Asymmetry in Old Age

Changes in intellectual and cognitive functions in old age are not uniform. For example, Wechsler (1958) reported that elderly persons often do poorly relative to younger adults on subtests (such as block-design and picture-arrangement tasks) of the WAIS used to measure what has been called "Performance IQ." By way of contrast, the same individuals often show little or no decline on a variety of verbal subtests (tests of vocabulary and language comprehension) of the WAIS used to measure what has been called "Verbal IQ." In addition, Horn and Cattell (1972; see also Horn, 1988) have found greater age-related declines in fluid intelligence (e.g., logic, reasoning) than in crystallized intelligence (e.g., culture-specific knowledge accumulated over time).

Horn and Cattell (1972) suggested that declines in intellectual functioning in the elderly might be the result of the accumulation of minor insults to the central nervous system. However, when the performance of elderly persons is compared with the performance of individuals with diffuse brain damage, the patterns are quite different. In particular, the pattern of impaired Performance IQ and spared Verbal IQ is more typical of neurological patients with right-hemisphere injury than with left-hemisphere injury. This has led to the suggestion that aging involves greater loss of right- than of left-hemisphere function and even to the hypothesis that the right hemisphere ages more rapidly than the left hemisphere (see Klisz, 1978; Ellis and Oscar-Berman, 1989; Schaie and Schaie, 1977). The remainder of this section considers this "right-hemisphere aging" hypothesis in more detail and examines how studies of hemispheric asymmetry in old age can benefit from considering the dimensions of individual variation outlined in Chapter 7.

Do the Hemispheres Age Differently?

The idea that normal aging might have different consequences on the two cerebral hemispheres and, consequently, on hemispheric asymmetry is plausible in view of certain age-related changes in other species. For example, in at least some species of rats certain portions of the right side of the brain are larger than homologous areas of the left side of the brain in young animals, but these asymmetries become smaller or disappear in old age (e.g., Diamond, 1985; see also Chapter 5). In view of evidence that certain areas of the human right hemisphere develop faster than homologous areas of the left hemisphere *in utero,* there would be an appealing symmetry if the aging process tended to be less kind to the right hemisphere than to the left hemisphere. However, despite early findings that seemed consistent with the hypothesis that the right hemisphere ages faster than the left hemisphere, the accumulated evidence does not support such a conclusion.

As noted earlier, the first suggestions of the right-hemisphere decline hypothesis came from the observation that psychometrically measured nonverbal, visuospatial functions tend to decline more rapidly than verbal functions in old age. Note that the hypothesis is based on the assumption that the right hemisphere is dominant for psychometrically measured nonverbal, visuospatial functions whereas the left hemisphere is dominant for verbal functions. While there is some truth to this assumption, we have seen in Chapters 2 and 3 that the pattern of hemispheric asymmetry is not so simple and that, when asymmetries exist, they tend to be relative rather than absolute. Furthermore, age-related differences in such things as Performance IQ relative to Verbal IQ can be explained in other ways. For example, the performance tests typically require greater ability to deal with novelty or complexity than do the verbal tests. This being the case, it is important to examine age-related changes for tests and tasks that are related more clearly to the integrity of one hemisphere or the other.

More recent studies of age-related changes in specific cognitive abilities do not support the conclusion that abilities associated with right-hemisphere injury decline more rapidly than abilities associated with left-hemisphere injury. For example, Goldstein and

Shelly (1981) tested 1,247 subjects ranging in age from the early twenties to the late seventies on several tests from the Halstead-Reitan neuropsychological battery. The battery consisted of a number of sensorimotor tasks that were developed to index the presence and location of brain injury in neurological patients. The test results were scored in a way that produced "points" associated with right- versus left-hemisphere function, such that a greater number of points is associated with poorer performance (i.e., possible deterioration). They found that the number of points associated with each hemisphere increased linearly across the age range tested: there was no hemisphere-by-age interaction. Thus, there is no clear evidence of significantly greater decline of right-hemisphere function. However, when Goldstein and Shelly examined which specific tasks contributed to the decline associated with each hemisphere, they found that declines associated with the left hemisphere involved primarily the decline of somatosensory functions for the right hand whereas declines associated with the right hemisphere involved primarily the decline of psychomotor problem solving with the left hand. On this basis Goldstein and Shelly suggest that the two hemispheres may decline to the same extent with advancing age, but the effects of aging are different in the two hemispheres (or, as they put it, one hemisphere ages in a different manner from the other).

Mittenberg, Seidenberg, O'Leary, and DiGiulio (1989) examined the right-hemisphere decline hypothesis in 68 volunteers who were free of systemic and neurological illness. One group consisted of individuals aged 20–35 years and the other consisted of individuals ranging in age from 55 to 75 years of age. The researchers administered a battery of neuropsychological tests that have been developed to localize brain injury, and the tests were chosen to control for such things as being timed versus untimed, familiar versus unfamiliar, and so forth. In general agreement with the results obtained by Goldstein and Shelly (1981), Mittenberg et al. found no evidence of greater decline for tests associated with right- rather than left-hemisphere injury. However, they did find some evidence of greater decline on tests associated with frontal-lobe function than with tests associated with other areas, such as the temporal, parietal, and occipital lobes.

Another approach to the study of possible changes in hemispheric asymmetry in old age involves the examination of perceptual and response asymmetries. For example, to the extent that the right hemisphere in some functional sense ages earlier or faster than the left hemisphere we might expect greater age-related declines when stimuli are presented directly to the right hemisphere (via the left ear, left visual field, or left hand) than when stimuli are presented directly to the left hemisphere (via the right ear, right visual field, or right hand). There is no clear, unequivocal indication that this is the case. For example, when individuals with peripheral hearing asymmetries were excluded, Ellis (1987, as described in Ellis and Oscar-Berman, 1989) found that both young and elderly right-handed men showed a right-ear advantage for a dichotic-listening task using verbal/phonemic stimuli and a left-ear advantage for a dichotic-listening task involving the discrimination of tonal patterns. Furthermore, the magnitude of the ear advantages were equivalent for both the young and old groups (for additional discussion of relevant dichotic-listening studies, see Ellis and Oscar-Berman, 1989).

Studies of visual-half-field asymmetry also suggest that hemispheric asymmetry is equivalent for young and elderly individuals. For example, in a study of visual masking, Byrd and Moscovitch (1984) required young and elderly adults to identify 3-letter words presented briefly to the LVF/right hemisphere or RVF/left hemisphere on each trial. Although the elderly adults identified fewer words overall than the young, both groups showed an RVF/left-hemisphere advantage of the same magnitude. Similar results have been reported by Obler, Woodward, and Albert (1984), who had young and elderly adults perform verbal (letter-name matching) and nonverbal (face matching) tasks. They found an RVF/left-hemisphere advantage for the verbal task and an LVF/right-hemisphere advantage for the face task, with the magnitude of the asymmetries being equivalent for the young and elderly (for additional review of relevant visual-half-field studies as well as studies of tactile asymmetry, see Ellis and Oscar-Berman, 1989). In addition, Levine and Levy (1986) examined perceptual asymmetry on the free-vision face task developed by Levy et al. (1983a,b; see Chapter 7) in right-handers ranging in age from 5 years to over 80 years. With the possible exception of 5-year-olds (whose data were very noisy), the magnitude of the expected left-side bias did not change across the life span. None of these results offers support for any straightforward hypothesis of differential-aging of the two hemispheres.

Although there is little behavioral or biological evidence to support any simple version of the differential aging hypothesis, Stern and Baldinger (1983) have reported results that suggest elderly people who prefer what might be termed right-hemisphere modes of processing are at something of a disadvantage relative to elderly people who prefer left-hemisphere modes of processing, whereas this distinction is less important in young adults. Stern and Baldinger classified individuals into two groups on the basis of whether they made more left or right lateral eye movements when being asked a series of questions face-to-face. It has been argued that "right-movers" (those who make more eye movements toward the right in this situation) tend to have more left- than right-hemisphere activation, whereas "left-movers" tend to have more right- than left-hemisphere activation (for discussion pro and con about the validity of this technique, see Ehrlichman and Weinberger, 1978; Kinsbourne, 1972). On this basis Stern and Baldinger suggest that right- and left-movers prefer to use modes of processing associated with the left and right hemispheres, respectively. For the elderly subjects, right-movers tended to outperform left-movers on both verbal and nonverbal measures of cognitive abilities, with statistically significant differences obtained for vocabulary, solving anagrams, and block design. Such differences did not occur for the young subjects. These results lead Stern and Baldinger to conclude that elderly individuals who rely more on the right hemisphere will show greater deficits in all aspects of intellectual functioning, relative to young adults and to elderly individuals who rely more on the left hemisphere. Although some of the links in their logical chain are weak, the hypothesis advanced by Stern and Baldinger is extremely interesting and merits additional investigation.

Aging and the Dimensions of Individual Variation

In Chapter 7, I argued that it is misleading to think about individual differences in something as multifaceted as "hemispheric asymmetry" because there are several dimensions related to functional hemispheric asymmetry and individuals have been shown to vary on a number of those dimensions. Among the dimensions known to produce individual variation are direction of hemispheric asymmetry, magnitude of hemispheric asymmetry, asymmetric arousal of the two hemispheres, complementarity of asymmetries, and interhemispheric communication. Part of the problem in trying to determine whether hemispheric asymmetry changes in some systematic way as people age comes from the fact that most of the available studies confound a number of these dimensions. Consequently, when age-related differences are found, they may be attributed to a change in the magnitude of hemispheric asymmetry, for example, when they actually reflect changes in interhemispheric communication. In addition, age-related changes in such things as the magnitude of hemispheric asymmetry can even be masked if there are concomitant changes in interhemispheric communication. Furthermore, it is not clear how the notion of right-hemisphere decline should be interpreted with respect to these different dimensions.

With these problems in mind, I think it is important in future studies of hemispheric asymmetry and aging to compare the performance of young and elderly adults on a variety of tasks chosen because they allow some separation of these various dimensions (see Chapter 7 for examples of such tasks). What evidence there already is suggests that the direction and magnitude of perceptual asymmetries probably does not change in the same way for most elderly people. With respect to possible changes in characteristic arousal asymmetries, the little evidence that exists is complex. It has been suggested that individual differences in left/right bias for the free-vision face task discussed in Chapter 7 reflect individual differences in characteristic arousal asymmetry. To the extent that this is so, the fact that the average left/right bias is the same for children, young adults, and elderly adults suggests that the pattern of characteristic arousal asymmetry does not change with age. If lateral eye movements are taken as a measure of characteristic arousal asymmetry, however, then the results reported by Stern and Baldinger (1983) suggest a different relationship between arousal asymmetry and intellectual abilities in elderly adults compared with younger adults. With respect to interhemispheric communication, very little is known or even hypothesized about the effects of aging, although there has been some suggestion that interhemispheric communication is not as efficient in elderly adults as in young adults (e.g., Goldstein and Braun, 1974).

Given all these points, it would seem prudent to keep an open mind about whether aging has any systematic effects whatsoever on "hemispheric asymmetry" until the various dimensions of individual variation have been separated adequately. Although it seems unlikely that the right hemisphere "declines" or "ages" earlier than the left hemisphere in a large majority of the population, there are other possibilities that remain to be investigated. In addition to such things as interhemispheric transfer, it is useful to consider the possibility that the two hemispheres do not decline at the same rate in most people (perhaps as the result of asymmetry for small, accumulated cerebral insults), with the direction of asymmetric decline varying randomly from person to person. As a result, the average asymmetries of younger and older adults might not differ, but the individual variation in measured asymmetry would be larger in the elderly than in the young. Although this possibility is admittedly speculative, it is consistent with the fact that overall performance on a variety of cognitive and behavioral tasks is more variable in elderly than in younger adults.

Summary and Conclusions

The seeds of functional hemispheric asymmetries are sown long before birth, and the various asymmetries unfold in an orderly fashion throughout the life span (theme 5 in Chapter 1). The ultimate ontogenetic origins of hemispheric asymmetry probably date back to formation of the first neural elements, to asymmetries of the ovum, or even to various molecular asymmetries. From these early origins, the eventual emergence of functional hemispheric asymmetry is determined by the complex interplay of a number of biological and environmental factors, beginning with the fetus as it develops *in utero* and continuing into old age.

During the course of fetal development, certain areas of the right hemisphere appear to mature earlier than homologous areas of the left hemisphere, and it has been hypothesized that for at least some areas this maturational asymmetry continues to be present for a short time after birth. Various scenarios are proposed to suggest how functional hemispheric asymmetries could arise from the interaction of these different rates of maturation and the changing nature of environmental information presented to the brain. In general, the right hemisphere is proposed to be more influenced by the sort of impoverished information that the developing brain receives first (e.g., noises in audition, low spatial frequencies in vision, very coarse sensorimotor feedback), whereas the later-developing left hemisphere seems to be saved to be more influenced by higher-quality information that the brain receives only later (e.g., language sounds from the mother's voice in audition, high spatial frequencies in vision, precise sensorimotor feedback). It has also been proposed that asymmetric growth spurts during childhood provide a mechanism for continued unfolding of functional hemispheric asymmetry. For example, during the period from 3 to 6 years of age both frontal and occipital areas of the left hemisphere develop more rapidly than homologous areas of the right hemisphere, and this is the period that is also critical for the acquisition of grammar (an aspect of language for which the left hemisphere is usually dominant).

Additional prenatal asymmetries have also been hypothesized to influence later functional hemispheric asymmetry. For example, various asymmetries in the skull and face begin to appear during the first trimester of pregnancy, and it has been proposed that these cranio-facial asymmetries constitute the prenatal origins of left-hemisphere dominance for perceiving and producing speech. The rationale is that for most individuals the direction of cranio-facial asymmetry leads to greater sensitivity of the right ear to sounds in the range critical for language and that this slight peripheral ear difference makes the left hemisphere more responsive to speech. In addition, asymmetries of fetal position, primarily during the last trimester of pregnancy, have been linked to such things as later handedness. On this basis it has been argued that fetal position is an important contributor to later functional hemispheric asymmetry. One of the most ambitious hypotheses is that the asymmetric fetal position favors development of the left otolith, and it is this asymmetry in the organs of balance that underlies later handedness and right-hemisphere dominance for spatial processing, the latter because of neural connections between primary vestibular cortex and the parietal lobe.

Once a specific functional hemispheric asymmetry first appears after birth, its direction and magnitude probably remains approximately the same until death. Of course, the emergence of any specific behavioral manifestation of hemispheric asymmetry depends on the level of overall cognitive development. That is, asymmetry for a specific cognitive ability cannot be seen until that ability is sufficiently developed. However, if the hemispheres will be asymmetric for that ability in adulthood, then they will be asymmetric for that ability from the time of its first occurrence and will continue to be asymmetric for that function in old age. In addition, many precursors of later asymmetries are present at very young ages and, in many cases, from birth. For example, neonates show postural biases that are correlated with later handedness, and their hemispheres are differentially responsive during speech-discrimination tasks (the left hemisphere is more responsive) and during face-discrimination tasks (the right hemisphere is more responsive). In addition, in young infants, the relative levels of left- and right-hemisphere activation are related to emotional approach versus withdrawal. Certain biological asymmetries that characterize the brains of adults also appear very early, in some cases prenatally. During childhood there are asymmetric growth spurts and continued maturation of the corpus callosum, however, and these biological changes may be related to the manner in which functional hemispheric asymmetries unfold.

Elderly people are more likely to show declines for a variety of nonverbal tasks (e.g., solving block-design problems) than for verbal tasks (e.g., vocabulary). On this basis it has been proposed that aging involves greater loss of right- than of left-hemisphere function, possibly because the right hemisphere ages faster or earlier than the left hemisphere. Despite some early suggestive evidence in its favor, this hypothesis is not supported by more recent studies. For example, when tasks are chosen that control for such things as familiarity, there is no evidence of greater age-related decline on tests associated with right-hemisphere dysfunction than on tests associated with left-hemisphere dysfunction. Furthermore, perceptual and response asymmetries do not differ for younger and elderly adults. However, there is some evidence that elderly people who rely primarily on right-hemisphere modes of processing show widespread intellectual impairment relative to elderly people who rely primarily on left-hemisphere modes of processing. This hypothesis merits additional investigation in studies that provide a clear separation of the several dimensions of individual variation discussed in Chapter 7. Of special importance are direction of hemispheric asymmetry, magnitude of hemispheric asymmetry, asymmetric arousal of the two hemispheres, and efficiency of interhemispheric communication.

9

The Evolution of Hemispheric Asymmetry

It is humbling to realize that our own species, *Homo sapiens sapiens,* has existed in its present form for only about 35,000 years, a brief moment in evolutionary time. It is even more humbling to realize that we have yet to demonstrate whether we have any more staying power than the products of earlier experiments of evolution, species that enjoyed a momentary flurry of success followed by the silence of extinction. Our capacity for humility, however, has not prevented us from being preoccupied with how and when we came to be what we are. To see that this is so, one need only consider the often heated exchanges between evolutionary theorists and those whose religious beliefs contain the idea that we are the result of a special act of creation. The preceding chapters review a small part of what we have learned about ourselves—that our brains are functionally asymmetric in a variety of ways. The purpose of the present chapter is to consider the manner in which such asymmetries could have evolved.

In evaluating possible scenarios for the evolution of hemispheric asymmetry, keep in mind that contemporary research has made it clear that functional and biological asymmetries are ubiquitous in nonhuman species (see Chapter 5), despite earlier views to the contrary. Furthermore, some of those asymmetries are hypothesized to be homologous to asymmetries in humans. At the same time, in humans as well as other species there are many functions for which there is no asymmetry or evidence of hemispheric dominance. Consequently, first I will review in a general way the evolutionary pressures that may have favored the emergence of symmetry versus asymmetry for different functions. Then I discuss the potential snowball effect caused by changing environmental demands on organisms that might have been only slightly asymmetric and the extent to which asymmetries might be considered continuous across species. These discussions provide a backdrop against which human evolution can be considered in more detail.

In order to assess scenarios specific to the evolution of hemispheric asymmetry in humans, one must know a bit about the time course of human evolution more generally. With this in mind, I present in a later section of the present chapter a brief review of the hypotheses about the divergence of hominids from the apes and about hominid evolution during the last 4 million years or so. Following this is a more detailed discussion of milestones of hominid evolution that are believed to be important for the continued evolution of hemispheric asymmetry, leading to its present form. These milestones include walking upright, tool use, language, and the prolonged period of postnatal immaturity that is characteristic of humans. For each of them I will discuss alternative views of how that milestone might have shaped the hemispheric asymmetry we see today.

Symmetry versus Asymmetry

The beauty of symmetry can be very compelling. Perhaps for that reason it is tempting to consider bilateral symmetry the natural (or at least original) order of living things and to consider the development of asymmetry as a somewhat deviant (or at least later) evolutionary adaptation. In fact, something close to this point of view has sometimes been implied in discussions of the evolution of hemispheric asymmetries in humans. To some extent this view was driven by the belief that hemispheric asymmetry is unique to humans, that its emergence was closely tied to the emergence of strong right-handedness at the population level, tool use, and language. With the discovery of so many asymmetries in the brains of other species, the notion that hemispheric asymmetry is a unique property of humans must be abandoned. Furthermore, despite a certain intuitive appeal, there is no reason to suppose that bilateral symmetry is somehow closer to the natural order of things than is asymmetry or that it is only the development of asymmetry that is the product of evolutionary adaptation. In fact, there is every reason to believe otherwise.

As noted earlier, the molecules and physical particles of which living things are constructed are themselves asymmetrical. In addition, asymmetry is characteristic of many single-celled organisms and of individual cells in more complex organisms. These observations, among others, have led to the realization that symmetry is in no sense more fundamental as a property of biological systems than is asymmetry. Instead, both symmetry and asymmetry are equally the result of evolutionary adaptation. This being the case, the remainder of this section considers evolutionary pressures that are likely to have favored symmetry versus asymmetry.

A compelling hypothesis about the evolutionary pressures favoring bilateral symmetry for sensorimotor processes has been advanced and summarized nicely by Corballis (1991, p. 81):

> To freely moving animals, the world is essentially indifferent with respect to left and right; there are no pressures or contingencies that should make us more responsive to one side or the other. Limbs are symmetrically placed so that movement, whether achieved by running, swimming, or flying, may proceed in a straight line. Any asymmetry in legs or wings would be likely to cause an animal or bird to proceed in fruitless (and perhaps meatless) circles. Sense organs, such as eyes, ears, and nostrils, are symmetrically placed so that we may be equally alert to either side. Those parts of the brain and central nervous system that are concerned with these functions mirror the symmetry of the functions themselves.

However, even symmetry in sensory or motor systems will be replaced by asymmetry when it is more adaptive. For example, in the flounder, which swims on one side along the ocean bottom, both eyes have migrated over evolutionary time to the side facing upward, presumably because the sensory capacities of an eye facing downward would be wasted. An interesting motor asymmetry is illustrated by the New Zealand wrybill plover, a bird with a beak that is curved to the right. As noted by Corballis (1991), this asymmetric adaptation makes it easier for the bird to use its beak to overturn stones to look for food and to use its beak as a sieve to extract tiny crustaceans from water.

There may be little or no pressure for symmetry for bodily and cortical systems that are not involved in direct sensory or motor responses to the environment. Instead, asymmetry would be just fine for these systems, especially if it offers the potential advantage of compact packaging. For example, most of our internal organs are asymmetric in shape and function and are asymmetrically placed within the body cavity, an arrangement that permits efficient use of limited space. It has been argued that, in an analogous way, hemispheric asymmetry allows homologous areas of the two hemispheres to play primary roles in different aspects of information processing, thereby increasing the range of processing that can be performed efficiently in a brain of limited size. In addition, we have seen that the corpus callosum, the major fiber tract that connects the two cerebral hemispheres, is both a channel by which the two hemispheres can share information and also something of an inhibitory barrier (see Chapter 6). With respect to this latter function, the corpus callosum seems to have evolved, in part, to reduce maladaptive cross-talk between homologous regions of the two hemispheres. To the extent that this is the case, it would add to the evolutionary pressure for *complementary* or *mutually inconsistent* processes to migrate toward opposite hemispheres.

To be sure, it does not seem likely that any single description or processing dimension can capture the full range of functional hemispheric asymmetries in present-day humans. It is remarkable, however, how many of the specific asymmetries have been described as complementary, at least at the population level. For example, with respect to visual cognition, the right and left hemispheres are predisposed toward global and local aspects of stimuli, respectively (see Chapter 3). Both aspects of the visual world are important for survival and the brain must be responsive to both, but it may be difficult or impossible for the same neural tissue to do so at the same time. As a result, evolutionary pressures may have favored segregation of the neural mechanisms that subserve global and local analysis of visual patterns. At the same time, this segregation would have to be accomplished in a manner that made coordination of information from the two levels relatively easy. From this point of view, the segregation of these specific components of visual processing into opposite hemispheres would seem to be an efficient solution.

Snowball Effects

The primary point of the preceding section is that both symmetry and asymmetry are the result of evolutionary adaptation. From this perspective, it is the case that a certain amount of asymmetry characterized living things from the time of their first moments on Earth. In the present section I consider how, once asymmetry was present, evolutionary adaptation could result in increased as well as decreased asymmetry, even when relevant environmental pressures did not favor asymmetry per se. Increasing asymmetry seems particularly likely for adaptations that do not involve the type of direct sensory or motor reactions to the environment discussed by Corballis (1991) as pressures that would serve to reduce asymmetry.

The hypothesis advanced here is analogous to the notion of a snowball mechanism discussed earlier for development of hemispheric asymmetry across the life span (see Chapter 8 and also Kosslyn, 1987). In organisms that are already functionally asymmetric, at least some of the adaptations favored by the environment are likely to be implemented more efficiently on one side than on the other. When this is the case, those adaptations are also likely to be asymmetric when they emerge and set the stage for even further asymmetries. For example, suppose that in the course of hominid evolution there was a time when the left hemisphere had already become dominant for producing sequential, communicative gestures of the hands but the vocal tract had not yet evolved in a way that permitted speech. Later in the present chapter I will discuss reasons why offspring that also had some ability for vocal communication would be more likely to survive and reproduce than offspring without such an ability. That is, the environment favored the eventual development of speech. Given that this new ability would have to be built upon already-existing abilities and given sufficient similarity between certain elements of speech and communication by gesture, it is reasonable to suppose that, when speech emerged, it would be in the context of left-hemisphere dominance. Notice that this would be the case even though there was nothing *in the environment* that necessarily favored hemispheric asymmetry for speech at all, much less left-hemisphere dominance in particular. The environment merely favored the development of speech and the already asymmetric state of the organism made it more likely that speech would emerge in the context of left-hemisphere dominance.

This interplay between the nature of the environment and the existing state of organisms is reminiscent of the interplay between environmental and biological factors in the unfolding of hemispheric asymmetry within an individual from conception through old age. While it would be unwise to make too much of this analogy, there is a sense in which the evolutionary seeds of hemispheric asymmetry were sown millions of years ago, and asymmetry has continued to unfold in response to changing environmental pressures. Later I will discuss hypotheses about the time course of this unfolding. To some extent, how one views the various scenarios that have been suggested depends on the extent to which various functional asymmetries are seen as continuous across species. It is to this issue that I now turn.

Continuity across Species?

An important issue that arises in considering alternative scenarios about the evolution of hemispheric asymmetry is the extent to which various functional asymmetries are continuous across species. In this section, I discuss what is meant by continuity and the kind of evidence that is relevant to argue for or against continuity. I then consider whether certain hemispheric asymmetries in humans might be continuous with those found in other species and the extent to which discontinuities in cognitive abilities might lead to discontinuities in hemispheric asymmetry that are more apparent than real.

In order to understand what is meant by continuity in an evolutionary context, let us review briefly differences of opinion about whether human language is unique or is continuous with earlier primate vocalization. The hypothesis that human language is unique has been advanced by Chomsky (1980), Jerison (1982), and Corballis (1991), among others. Among the kinds of evidence cited for this point of view are several observations about qualitative differences between language in present-day humans and vocalizations in present-day primates. For example, primate vocalizations tend to occur as *reactions* to environmental events (e.g., the approach of a predator) and lack both the sophisticated syntax and the *generative* capacity of human language. From this perspective of discontinuity, it is unlikely that human language is a direct descendant of earlier primate vocalizations. Consequently, we are encouraged to look elsewhere in building a scenario about the evolution of language (and about hemispheric asymmetry for language).

The alternative hypothesis, that human language is continuous with (that is, evolved gradually from) earlier primate vocalization is suggested by Bradshaw (1988, 1989), among others. Evidence for this hypothesis emphasizes similarities between human language and primate vocalization. For example, Bradshaw notes how communicatively subtle monkey calls can be: they can provide information about sex, social relationships, personal identity, and so forth (see also Chapter 5). This is not meant to deny that present-day human language is very different from present-day primate vocalization or from its evolutionary precursors. It *is* meant to argue that the precursors of language were not totally unique to humans and that human language may have evolved in a continuous and gradual fashion from earlier primate vocalizations.

As with many issues that arise in considering the evolution of species, it is difficult to provide clear-cut tests of alternative hypotheses about continuity. As indicated above, one way is to consider how similar potentially continuous properties are in present-day species. The general notion is: the greater the present-day similarity between two species for a certain property, the greater the likelihood that a common ancestor possessed a property that was similar to those elements that the two present-day species have in common. For some biological properties, the evidence in favor of continuity is relatively clear-cut. For example, the physical structure and functional properties of human and monkey hearts are so similar that it makes sense to treat them as continuous and to suppose that the common ancestor of monkeys and humans had a heart that was substantially similar to those of present-day species in both structure and function. Problems arise, however, when considering properties for which there is a great deal of room for disagreement about present-day similarity. This is typical of "cognitive" properties like language and properties like hemispheric asymmetry that are associated with them.

As the foregoing discussion indicates, it is unlikely that the issue of whether hemispheric asymmetry is or is not continuous across species will be resolved soon. However, it is useful to consider which aspects of hemispheric asymmetry seem most similar across present-day species and have the greatest potential for being continuous. In order to do this, one might look for instances in which hemispheric asymmetry for some skill or component of processing is in the same direction across species. A problem in doing this is that research with nonhuman species is still sparse and the specific tasks that can be tested vary widely from species to species. Nevertheless, attempts of this sort can be instructive.

For example, the right hemisphere seems more involved than the left hemisphere in producing affective responses in several species (such as humans, rats, and chicks; see Chapter 5), and the left hemisphere perhaps serves to inhibit the affective responses activated by the right hemisphere. This raises the possibility that right-hemisphere dominance for affective processing arose sufficiently early in evolution to be continuous across a number of species.

In several species the right hemisphere seems more involved than the left in certain aspects of visuospatial processing. For example, in chimpanzees there is right-hemisphere dominance for a line-location task and for processing meaningless shapes of the sort that also produce right-hemisphere dominance in humans (see Chapter 5). This raises the possibility that this specific aspect of hemispheric asymmetry is continuous between chimpanzees and humans. By way of contrast, studies with monkeys (which split earlier than chimpanzees from the line that eventually led to humans) have suggested left-hemisphere dominance for processing meaningless shapes and for certain other visuospatial processes. This could suggest that there is an important discontinuity between humans (and chimpanzees) and monkeys. It must be noted, however, that some of the tasks used with monkeys could be classified as "categorical" spatial tasks, for which there may also be left-hemisphere dominance in humans (see Chapter 3). Furthermore, the right hemispheres of both rhesus monkeys and humans are dominant for processing certain aspects of faces, suggesting a continuity that goes back at least to the time when rhesus monkeys split from the line leading to humans. Beyond that, it has even been argued that the right hemisphere of rats is dominant for aspects of spatial processing.

As discussed in Chapter 5, it has been proposed that certain motor asymmetries in other species were precursors of handedness in humans. For example, it has been hypothesized that a left-hand/right-hemisphere advantage for visually guided reaching evolved in early primates in conjunction with right-arm and right-hand dominance for postural support. Later, with the emergence of an upright posture, the right hand was freed from its burden of postural support to gradually become dominant for manipulation and bimanual coordination. To be sure, proposals about continuity between human handedness and motor asymmetries in other species have generated considerable controversy; they will be discussed in more detail later in the present chapter.

Despite disagreements noted earlier about continuity of human language and primate vocalizations, there is growing evidence that asymmetries in the production and perception of vocalizations are characteristic of a number of present-day nonhuman species, including various species of primates, songbirds, and rats (see Chapter 5). This is at least consistent with the possibility that, when something close to present-day human language finally emerged, it emerged in organisms whose cerebral hemispheres were not equally hospitable to at least some components of vocalization and communication. Note that this could be true even if it is the case that human language is not completely continuous with primate vocalization. That is, hemispheric asymmetry for any function or processing component is not possible until that function or component has emerged during the course of evolution. This is analogous to the fact that in the life span of an individual there can be no hemispheric asymmetry for a specific cognitive ability until that ability is present. From this perspective, genuine discontinuities can occur in terms of cognitive abilities (like language), but any apparent discontinuity in hemispheric asymmetry is only a by-product that has no existence independent of the cognitive discontinuity.

From Monkeys to Humans

This section outlines critical events in the evolution of primates and in the eventual emergence of hominid species leading to present-day humans. Of course, the dates when certain species split from each other are not fixed with precision, and there is a good deal of controversy about the manner in which certain hominid species were related to each other. Some of these details are relevant for a discussion of the evolution of hemispheric asymmetry and some are not. Despite controversy about certain issues, there is sufficient agreement to present the following general picture of our ancestry over the last 60 million years or so. The time course presented here is based largely on recent reviews by Bradshaw (1988, 1989) and by Corballis (1991), which can be consulted for additional discussion of the evidence (e.g., fossil record) used to reconstruct the sequence of events so long ago. As these events are reviewed, asymmetries found in present-day species will be noted as a way of suggesting hypotheses about the sorts of asymmetries that may have been present at different moments in evolution.

The Evolution of Primates

Figure 9.1 illustrates some of the "branching events" and their approximate dates in the evolution of the primates, the line that led eventually to present-day humans. New World and Old World monkeys are thought to have evolved from a common ancestor and split from each other approximately 50 million years ago (mya). The group of primates known as the *prosimians* are more primitive than either of these groups of monkeys and are thought to have changed little since the primates first branched off from the insectivores more than 50 mya. (Examples of this suborder are the lemurs and tarsiers.) Consequently, it is interesting that a number of present-day prosimian species show a left-hand preference for visually guided reaching, although the preference may be restricted to what has been referred to as high-level manual activity (e.g., Fagot and Vauclair, 1991) and may be moderated by sex and age (e.g., Ward, 1991; see also Chapter 5). Such results suggest that at least some asymmetry was already present at the earliest time illustrated (at label *a*) in Figure 9.1.

Additional asymmetries have also been reported for a number of present-day species of Old World monkeys, which split off from the line leading to humans approximately 40 mya. As noted earlier, the right hemisphere of rhesus monkeys is dominant over the left for processing certain aspects of faces, whereas there is some evidence of left-hemisphere dominance for other visuospatial tasks. In addition, the left hemisphere is dominant for the perception of communicatively relevant species-specific vocalizations in Japanese macaques, but there is in this species no hemispheric asymmetry for the processing of features of vocalization

Figure 9.1. A schematic "time line" indicating when other primates split off from the lineage that eventually led to humans. The existence of important asymmetries, as discussed in the text, are noted by the letters *a–c*.

that are not communicatively relevant (see Chapter 5). It is also the case that present-day rhesus monkeys show a left-hand preference for so-called high-level manual tasks, such as making a somatosensory discrimination (for review, see Fagot and Vauclair, 1991). Of course, the evolution of these species has undoubtedly continued over the last 40 million years, which means that some of the present-day asymmetries may not have been characteristic of the ancestors common to humans and Old World monkeys. Nevertheless, the discovery of these present-day asymmetries that resemble certain asymmetries in present-day humans makes it likely that at least some of them were present in our common ancestors (at point *b* in Figure 9.1).

Various species of apes began to split off from the line leading to humans approximately 16–23 mya. As illustrated in Figure 9.1, the gibbons, orangutans, gorillas, and chimpanzees split off in that order. The chimpanzees, our closest present-day primate relatives, are thought to have split off from the line leading to humans approximately 5–10 million years ago. The details of that split and of the emergence of various hominid species have been difficult to determine, in part because there is a gap in the fossil record covering the critical period from approximately 4 to 8 mya. Investigations of several species of present-day apes indicate at least some amount of manual asymmetry for high-level tasks (see Fagot and Vauclair, 1991), with right-hand dominance being reported in recent studies with chimpanzees. As noted earlier, language-trained chimpanzees also show an RVF/left-hemisphere advantage for processing communicatively relevant symbols but not for other symbols. These same chimpanzees showed an LVF/right-hemisphere advantage for various visuospatial tasks that tend to show similar asymmetries in humans. These findings raise the possibility that, by the time illustrated by point *c* on Figure 9.1, an ancestor common to chimpanzees and humans had several asymmetries similar to those of both present-day species.

It should also be noted that certain biological asymmetries that are characteristic of the brains of present-day humans are also present in less dramatic (and sometimes less frequent) form in other present-day primates (see Chapter 5). For example, in both humans and apes the brain has a counterclockwise torque, and the brains of both Old (and, perhaps, New) World monkeys show a smaller asymmetry of this sort. In addition, the sylvian fissure tends to be longer on the left than on the right side in both humans and chimpanzees but not in rhesus monkeys.

The Emergence of Hominids and Humans

From the preceding outline it seems likely that certain hemispheric asymmetries were already present before the emergence of the first hominids. However, there is no doubt that such asymmetries continued to evolve with the emergence of new skills and new cognitive abilities. This section outlines the emergence of hominid and human species during the last 4 million years or so and notes the approximate time that certain critical milestones began to appear and evolve into what we see today. The manner in which each of those milestones might relate to hemispheric asymmetry will be discussed in more detail in the following section.

Figure 9.2 illustrates the approximate time span that various

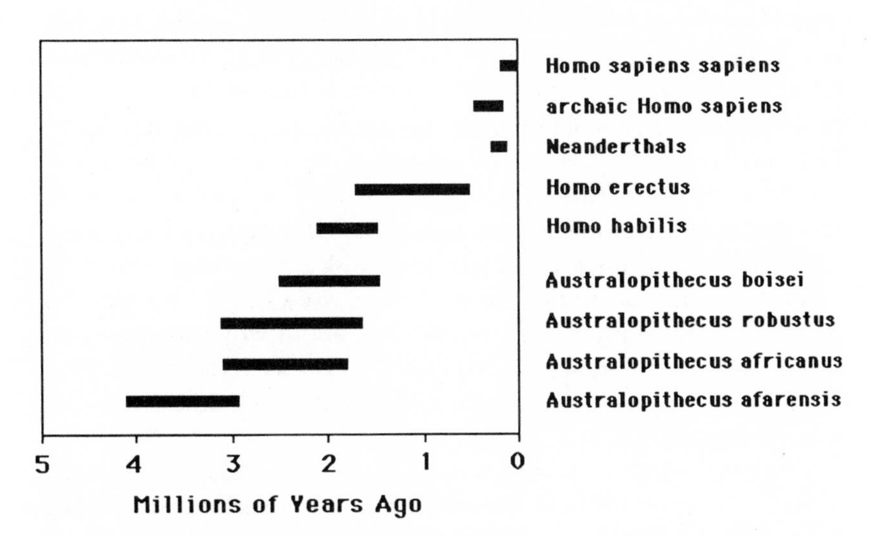

Figure 9.2. The emergence of hominids during the last 4 million years. Each bar in the figure indicates the approximate time of existence of the hominid group whose name appears to the right.

species of hominids and humans are thought to have inhabited the Earth. Given the great deal of uncertainty surrounding the specific lineage from the early species of *Australopithecus* to *Homo sapiens sapiens,* no specific line of descent is illustrated. As shown in Figure 9.2, the earliest form of hominid *(A. afarensis)* is thought to have emerged approximately 4 mya and remained for a million years or so. As also shown in Figure 9.2, at least three other species of australopithecines are thought to have emerged later and co-existed with each other until sometime between 1 and 2 mya. As discussed by Bradshaw (1988) and Corballis (1991), the australopithecines were divided into robust (*A. robustus* and *A. boisei*) and gracile *(A. africanus)* forms. The robust forms are thought to have lived mainly on coarse vegetation whereas the gracile form is thought to have scavenged for meat. The species *Homo* is thought to have evolved from the gracile australopithecines, whose scavenging life-style may have favored the emergence of increased brain size and the manufacture of tools.

The brains of australopithecines are thought to have been about the size of the brains of present-day apes, significantly smaller than the brains of present-day humans. Endocasts of the skull remains of australopithecines show elements of the counterclockwise torque that is characteristic of the brains of present-day humans and of at least some apes and monkeys (e.g., Holloway and de la Coste-Lareymondie, 1982). As would be expected, similar findings have been obtained from endocasts of later hominids, including *H. erectus,* archaic *H. sapiens,* early *H. sapiens sapiens,* and the Neanderthals. In addition, there is fossil evidence that even the earliest australopithecines walked upright much of the time, a dramatic difference from the apes and a habit that was to be accentuated in the species *Homo.* As discussed in the next section, there are several interesting hypotheses about why this transition to an upright, bipedal posture might have occurred and why it seems to have been associated with dramatic changes in hemispheric asymmetry.

The first members of the species *Homo (H. habilis)* are believed to have emerged between 1.5 and 2 mya and overlapped with the later australopithecines. Although the line of descent has not been determined with precision, it has been suggested that *H. habilis* descended from a gracile australopithecine (see Corballis, 1991).

The brain of *H. habilis* was larger than that of the australopithecines and of present-day apes, but still much smaller than the brains of present-day humans. From the study of endocasts of fossil skulls it has been determined that certain additional asymmetries that are characteristic of the brains of present-day humans were also present in the brains of *H. habilis* (for review, see Corballis, 1991; Tobias, 1987). For example, the sylvian fissure was longer and straighter on the left side than on the right side and, as discussed by Tobias, there was an increase in the size of both the inferior parietal lobe and the motor speech area (Broca's area). Whether *H. habilis* "spoke" in a manner that was different from the vocalizations of other primates is unclear, but it does appear that an asymmetric substrate was emerging that would make language production and perception asymmetric when it did finally appear.

There is very clear evidence that *H. habilis* walked upright and manufactured primitive tools, indicating the emergence of another important milestone to be discussed later. In fact, from an analysis of the flaking patterns on the primitive stone tools manufactured by *H. habilis,* Toth (1985) has suggested that the majority of the population was already right-handed, perhaps in a ratio slightly less than 2:1.

The more advanced *H. erectus* is thought to have emerged (possibly descended from *H. habilis*) approximately 1.5 mya. Relative to *H. habilis, H. erectus* had a still larger brain with many of the same anatomical asymmetries (e.g., a longer, straighter sylvian fissure on the left side). *H. erectus* used more sophisticated tools, including stone scrapers, choppers, and hand axes (e.g., Corballis, 1991), and the flaking patterns on stones again indicate a population proportion of right-handedness that approximates a 2:1 ratio (Toth, 1985). There is even some evidence of fire use and hunting. Corballis (1991) suggests that the prolonged period of postnatal immaturity that is unique to humans probably also emerged during the transition from *H. habilis* to *H. erectus,* given such indications as the size of the birth canal relative to the size of the fully grown adult.

The later-occurring archaic *H. sapiens* may have evolved from *H. erectus* and may represent a transition from *H. erectus* to present-day humans. For example, archaic *H. sapiens* was characterized by a brain that was approximately the size of the brains of present-day humans and showed increasing sophistication in tool manufacture and use relative to *H. erectus.*

The Neanderthals emerged approximately 100,000 years ago and disappeared approximately 35,000 years ago. They were clearly a distinct subspecies of *H. sapiens* and, for a time, they were thought to be ancestors of present-day humans. There is still controversy surrounding this, but the predominant contemporary view seems to be that they were not. Relative to present-day humans, the Neanderthals had flattened foreheads, receding chins, and large noses. Their brains tended to be asymmetric in the ways described earlier for other hominids, they were adept at tool manufacture and use, and they practiced rituals associated with death. Their disappearance may have been caused by competition from the superior *H. sapiens sapiens,* with whom they coexisted for a time.

The exact time of the appearance of the first anatomically modern humans *(H. sapiens sapiens)* is a matter of great speculation, but it appears to have been between 100,000 and 200,000 years ago (for discussion of the evidence, see Bradshaw, 1988; Corballis, 1991). Almost certainly, *H. sapiens sapiens* first emerged in Africa and then migrated to other locations. Regardless of when the first members of the species appeared, artifacts from Europe and Asia suggest an evolutionary explosion that occurred from approximately 10,000 to 35,000 years ago (e.g., Pfeiffer, 1985) as these new humans began to assert their dominance over the planet. For example, this time period saw a dramatic increase in the sophistication of tools, the emergence of cave drawings and other types of symbolic and metaphoric representation, and the development of language.

It is difficult to say exactly when in the history of *H. sapiens sapiens* language first existed with anything like the generative capacity and grammatical complexity of present-day human language, although most estimates suggest that the common ancestor of modern languages was present by approximately 40,000 years ago. The extent to which language evolved gradually from earlier forms of vocalization or by occasional sudden jumps is a matter of considerable debate. For example, on one view (e.g., Jaynes, 1977), the sudden increase in art and artifacts that began to appear between 30,000 and 50,000 years ago reflects the relatively sudden appearance of language, and truly modern language (and left-hemisphere dominance for that language) did not emerge fully until the last 5,000 years or so—bringing with it new cognitive abilities, including self-consciousness. Another view is presented by Bradshaw (1988), who suggests that by approximately 50,000 years ago the coevolution of several factors (including increasing brain capacity, the technology of tool manufacture and use, social interaction and parenting) reached a critical mass that precipitated the relatively sudden appearance of something close to modern language. According to Bradshaw's view, prior to this time language had a long evolutionary history and was continuous with primate vocalizations. Still a third view (discussed in more detail in the next section) has been suggested by Corballis (1991), who considers how the relatively sudden emergence of language as a form of communication could have been based on the prior use of gestures to communicate and how both might be related to what he calls the "generative mind" and to hemispheric asymmetry.

Many earlier accounts of the evolution of hemispheric asymmetry have argued that it emerged first with the species *Homo* (or, possibly, with the hominids that appeared slightly earlier) and was tied to the emergence of language and tool manufacture. These accounts share several common characteristics. For example, they include the idea that there was evolutionary pressure for one hemisphere to control speech, possibly in order to reduce competition for control of the single vocal apparatus. From this pressure right-handedness could have emerged because it was more efficient to have the same hemisphere that was dominant for programming the rapid, sequential movements associated with speech also become dominant for well-coordinated, sequential movements of the fingers. It is also usually suggested that right-hemisphere dominance for such things as visuospatial processing or for producing and perceiving affect emerged later, possibly by default because the left hemisphere became so devoted to language that its capacity to do other things diminished.

Several aspects of this popular scenario seem unlikely. The emergence of language seems too late and was preceded by too many prior asymmetries to have been the single driving force behind the evolution of hemispheric asymmetry. In fact, several asymmetries seem to have been present prior to both the manufacture of tools and language and, if anything, they suggest very early precursors of certain aspects of right-hemisphere dominance for certain functions in present-day humans. For example, in rats the right hemisphere tends to be larger than the left and has been hypothesized to be dominant for both generating emotion and for aspects of spatial processing. At the same time, there is evidence that the left hemisphere of at least some species of rats is dominant for the perception of communicatively relevant vocalizations. In addition, as noted throughout the present section, there seem to have been several precursors of both right- and left-hemisphere dominance in our primate ancestors. Consequently, it does not seem likely that hemispheric asymmetry emerged for the first time in hominids or that its original appearance was driven completely by the requirements of either language or tool use.

The preceding discussion is not meant to deny or to minimize the tremendous physical and intellectual differences between modern-day humans and other primates. The fact is that the last few hundred thousand years have seen dramatic responses to environmental pressures. No species comes close to the ability of present-day humans to communicate, to manufacture tools, to think, and to inhabit the entire planet. It is not surprising that some aspects of hemispheric asymmetry are different for humans than for other species, and the next section discusses how several important milestones in hominid evolution may have led to such differences. However, an important message of the present section is that hemispheric asymmetry almost certainly did not emerge for the first time with our species or in conjunction with abilities that are unique to us.

Milestones in Hominid Evolution

I have already mentioned four milestones that figure prominently in various scenarios about the evolution of hemispheric asymmetry: walking upright, tool manufacture and use, language, and prolonged postnatal immaturity. The emergence and continued development of these milestones did not take place in a strictly sequential manner. Instead, their evolutionary histories are intertwined in ways that remain a matter of considerable speculation. So are their contributions to the evolution of hemispheric asymmetry. In other words, it is not the case that one milestone appeared and had its full effect on hemispheric asymmetry before another appeared and built on the first, and so forth. Instead, it appears that these factors interacted to produce the pattern of hemispheric asymmetry that is characteristic of present-day humans. It is instructive to consider what each of them contributed that may have moved our ancestors in the direction of present-day asymmetries and how the various pressures for asymmetry may have been mutually reinforcing.

Walking Upright

Upright posture and bipedal locomotion were probably characteristic of the first hominids and were clearly well established from the time of *H. habilis* on. Although other primates occasionally stand upright, and some can go short distances on two legs, humans are unique in that walking upright is the prototypical means of locomotion. Several scenarios have been proposed to explain what aspects of the environment might have favored the development of bipedalism (for discussion, see Bradshaw, 1988, 1989; Corballis, 1991; Lovejoy, 1981; Morgan, 1984). According to most conventional views, several geographic events conspired to increase the diversity of climate and terrain in East Africa between 8 and 12 mya, and bipedalism evolved as a consequence of these changes. For example, standing upright would have enabled the hominids to see farther across the open savanna, providing earlier warning of predators and better detection of potential sources of food and shelter. Bipedalism would also have freed the hands from the need to provide postural support or aid in locomotion. Consequently, the hands could become adapted for carrying such things as food and infants—activities that would be important for the survival of species that lived a nomadic life and that scavenged for food—and for wielding weapons and tools.

An interesting alternative to these conventional views has been provided by Morgan (1984), who argues that the ancestral line leading to hominids passed through an aquatic phase. This sojourn led to a variety of characteristics that are typical of aquatic mammals but that set humans apart from other primates. Among these characteristics are the loss of most body hair, the downward direction of nostrils, weeping of salt-water tears, and face-to-face copulation. On this view, bipedalism arose because an upright posture would be favored for wading and treading water, and the posture appropriate for swimming (e.g., the legs in line with the body) is also appropriate for walking upright. Thus, when the hominids eventually emerged from the sea, the stage would have been set for bipedalism. Despite a certain intuitive appeal, this scenario has been greeted with great skepticism, partly because there is no converging evidence of climatic changes that could have led to an aquatic phase and partly because there are other plausible explanations, such as those listed above, for the evolution of bipedalism.

Regardless of why bipedalism emerged in the hominids, it is generally regarded as a critical event in setting humans apart from other species—in part because it influenced the evolution of hemispheric asymmetry. Earlier in the present chapter I discussed the idea that evolutionary pressures favor bilateral symmetry for parts of the brain and central nervous system associated with locomotion. Even though there is some evidence for precursors of handedness as far back as 60 mya, the need to use the forelimbs for locomotion may have discouraged the development of more dramatic motor asymmetries. Once the need to participate in locomotion was decreased dramatically with the emergence of bipedalism, however, there would be far less pressure for bilateral symmetry to be maintained. In addition, for the efficient performance of many manual activities (e.g., peeling a banana or opening the lid of a small box), the two hands must play complementary roles—one hand holds an object while the other manipulates it. This is likely to add to the pressure for asymmetry. Given potential precursors of handedness in other primates, it is conceivable that sufficient asymmetry already existed in monkeys and apes to set the stage for asymmetric use and control of the two hands and even to create a bias toward the right-handedness which seems to have been characteristic of the first hominids. On this view, handedness did not appear for the first time in hominids, but their upright posture encouraged it to develop more far more fully than in other species.

Another interesting scenario has been advanced by Corballis (1991), who emphasizes the development of left-hemisphere dominance for *praxis.* Corballis defines praxis as the organization of purposeful, sequential actions in which spatial constraints imposed by the environment are minimal. In many respects, vocalization for the purpose of communication is a prototypical example of movements resulting from praxis. Whereas movements of the hands that represent *reactions to* the environment are not good examples of praxis, many *manipulations of* the environment by the hands are good examples. As we have seen earlier in the present chapter, there is reason to believe that the left hemisphere had already become at least somewhat dominant for vocalization before the emergence of the first hominids. Corballis suggests that this made the left hemisphere a better neural substrate for *manual* praxis when the hands were freed to engage in manipulative activities, which would further encourage the emergence of strong right-handedness.

A very different way in which the emergence of an upright posture might have influenced hemispheric asymmetry has been proposed by Previc (1991). As discussed in Chapter 8, Previc argues that both right-handedness and right-hemisphere dominance for visuospatial processing in present-day humans come about because the otolith organs of the inner ear are subjected to asymmetric shearing forces during fetal development. The asymmetric shearing comes about as the mother walks and is related to the typical asymmetric position of the fetus during the last trimester of pregnancy. As Previc notes, the asymmetric shearing forces are present when the mother walks upright but would be much smaller or nonexistent if the mother were to walk on all fours. This may explain why handedness and certain other asymmetries are weaker in other present-day primates than they are in humans. In addition, this may also explain why the emergence of an upright posture and bipedal locomotion led to increasing right-handedness as well as to other asymmetries in hominids.

Tool Manufacture and Use

Among the most important praxis-governed activities performed by our hands are tool manufacture and tool use. Of course, humans are not the only present-day species to use tools as an extension of their own limbs or mouth or even to shape objects to do so. For example, chimpanzees strip bark from twigs and use the resulting "tool" to extract termites from their holes (e.g., Goodall, 1970). But no other present-day species even begins to approach our sophistication in the manufacture and use of tools. As outlined earlier in the present chapter, primitive tools were manufactured by *H. habilis,* but the sophistication of tools changed only gradually until the advent of *H. sapiens sapiens.* It has been argued that bipedalism helped create the conditions that favored increasing use of tools and that tool use and manufacture favored the development of both handedness and hemispheric asymmetry. In addition to factors discussed earlier, it has been suggested that the need to communicate the method of tool manufacture and use from individual to individual may have added to the pressure for the direction of handedness to become more standardized across the population (for discussion, see Bradshaw, 1988).

Corballis (1991) has pointed out several interesting parallels between the manufacture of tools and the use of language for the purpose of communication. These parallels suggest that it would be efficient for hemispheric dominance for handedness and many aspects of language to be in the same direction. He argues, for example, that the fact that one tool can be used to make another tool is an elementary example of the property of *recursion,* which is an important characteristic of full-fledged human language. Furthermore, there is a sense in which a potential tool (e.g., a stone scraper) is *embedded* in the stone from which it is extracted. In human language, one structure (e.g., a subordinate clause) is frequently embedded in another (e.g., a sentence). In addition, both tool manufacture and language have the capacity of *generativity,* defined by Corballis as "the ability to construct an unlimited number of different forms from a finite number of elementary parts" (Corballis, 1991, p. 65). For example, the same collection of raw materials can be used to manufacture a variety of tools to be used for different purposes, just as an almost unlimited number of words can be derived from a small set of phonemes and an unlimited number of sentences can be created from a finite vocabulary.

Given that both tool manufacture and language exhibit properties of recursion, embeddedness, and generativity, it is possible that emerging hemispheric asymmetry for either reinforced a similar pattern of hemispheric asymmetry for the other. For example, increasing sophistication in tool use may have helped set the stage for increasing sophistication in language and vice versa. The possibility that these abilities may have interacted in a synergistic fashion is particularly plausible in view of the fact that both seem to have increased dramatically in sophistication during the period from about 10,000 to 30,000 years ago.

Language

Left-hemisphere dominance for the production and perception of speech and other components of language, such as syntax, are among the most well-established hemispheric asymmetries in present-day humans (see Chapters 2 and 3). Consequently, it should come as no surprise that the emergence of language is an important milestone in the evolution of hemispheric asymmetry. As I have argued earlier, however, it does not seem likely that the hemispheres were functionally symmetric until modern language appeared and only then began to become asymmetric or that all other functional asymmetries arose as a by-product of hemispheric asymmetry for language. Instead, several alternative scenarios have been sketched, including the idea that hemispheric asymmetry for human language is continuous with asymmetry for earlier primate vocalizations and was influenced by emerging left-hemisphere asymmetry for praxis.

In addition to issues that have already been considered, it is useful to consider whether the prior use of communication by gesturing with the hands was a precursor of vocal language or of asymmetry for the production and perception of vocal language. Both modes of communication require timed sequences of movements (see Bradshaw, 1988; Bradshaw and Nettleton, 1981) and, in present-day humans, there is evidence of left-hemisphere dominance for communication by gesture in the deaf as well as for vocal communication in the hearing (see Chapters 2 and 3). Despite this kind of evidence, Bradshaw (1988) is skeptical about whether gesture was truly ancestral to spoken language. In part, his doubt stems from his willingness to consider spoken language continuous with earlier primate vocalizations and from skepticism about whether there was a strictly linear sequence from an upright posture leading to tool use and gesture and then, finally, to language.

A somewhat different position is taken by Corballis (1991), who suggests that an emerging left-hemisphere dominance for praxis (which may have come about, in part, because of prior left-hemisphere dominance for primate vocalizations) made the left hemisphere a preferred neural substrate for sequential, timed movements of the hands and arms and for factors such as recursion, embeddedness, and generativity. It is at least possible that such factors continued to evolve in sophistication during a time when the primary mode of communication was by gesture. The eventual emergence of spoken language probably took place because of the advantages of spoken language over gesture. For example, spoken language permits communication in the dark or when the individuals communicating are not visible to each other for other reasons, and it would permit communication to continue while the hands were used for other things. To the extent that the left hemisphere had already become dominant for important aspects of gestural language, it would provide a welcome substrate for similar aspects of vocal language. In this way, communication by gesture may have been an important step on the way to hemispheric asymmetry for language.

Prolonged Immaturity

An additional factor of some importance might well be the prolonged period of postnatal immaturity that characterizes present-day humans and that probably emerged in the transition from *H. habilis* to *H. erectus*. As the size of the brain (and surrounding skull) increased, it would have become impossible for the fetus to traverse the birth canal without some type of evolutionary adaptation. The solution that seems to have been favored was for the fetus to emerge before its head grew too large, which meant that it was entering the world considerably less mature than was the case in earlier primates. Thus, the nervous system of the human infant and child undergoes much more maturation after birth than do the nervous systems of other primates. One can only speculate about why this solution was favored over other possibilities, such as increasing the size of the birth canal. One possibility is that there are advantages to having the developing nervous system shaped as much as possible by the external environment instead of by the relatively uninformative uterine environment.

A prolonged period of postnatal immaturity could have had important consequences for hemispheric asymmetry. There is evidence that homologous areas of the two cerebral hemispheres do not mature at the same rate, either before birth or after birth (see Chapter 8). There is no way to know for certain if this was also true during the gestational period of our primate ancestors, but the possibility is likely in view of recent studies with present-day nonhuman species (see Geschwind and Galaburda, 1987). Because the uterine environment presents the brain with a very restricted range of information, cerebral asymmetries in the rate of maturation may have had only modest consequences in earlier primates. When the fetus is born less fully developed and there is a prolonged period of postnatal immaturity, however, a greater proportion of asymmetries occur either very late in fetal development or after birth and the consequences for functional hemispheric asymmetry may be much more marked. As discussed in Chapter 8, in humans the sensorimotor capacities of newborns change dramatically over the first few days, weeks, and months of life, which means that the brain is continually being presented with better-quality information. The rapidly changing nature of information presented to the brain may interact with any asymmetries in maturation to sow the seeds of functional asymmetries that will continue to unfold throughout the life span (see Chapter 8). From this perspective, hemispheric asymmetries may be more marked in humans than in other primates partly as an inadvertent byproduct of prolonged immaturity.

There may even have been evolutionary pressures that favored the existence and enhancement of asymmetric maturational gradients. On one hand, it would seem to be advantageous for survival to have an organism's brain begin learning as much as possible about its environment from the moment of birth (or even before), thereby favoring development of a brain that is ready to learn as much as possible as soon as possible. On the other hand, if other pressures favor the organism being born in an extremely immature state, it would also seem to be advantageous to have its brain learn as much about the world as its fully mature sensory and motor systems will allow. There is a potential conflict between these two goals. If the brain is modified permanently by the degraded sensorimotor information that it receives initially, it may no longer be the best possible substrate for handling the higher-quality information that it will receive later. If the brain does not begin to modify itself immediately, however, the organism may not survive until later. One solution would be to give the organism two "brains," largely identical, except that homologous areas are on different maturational trajectories, with one being modified a great deal initially and the other being more sensitive to modification somewhat later. Of course, eventually these two brains would somehow have to integrate their activities to minimize conflict, favoring the development of a sophisticated network of connecting fibers. Although this scenario is highly speculative, asymmetrical maturational gradients characterize the brains of present-day humans and may well have become more adaptive with significant decreases in the maturity of hominid newborns.

The explanations discussed in this section of how each of four evolutionary milestones might have contributed to the hemispheric asymmetry of present-day humans are, of necessity, speculative and extremely difficult to test. For the most part, it is impossible to go into the laboratory and resolve the critical issues by studying present-day species. Instead, we must continue to look for clues in the fossil records of earlier species and among the ruins of ancient civilizations. In addition, most of the ideas presented here are not mutually exclusive: there is every reason to suppose that a number of these factors (as well as others) interacted in ways that we are only beginning to understand. Given the inherent difficulties of studying the evolution of brains and behaviors, one might be tempted to throw up both hands in frustration (one of our remaining symmetrical behaviors) and accept the fact that we may never know the evolutionary history of hemispheric asymmetry. Despite the inevitable frustrations, however, the quest for answers to evolutionary riddles will continue. Our human nature will not allow us to ignore them. With respect to hemispheric asymmetry, the quest is especially important because any links between those things for which the same hemisphere tends to be dominant are likely to be found buried in our evolutionary past.

Summary and Conclusions

The seeds of hemispheric asymmetry in present-day humans were sown in our distant evolutionary past, perhaps even in the very molecules and physical particles of which all living things are constructed (theme 5 in Chapter 1). Despite its inherent beauty, bilateral symmetry is no more fundamental a property of biological systems than is bilateral asymmetry. Instead, both are equally the result of evolutionary adaptation. Because freely moving animals must be equally attentive to both the left and right sides of space and because they must be able to move in a straight line, there are strong evolutionary pressures for bilateral symmetry in the placement of sense organs and limbs and in those parts of the brain and nervous system that are associated with them. In contrast, there is little pressure for symmetry for bodily and cortical systems that are not involved in direct sensory or motor responses to the environment. Indeed, in those cases, asymmetry is advantageous because it allows for more compact packaging of organs into a limited body space and cognitive functions into a limited cortical space. Furthermore, the development of a corpus callosum, which, in part, serves to reduce maladaptive cross-talk between functions carried out in homologous regions of the two hemispheres, adds to the evolutionary pressure for complementary or mutually inconsistent processes to migrate toward opposite hemispheres.

In organisms that are already functionally asymmetric, at least some of the additional evolutionary adaptations that are favored by the environment are likely to be implemented more efficiently on one side than on the other. When this is the case, the adaptations are also likely to be asymmetric when they emerge. Over the course of several such adaptations, this snowball mechanism can greatly increase the extent of functional hemispheric asymmetry. This is true even when there is nothing in the environment that necessarily demands hemispheric asymmetry for processing abilities that emerged relatively recently.

At least some of the hemispheric asymmetries found in present-day humans seem continuous with similar asymmetries found in other present-day species. For example, in several species the right hemisphere seems to be more involved than the left in producing affective responses and in several aspects of verbal processing. Although no present-day species shows a pattern of population-level handedness that approaches the strength of right-handedness in the human population, there are motor asymmetries in a variety of other species that may be precursors of handedness in humans. In addition, there are indications of left-hemisphere dominance for the production and perception of vocalizations in a number of present-day species, which is at least consistent with the possibility that human language emerged in brains whose hemispheres were already asymmetric for at least some of the relevant processing components. In general, the asymmetries found in present-day nonhuman species can be used to generate hypotheses about the sorts of asymmetries that may have been present at different moments in the evolutionary chain that led eventually to present-day humans.

The existence of a left-hand preference for visually guided reaching in present-day prosimians suggests that at least some motor asymmetry was present in our ancestral line by 60 mya. The left-hand advantage for high-level manual tasks extends to some species of Old World monkeys. In addition, for some species of Old World monkeys the right hemisphere is dominant over the left for processing certain aspects of faces and the left hemisphere is dominant over the right for the perception of communicatively relevant species-specific vocalizations. The similarity of these asymmetries to those of present-day humans makes it likely that at least some of them were present in our common ancestors, approximately 40 mya. Our closest present-day primate relatives, the chimpanzees, show right-hand dominance for certain high-level tasks, an RVF/left-hemisphere advantage for processing communicatively relevant symbols, and an LVF/right-hemisphere advantage for visuospatial tasks. This raises the possibility that asymmetries of this sort were present in an ancestor common to present-day humans and chimpanzees, approximately 5 mya.

The preceding summary is consistent with the possibility that certain hemispheric asymmetries were already present before the emergence of the first hominids approximately 4 mya. If hemispheric asymmetry did not emerge for the first time in hominids, then its initial appearance was not driven completely by either language or tool use, as some have argued. This is not meant to deny or to minimize the tremendous physical and intellectual differences between present-day humans and other primates. In fact, the last few hundred thousand years have seen dramatic responses to environmental pressures and the emergence of several milestones that have set humans apart from others. Some of these milestones are likely to have had important consequences for the continued evolution of hemispheric asymmetry, and there are several alternative hypotheses about the reasons for these consequences.

From approximately 2 to 4 mya, hominids mastered the art of standing and walking upright. Among other things, an upright stance freed the hands from the need to provide postural support and minimized their use in locomotion. Thus freed, the hands would be under far less pressure to be controlled by a bilaterally symmetric nervous system. In fact, for efficient performance it is important for the hands to be used in a complementary fashion (e.g., one hand holds a stone while the other chips away at it to fashion a tool). Given precursors of handedness in other primates, it is conceivable that sufficient asymmetry already existed to create the bias toward right-handedness that seems to have been characteristic of even the first hominids. It is also possible that prior left-hemisphere dominance for vocalization made the left hemisphere a better neural substrate than the right for manual *praxis* —the organization of purposeful, sequential actions in which spatial constraints imposed by the environment are minimal. In addition, the adoption of an upright posture and bipedal locomotion may have created asymmetric shearing forces on the otolith organs during fetal development, which have been hypothesized to contribute to right-handedness and to right-hemisphere dominance for visuospatial processing.

Among the most important praxis-governed activities performed by human hands are tool manufacture and use. Perhaps the need to communicate the method of tool manufacture added to the pressure for the direction of handedness to become more uniform across the population. Both manufacturing tools and producing language involve properties such as recursion, embeddedness, and generativity. Furthermore, communicating both by hand gesture and by vocalizing requires sequential, timed movements. It would therefore seem efficient to have the same hemisphere be dominant for both hand use and for at least those aspects of vocalization and language that are shared with tool manufacture and communicating by gesture. Thus, tool use and language may have interacted in a synergistic fashion to produce a dramatic increase in the sophistication of both skills during the period from about 10,000 to 30,000 years ago.

Present-day humans are characterized by a prolonged period of postnatal immaturity, a characteristic that is believed to have emerged approximately 2 mya. To the extent that homologous areas of the two hemispheres mature at different rates, the consequences for functional hemispheric asymmetry are likely to become more dramatic when the brain undergoes more of its development after birth and less in the impoverished uterine environment. In part, this is attributable to the fact that the sensorimotor capacities of newborns change dramatically over the first few days, weeks, and months after birth. In fact, there may even be evolutionary pressures that favor the enhancement of asymmetric maturational gradients as a way of allowing areas of one hemisphere to be modified initially by the relatively impoverished sensorimotor information presented to the brain while at the same time saving homologous areas of the other hemisphere to be more sensitive to modification at a later time, when the brain is presented with much better sensorimotor information. In this way, nature could create a brain that learned as much as possible as soon as possible and eventually learned as much about the world as its fully mature sensory and motor systems would allow.

Scenarios about the evolutionary history of hemispheric asymmetry are, of necessity, speculative and difficult to test. Most of the hypotheses considered here are less complete than one would like and most will turn out to be wrong in part or *in toto*. Nevertheless, as the fossil record becomes more complete and as we uncover the ruins of more ancient civilizations it will be important to consider what new information they provide about the emergence and evolution of functional hemispheric asymmetry. In earlier chapters I have argued that there is little evidence that a single information-processing dichotomy underlies or can describe the full range of hemispheric asymmetries that characterize present-day humans. At the same time, it is possible that there are important links of another kind among the things for which the same hemisphere is typically dominant, but evidence of those links are likely to be buried deep within our evolutionary past.

10

Epilogue

The two hemispheres of the human cerebral cortex have different information-processing abilities and propensities. In the foregoing chapters I have reviewed much of what is currently known about hemispheric asymmetry and about the ways in which the two hemispheres interact to produce a unified information-processing system. Throughout that review I have tried to present an accurate and objective picture of the accumulated empirical findings and of the sometimes contradictory theoretical positions that have been taken by different authors. At the same time, I have tried to look for relationships among findings that have heretofore been treated as unrelated and to develop a coherent story about hemispheric asymmetry. The purpose of this epilogue is to review the major conclusions that have been reached along the way and to summarize the story that has emerged. Although the story stops short of what I would expect of a truly comprehensive model of hemispheric asymmetry, it at least illustrates the direction in which I think such a model must go.

Most of the major conclusions that I have reached are related to the five themes that have come up repeatedly in the foregoing chapters. Furthermore, the empirical findings that are consistent with those themes place important constraints on the development of a comprehensive model of hemispheric asymmetry. Accordingly, I begin with a brief review and elaboration of those five themes, emphasizing the broad implications of each. This is followed by a sketch of the story about hemispheric asymmetry that has emerged from the present review.

The Five Themes Revisited

I began Chapter 1 by previewing the following five themes that would recur throughout this book. (1) There are many cognitive and behavioral asymmetries in humans that can be attributed to asymmetries in the brain. (2) The two hemispheres are parts of a much larger, anatomically extensive information-processing system. (3) Some asymmetries in humans have behavioral and biological parallels in other species. (4) Individuals differ from each other in patterns of hemispheric asymmetry and in the ways that the two hemispheres interact. (5) Questions about the unfolding of hemispheric asymmetry over the life span and about the evolution of hemispheric asymmetry have important implications for understanding the nature of hemispheric asymmetry and the mechanisms that create it.

At first glance, theme 1 is quite simple. There are a great many behavioral asymmetries in humans and many of those asymmetries can be attributed to cerebral asymmetries. However, along the way we have learned many specific details about the nature and extent of such asymmetries, and they provide important guidelines for models of hemispheric asymmetry. Certainly, models of hemispheric asymmetry must take account of the specific asymmetries that have been discovered. In addition, such models must also take account of more general aspects of behavioral and brain asymmetry. It is particularly interesting to consider both the *subtlety* and the *breadth* of hemispheric asymmetry.

In the intact brain, it is rarely the case that one hemisphere can perform a task normally whereas the other hemisphere is completely unable to perform the task at all. Instead, both hemispheres often have considerable ability to perform a task, even though they may go about it in different ways. That is, hemispheric differences tend to be smaller and more subtle than popularized views have suggested. A possible exception is overt speech, which the normal left hemisphere can typically produce whereas the normal right hemisphere cannot. However, even individuals who are born without the left hemisphere learn to speak without any great effort, although their speech and language may be deficient in subtle ways. In general, then, it seems that each hemisphere provides a neural substrate that is capable of supporting a wide range of cognitive abilities and propensities to some extent. In this sense, having two cerebral hemispheres is akin to having two reasonably complete "brains" whose differences, compared with their many similarities, are likely to start out being subtle.

Although many hemispheric asymmetries are very subtle, the range of tasks showing hemispheric asymmetry is quite broad and spans such diverse domains as motor performance, language, spatial processing, and emotion. Thus far, it has not been possible to identify any single information-processing dichotomy that could account for anything close to this entire range of hemispheric asymmetries. In fact, there is sufficient independence among the various asymmetries to make it unlikely that such a dimension exists or that models of hemispheric asymmetry that are based on a single dimension can ever be very comprehensive. Whatever links there might be between the various hemispheric asymmetries, they would seem to be determined in some other way or according to some other principle.

We have seen that even simple tasks consist of a number of specific subprocesses or components and that hemispheric asymmetry is not always the same for all components of a task. In fact, from a componential perspective it appears that oftentimes neither hemisphere is uniformly superior to the other or specialized for an entire multicomponent task. Instead, both hemispheres often play a role in complex tasks, and one could describe these roles as complementary. Complementarity may provide important clues for developing a comprehensive model of hemispheric asymmetry.

According to theme 2, the two hemispheres are part of a much larger, anatomically extensive information-processing system. Hemispheric asymmetry cannot be understood without taking into account the unity of the brain. Because both cerebral hemispheres are involved in virtually everything we do, there is potential for conflicts in perception, cognition, emotion, and action. Despite this potential, true conflicts are rare because the two hemispheres in the intact brain interact with each other in a variety of ways. It is particularly important to note that the corpus callosum seems to play two somewhat contradictory roles. On the one hand, the corpus callosum (along with various subcortical structures) permits the transfer of *information* from one hemisphere to the other so that the two hemispheres can coordinate their activities by sharing the results of their computations. On the other hand, it also provides something of a barrier between the hemispheres, reducing maladaptive cross-talk between the sometimes mutually inconsistent (and complementary) *processes* being carried out in the different hemispheres. Both of these roles are important if the two hemispheres are to work together in an efficient manner, and they may help to explain why the functions that have tended to migrate to opposite hemispheres are so often described as complementary.

As stated in theme 3, there are many reliable behavioral and biological asymmetries in nonhuman species, and some of those asymmetries are analogous or even homologous to asymmetries in humans. Asymmetries in other species indicate that whatever evolutionary pressures led to hemispheric asymmetry were not unique to humans and must have included many things besides language and tool use. In addition, some of the asymmetries in other species provide clues about the sorts of asymmetries that may have been present at different moments in the evolutionary chain that led eventually to present-day humans. In some cases, the existence of asymmetries in our evolutionary ancestors has been confirmed by fossil evidence and by the careful examination of artifacts produced by those ancestors.

Individuals differ from each other in patterns of hemispheric asymmetry and in the ways that the two hemispheres interact, as noted in theme 4. Although the existence of individual variation does little to constrain models of hemispheric asymmetry, a comprehensive model should, at the very least, provide clues about the mechanisms that may be responsible for individual variation. For example, such a model should consider why some individuals are right-handed and some are left-handed or why the rate of left-handedness is higher for men than for women.

Theme 5 emphasizes the importance of understanding how hemispheric asymmetry unfolds over time. There are two parts of this theme: hemispheric asymmetry across the life span of an individual and the evolution of hemispheric asymmetry. A comprehensive model of hemispheric asymmetry should consider both.

As we consider the unfolding of hemispheric asymmetry across the life span, we must acknowledge that the seeds of asymmetry are sown long before birth. The ultimate ontogenetic origins of hemispheric asymmetry may date back to the formation of the first neural elements, to asymmetries of the ovum, or even to various molecular asymmetries. The eventual emergence of functional hemispheric asymmetry is determined by the interplay of a number of biological and environmental factors, beginning as the fetus changes *in utero*. Relevant factors may include asymmetric rates of maturation of the two cerebral hemispheres, asymmetric development of the skull and face, and asymmetric fetal position. Once a specific functional hemispheric asymmetry first appears, its direction and magnitude probably remains approximately the same until death. It is noteworthy that many precursors of later asymmetries are present at very young ages and, in many cases, from birth.

With respect to the evolution of hemispheric asymmetry, it appears that the seeds of asymmetry in present-day humans were sown in our distant evolutionary past, perhaps even in the very molecules and physical particles of which all living things are constructed. In organisms that are already functionally asymmetric to some extent, at least some of the additional evolutionary adaptations that are favored by the environment are likely to be implemented more efficiently on one side or the other, and therefore the adaptations are also likely to be asymmetric when they emerge. Over the course of several such adaptations, this snowball mechanism can greatly increase the extent of functional hemispheric asymmetry. At least some of the hemispheric asymmetries found in present-day humans seem continuous with similar asymmetries found in other present-day species. This suggests that some asymmetries were present very far back in the ancestral line that eventually led to humans. Thus, when important evolutionary adaptations emerged in hominids they did so in beings whose brains were already asymmetric in many respects.

In this brief review of the five themes of the book I have tried to highlight the range of things with which a comprehensive model of hemispheric asymmetry must deal. I would like nothing better than to end this book with a precise and comprehensive model of hemispheric asymmetry that addresses all of the points that have been raised with clarity, elegance, specificity, and confidence. Unfortunately, I do not think that I can devise such a definitive model and, at the same time, be faithful to the entire corpus of empirical results. I can, however, sketch the story that I see emerging from the present review as a way of at least moving us toward a comprehensive model of hemispheric asymmetry.

Toward a Model of Hemispheric Asymmetry

A comprehensive model of hemispheric asymmetry must deal with all of the issues contained in the preceding section. It should explain how and why hemispheric asymmetry emerged during the course of evolution and identify the evolutionary milestones that shaped it into what we see today. It should consider how it is that hemispheric asymmetry emerges and changes over the life span of an individual and how asymmetry is shaped by biological and environmental factors. Ideally, the factors that influence the ontogenetic emergence of hemispheric asymmetry should help explain why there is individual variation in aspects of asymmetry. In doing these things, a comprehensive model should also be consistent with the subtlety and breadth of hemispheric asymmetry and should contain principles that predict which computational processes tend to be better implemented on the right side and which tend to be better implemented on the left side.

Here I will review the suggestions that have been made throughout the book about how and why hemispheric asymmetry has come to characterize the brains of humans, beginning with the emergence of asymmetry during the course of evolution and ending with the sort of generalizations about present-day asymmetries that have been reviewed earlier. Many of the specific suggestions made here are speculative and need to be tested in future empirical studies, but I include them because they seem plausible in view of the research that has accumulated and because they serve as a reasonable point of departure for moving toward a comprehensive model of hemispheric asymmetry.

Chapter 9 ended with several suggestions about the emergence of hemispheric asymmetry during the course of evolution. To the extent that those suggestions are accurate, they have important implications for the mechanisms that determine aspects of hemispheric asymmetry in present-day humans. Therefore, at the risk of being redundant, I begin by summarizing what seems to be a plausible evolutionary scenario.

Freely moving animals must be equally attentive to both the left and right sides of space and they must be able to move in a straight line through their environments. Consequently, there are strong evolutionary pressures for bilateral symmetry in the placement of sense organs and limbs and in those parts of the brain and nervous system that are associated with sensation and locomotion. On this view, it comes as no surprise that both hemispheres of present-day humans have excellent (and, perhaps, equal) sensory capabilities and that both are equally involved in locomotion. Because much of the cortex is involved in sensory or motor processes at least indirectly, it should also come as no surprise that most functional hemispheric asymmetries are subtle, involving small, quantitative differences in ability rather than dramatic, all-or-none differences.

At the same time, there is little evolutionary pressure for symmetry for many functions that are not involved in direct sensory or motor responses to the environment. Indeed, in those cases, asymmetry is advantageous because it allows for more compact packaging of functions into a limited cortical space. Furthermore, the simultaneous development of the corpus callosum, which, in part, serves to reduce maladaptive cross-talk between functions carried out in homologous regions of the two hemispheres, makes it advantageous for complementary or mutually inconsistent processes to migrate toward opposite hemispheres. In view of all that has been said, pressures toward asymmetry would be expected primarily for perceptual and cognitive functions beyond the earliest levels of sensory analysis and for motor activity that is not a direct response to environmental stimulation.

Research reviewed in Chapters 5 and 9 suggests that certain brain asymmetries were already present in our evolutionary ancestors before the emergence of the first hominids approximately 4 million years ago. Thus, subsequent adaptations to environmental pressures took place in the context of a brain whose hemispheres were already functionally asymmetric for at least some of the relevant processing components. Given the asymmetry that already existed, it is not surprising that some of the subsequent adaptations were implemented more efficiently in one hemisphere than in the other. This sort of snowball mechanism could account for increases in the extent of functional hemispheric asymmetry, even when there is nothing in the environment that necessarily favors hemispheric asymmetry for the processing abilities that have emerged recently. The last few hundred thousand years have seen especially dramatic adaptations to environmental pressures and the emergence of several milestones that have set humans apart from other species. Some of these milestones are likely to have had important consequences for the continued evolution of hemispheric asymmetry.

One important milestone is standing and walking upright. Among other things, this adaptation freed the hands from the need to provide postural support and minimized their use in locomotion. There was thus far less pressure for control of the hands to remain bilaterally symmetric. In fact, for efficient performance it is important for the hands to be used in a complementary fashion (e.g., one hand holds a stone while the other chips away at it to fashion a tool). As noted in Chapter 9, it is conceivable that prior to the emergence of an upright posture, sufficient asymmetry in control of the hands already existed to create the bias toward right-handedness that seems to have been characteristic of even the first hominids. It is also possible that prior left-hemisphere dominance for vocalization made the left hemisphere a better neural substrate than the right for manual praxis. In addition, the adoption of an upright posture and bipedal locomotion created asymmetric shearing forces on the otolith organs during fetal development, which has been hypothesized to contribute to right-handedness and to right-hemisphere dominance for visuospatial processing in present-day humans (see Chapter 8).

Additional evolutionary milestones discussed in Chapter 9 include the manufacture and use of tools and the emergence of language. Whatever specific effects each of these may have had on the evolution of hemispheric asymmetry, they seem to have interacted in a synergistic fashion to produce a dramatic increase in the sophistication of both tools and language during the period from about 10,000 to 30,000 years ago. It has been argued that manufacturing tools and producing oral language share certain processing needs (e.g., the need for sequential, timed movements), making it efficient for the same hemisphere to become dominant for both hand use and for certain aspects of vocalization and language.

An additional evolutionary milestone is the prolonged period of postnatal immaturity that characterizes present-day humans. The possibility that this adaptation may have contributed to enhanced hemispheric asymmetry has not received much attention in earlier accounts of the evolution of hemispheric asymmetry. In view of the novelty of some of the suggestions made in Chapter 9, a review of the potential consequences of postnatal immaturity would seem to be worthwhile.

As the size of the brain and skull increased during the course of evolution, it would have become impossible for the fetus to traverse the birth canal were it not for the fact that the fetus also came to be born earlier and earlier—before its head grew too large. As a result, the human fetus enters the world in considerably less mature form than the newborns of other contemporary primates or of our primate ancestors (before *H. erectus*). In Chapter 9, I speculated that this adaptation was favored because there are advantages of having the developing nervous system shaped as much as possible by the external environment instead of by the relatively impoverished intrauterine environment. A prolonged period of postnatal immaturity could have consequences for enhancing functional hemispheric asymmetry in the following way.

In several present-day species (including humans), there is evidence that homologous areas of the two cerebral hemispheres do not mature at the same rate, either before or after birth (Chapter 8). I have argued that asymmetries in the rate of maturation are likely to have greater consequences when they occur either late in fetal development or after birth than when they occur earlier. As noted in Chapter 8, in humans the sensorimotor capacities of newborns change dramatically over the first few days, weeks, and months of life, and therefore the brain is continually being presented with better-quality information. The rapidly changing nature of this information may interact with asymmetries in maturation to sow the seeds of functional hemispheric asymmetry. On this view, hemispheric asymmetries may be more marked in present-day humans than in either present-day primates or in our primate ancestors partly as a by-product of prolonged immaturity.

I have also suggested that two opposing evolutionary pressures may have favored the emergence and enhancement of asymmetric maturational gradients. One pressure favors the development of a brain that is ready to learn as much as possible as soon as possible (from the moment of birth or even before). The opposing pressure favors the development of a brain that is "saved" to learn as much about the world as the organism's fully mature sensory and motor systems will allow. Chapter 9 discusses the potential conflict between these two goals when an organism is born in an extremely immature state. One solution would be to give the organism "two brains," largely identical except that homologous areas are on different maturational trajectories (in other words, one is modified a great deal initially and the other is most sensitive to modification somewhat later). From this evolutionary perspective, functional hemispheric asymmetry in present-day humans emerges as the result of the interaction of a potentially large number of biological and environmental factors, beginning with the fetus as it develops *in utero*.

Sorting out all of the biological and environmental factors is likely to be time-consuming and difficult. Indeed, the complexity of these factors is a major obstacle to the development of a truly complete and comprehensive model of hemispheric asymmetry. While there is much left to do, progress is being made. For example, in Chapter 8 I discussed several factors that have been proposed to account for the emergence of functional hemispheric asymmetries across the life span of a contemporary human. Although more research is clearly needed to refine our understanding of those factors and to identify others, they serve to illustrate what I believe will be a fruitful approach to understanding functional hemispheric asymmetry. In addition, individual variation in the various biological and environmental factors may account for why individuals come to differ from each other in several aspects of hemispheric asymmetry and interhemispheric interaction.

During the course of fetal development, certain areas of the right hemisphere appear to mature earlier than homologous areas of the left hemisphere (see Chapter 8). Furthermore, this maturational asymmetry continues to be present for a short time after birth. Various scenarios have been proposed to suggest how functional hemispheric asymmetries could arise from the interaction of these different rates of maturation and the changing nature of environmental information presented to the brain. What these different scenarios have in common is that the right hemisphere is proposed to be more influenced than the left hemisphere by the sort of impoverished sensorimotor information that the developing brain receives first. This information has been hypothesized to include such diverse things as noises in audition, low spatial frequencies in vision, and very coarse sensorimotor feedback. By way of contrast, the later-developing left hemisphere is in some sense "saved" to be more influenced than the right hemisphere by the higher-quality sensorimotor information that the brain receives only later (e.g., language sounds from the mother's voice in audition, high spatial frequencies in vision, and more precise sensorimotor feedback). Note that this scenario links various components of processing together in the same hemisphere on the basis of early changes in the sensorimotor information presented to the brain rather than on the basis of there being a single, fundamental information-processing dichotomy that characterizes all hemispheric asymmetries. This is at least consistent with the inability to reduce various hemispheric asymmetries to a single, dichotomous information-processing dimension. It is also consistent with the subtlety and breadth that characterize hemispheric asymmetry.

It has also been proposed that asymmetric growth spurts during childhood provide a mechanism for continued unfolding of functional hemispheric asymmetry. For example, during the period from 3 to 6 years of age both frontal and occipital areas of the left hemisphere appear to develop more rapidly than homologous areas of the right hemisphere. It is interesting that this coincides with a period that is also critical for the acquisition of grammar, an aspect of language for which the left hemisphere is usually dominant.

Additional prenatal asymmetries have also been hypothesized to influence later functional hemispheric asymmetry. For example, various asymmetries in the skull and face begin to appear during the first trimester of pregnancy. It has been proposed that these cranio-facial asymmetries constitute the prenatal origins of left-hemisphere dominance for perceiving and producing speech. The rationale is that for most individuals the direction of cranio-facial asymmetry leads to greater sensitivity of the right ear to sounds in the range critical for language and that this slight peripheral ear difference makes the left hemisphere more responsive to speech. On this view, the eventual emergence of left-hemisphere dominance for speech production is a consequence of the earlier left-hemisphere dominance for speech perception. In addition, asymmetries of fetal position, primarily during the last trimester of pregnancy, have been linked to such things as later handedness. On this basis it has been argued that fetal position is an important contributor to later functional hemispheric asymmetry. One of the most ambitious hypotheses is that the asymmetric fetal position favors development of the left otolith and that it is this asymmetry in the organs of balance that underlies later handedness and right-hemisphere dominance for spatial processing, the latter because of neural connections between the primary vestibular cortex and the parietal lobe.

From this ontogenetic perspective, individual differences in aspects of functional hemispheric asymmetry can be traced to individual variation in the sorts of factors just considered. For example, individual brains are likely to differ in the rates at which homologous areas of the two hemispheres mature. Furthermore, it has been hypothesized that some of these differences are influenced by such things as the level of testosterone and other hormones during certain critical stages of fetal development. In addition, there is individual variation in the position of the fetus *in utero* and in the extent of asymmetric cranio-facial development. There are also individual variations in the pattern of environmental stimulation that an infant experiences over the first few days, weeks, and months of life. While it remains to be determined just how all of this might produce individual differences in functional hemispheric asymmetry, these sorts of individual variations seem sufficiently rich to produce the range of individual variation that is seen in adults (see Chapter 7).

It should also be noted that individual variation in one of these ontogenetic factors may be independent of individual variation in some of the others. For example, larger cranio-facial asymmetry is not necessarily related to fetal position during the last trimester of pregnancy. To the extent that this is the case, it means that specific processing components may become lateralized to one hemisphere or the other more or less independently. This is consistent with the fact that hemispheric advantages are not always correlated strongly across tasks, either in humans (Chapter 2) or in other species (Chapter 5). It also suggests that important insights about the ontogenetic mechanisms that lead to functional hemispheric asymmetry can be gained from future studies that examine the extent to which various asymmetries in adults are related to each other.

An important aspect of the view presented here is that asymmetries that are very small and subtle when they first emerge can eventually have more profound consequences for functional hemispheric asymmetry. As discussed in Chapter 8, the effect of small asymmetries at one moment in ontological development can snowball into larger and more complex asymmetries at later moments. As the fetus (and, later, the infant) is presented with new and richer aspects of sensorimotor information, the extent to which neural networks in the two hemispheres are responsive is likely to depend on how those networks have already been modified by input that has come before. If previous input has led to subtle differences in the neural networks contained in the two hemispheres, these differences may make one hemisphere's networks slightly more hospitable as a neural substrate for the new and richer information that is now available to the brain. Even a slight difference in hospitality may be sufficient for the neural networks in the two hemispheres to be modified differently by the new information. When this process is repeated over and over, the functional differences between the hemispheres grow into what we see in adulthood.

In Chapter 3, I argued in favor of what I call a componential approach to the study of hemispheric asymmetry. Briefly, this approach acknowledges that (1) even simple tasks require the coordination of a number of information-processing subsystems, components, or modules and (2) hemispheric asymmetry can vary from component to component even within what we would regard as a single task. Differences among components make it necessary to examine hemispheric asymmetry for specific components of processing or for specific types of neural computation. The story about the emergence of hemispheric asymmetry that is sketched here is consistent with this componential approach in the following way. According to the view taken here, subtle asymmetries begin to emerge very early and create small differences between the two hemispheres. These small differences eventually predispose homologous areas of the two hemispheres toward somewhat different forms of computation. As a consequence, one hemisphere may become superior to the other (usually in a relative rather than in an all-or-none sense) for a particular type of neural computation. Presumably, this hemispheric asymmetry is sufficiently stable that it contributes to virtually all tasks that involve that particular neural computation or processing component. This being the case, a more analytic picture of hemispheric asymmetry will emerge from a componential approach than from an approach that emphasizes tasks without dissecting them into the relevant processing components.

By way of illustration, consider the following hypothesis about the emergence of hemispheric asymmetry for certain aspects of visual processing. As noted earlier (see Chapter 8), there is some evidence that the development of brain areas important for visual cognition are more advanced in the right hemisphere than in the left hemisphere at the time of birth. The visual sensory system of newborns presents the brain with highly degraded visual input. In particular, the visual system of neonates is especially limited in its transmission of visual information carried by relatively high spatial frequencies. The more advanced right hemisphere may have priority in dealing with this highly degraded visual input shortly after birth, resulting in important modifications of its relevant neural networks. Later (perhaps only a few weeks or months), when relevant areas of the left hemisphere are now more developed than they were at birth, the visual sensory system is more fully developed and presents the brain with visual information conveyed by a higher range of spatial frequencies in addition to the low frequencies that were transmitted earlier. The right hemisphere might continue to be more responsive than the left hemisphere to information conveyed by relatively low spatial frequencies, which means that the right hemisphere is less available than the left for processing information conveyed by relatively high spatial frequencies. At the same time, the left hemisphere is now more advanced in development than it was earlier, which means that its relevant neural networks are modified at least as much by high- as by low-frequency information. As a result of all of this, the hemispheres become asymmetric for processing different aspects of visual information.

The kind of hemispheric asymmetry just considered could underlie a number of asymmetries with respect to visual information processing—asymmetries that might otherwise be viewed as unrelated to each other. In Chapter 3, I used the componential approach to suggest the following relationships among three aspects of vision. It is obvious how the mechanism outlined earlier could account for hemispheric asymmetries in processing high versus low spatial frequencies. Specifically, it appears that, at some level of processing beyond the sensory level, the left and right hemispheres are biased toward efficient use of higher and lower visual-spatial frequencies, respectively. Whatever computational or componential hemispheric differences account for this bias are also likely to account for the finding that the right hemisphere is well suited or predisposed for the processing of global aspects of the visual world whereas the left hemisphere is well suited or predisposed for the processing of local aspects of the visual world. For example, it is generally the case that information about global aspects of a visual stimulus is carried by a lower range of visual-spatial frequencies than is information about local aspects. I also noted in Chapter 3 how the same computational or componential hemispheric differences could also account for hemispheric asymmetry in processing two different types of spatial relations—categorical and coordinate. For example, it appears that information about categorical spatial relations may be carried more effectively by a higher range of spatial frequencies than is information about coordinate spatial relations.

The foregoing examples (and others contained in Chapters 2 and 3) illustrate the value of the componential approach for discovering possible links among various functional hemispheric asymmetries. For continued progress in understanding hemispheric asymmetry it is essential that larger tasks continue to be broken down into relevant processing components. As a starting point, it has proven worthwhile to consider the components that have been postulated by contemporary models of perception, cognition, and action. Discovering hemispheric asymmetries for specific components of this type refines our understanding of hemispheric asymmetry and also provides converging evidence that the decomposition of tasks into those particular components is on the right track.

Concluding Comments

The emerging field of cognitive neuroscience has set for itself a very large goal—understanding how our brains produce the range of abilities and propensities characteristic of our species. If we are to succeed in this endeavor, we must confront a great many mysteries about the structure of the brain and about the architecture of perception, cognition, emotion, and action. An important characteristic of our cortex is that it is divided anatomically into two cerebral hemispheres that have somewhat different abilities and propensities. The existence of functional hemispheric asymmetries has captured the imagination of scientists and of the lay public, especially during the last thirty years or so. As a result of this interest and of the development of new technologies, the past three decades have also seen great strides toward clarifying the nature of hemispheric asymmetries, the varieties of interhemispheric interaction, and the implications for cognitive processing in neurologically intact individuals. Although mysteries still abound, the progress of recent years suggests that we will eventually have a comprehensive model of this important aspect of cognitive neuroscience.

As I noted at the beginning of this epilogue, the story sketched here is not yet a comprehensive model of hemispheric asymmetry. In particular, there are many specific details that need to be determined by future empirical research and many specific hypotheses that require additional testing. I have mentioned many of those details and hypotheses throughout this book. At the same time, I do believe that any comprehensive model is likely to take the general form of the story that has emerged here and must integrate the evolution of hemispheric asymmetry with the ontogenetic emergence of hemispheric asymmetry for components of processing. Furthermore, solving the mysteries that still surround the emergence of hemispheric asymmetry will be an important part of coming to understand how our brains produce our perceptions, cognitions, emotions, and actions.

Bibliography

Adams, C. L., D. L. Molfese, and J. C. Betz. 1987. Electrophysiological correlates of categorical speech perception for voicing contrasts in dogs. *Developmental Neuropsychology, 3,* 175–189.

Allen, M. 1983. Models of hemispheric specialization. *Psychological Bulletin, 93,* 73–104.

Alivisatos, B., and J. Wilding. 1982. Hemispheric differences in matching Stroop-type letter stimuli. *Cortex, 18,* 5–21.

Alwitt, L. F. 1982. Two neural mechanisms related to modes of selective attention. *Journal of Experimental Psychology: Human Perception and Performance, 8,* 253–272.

Amaducci, L., S. Sorbi, A. Albanese, and G. Gainotti. 1981. Choline acetyltransferase (ChAT) activity differs in right and left human temporal lobes. *Neurology, 31,* 799–805.

Anderson, J. R. 1990. *Cognitive Psychology and Its Implications.* New York: Freeman.

Annett, M. 1985. *Left, Right, Hand and Brain: The Right Shift Theory.* Hillsdale, NJ: Erlbaum.

Arnold, A. P., and S. W. Bottjer. 1985. Cerebral lateralization in birds. In *Cerebral Lateralization in Nonhuman Species,* ed. S. D. Glick, 11–40. New York: Academic Press.

Badian, N. A., and P. H. Wolff. 1977. Manual asymmetries of motor sequencing in boys with reading disability. *Cortex, 13,* 343–349.

Bagnara, S., D. B. Boles, F. Simion, and C. Umilta. 1983. Symmetry and similarity effects in the comparison of visual patterns. *Perception and Psychophysics, 34,* 578–584.

Bakan, P. 1977. Left-handedness and birth order revisited. *Neuropsychologia, 15,* 837–839.

Banich, M. T., and A. Belger. 1990. Interhemispheric interaction: How do the hemispheres divide and conquer a task? *Cortex, 26,* 77–94.

Banich, M. T., S. Goering, N. Stolar, and A. Belger. 1990. Interhemispheric processing in left- and right-handers. *International Journal of Neuroscience, 54,* 197–208.

Banks, M. S., and J. L. Dannemiller. 1987. Infant visual psychophysics. In *Handbook of Infant Perception,* vol. 1, ed. P. Salapatek and L. Cohen, 115–184. Orlando: Academic Press.

Baynes, K., M. J. Tramo, and M. S. Gazzaniga. 1992. Reading with a limited lexicon in the right hemisphere of a callosotomy patient. *Neuropsychologia, 30,* 187–200.

Beaumont, J. G. 1982. *Divided Visual Field Studies of Cerebral Organization.* New York: Academic Press.

Beaumont, J. G., A. W. Young, and I. C. McManus. 1984. Hemisphericity: A critical review. *Cognitive Neuropsychology, 1,* 191–212.

Behrmann, M., and M. Moscovitch. 1992. Frames of reference and spatial attention in neglect. Paper presented at the Annual Convention of the International Neuropsychological Society, San Diego, CA.

Bellugi, U., H. Poizner, and E. S. Klima. 1983. Brain organization for language: Clues from sign aphasia. *Human Neurobiology, 2,* 155–170.

Benbow, C. P. 1986. Physiological correlates of extreme intellectual precocity. *Neuropsychologia, 24,* 719–725.

Bentin, S., A. Sahar, and M. Moscovitch. 1984. Intermanual information transfer in patients with lesions in the trunk of the corpus callosum. *Neuropsychologia, 22,* 601–612.

Benton, A. L. 1980. The neuropsychology of face recognition. *American Psychologist, 35,* 176–186.

Benton, A. L., H. J. Hannay, and N. R. Varney. 1975. Visual perception of line orientation in patients with unilateral brain disease. *Neurology, 25,* 907–910.

Berndt, R. S., A. Caramazza, and E. Zurif. 1983. Language functions: Syntax and semantics. In *Language Function and Brain Organization,* ed. S. J. Segalowitz, 5–28. New York: Academic Press.

Berrebi, A. S., R. H. Fitch, D. L. Ralphe, J. O. Denenberg, V. L. Friedrich, and V. H. Denenberg. 1988. Corpus callosum: Region-specific effects of sex, early experience and age. *Brain Research, 438,* 216–224.

Berryman, M. L., and K. J. Kennelly. 1992. Letter memory loads change more than visual-field advantage: Interhemispheric coupling effects. *Brain and Cognition, 18,* 152–168.

Bertelson, P. 1981. The nature of hemispheric specialization: Why should there be a single principle? *Behavioral and Brain Sciences, 4,* 63–64.

Best, C. T., H. Hoffman, and B. B. Glanville. 1982. Development of ear asymmetries for speech and music. *Perception and Psychophysics, 31,* 75–85.

Bever, T. G., C. Carrithers, W. Cowart, and D. J. Townsend. 1989. Language processing and familial handedness. In *From Reading to Neurons,* ed. A. Galaburda, 351–357. Cambridge, MA: MIT Press.

Bisiach, E., E. Capitani, C. Luzzatti, and D. Perani. 1981. Brain and conscious representation of reality. *Neuropsychologia, 19,* 543–552.

Bisiach, E., and C. Luzzatti. 1978. Unilateral neglect of representational space. *Cortex, 14,* 129–133.

Bloch, M. 1989. A test of four neuropsychological models of dyslexia. Unpublished Ph.D. dissertation, Department of Psychology, University of Southern California.

Bogen, J. E., and M. S. Gazzaniga. 1965. Cerebral commissurotomy in man: Minor hemisphere dominance for certain visuo-spatial functions. *Journal of Neurosurgery, 23,* 394–399.

Boles, D. B. 1979. Laterally biased attention with concurrent verbal memory load: Multiple failures to replicate. *Neuropsychologia, 17,* 353–361.

——— 1984. Global versus local processing: Is there a hemispheric dichotomy? *Neuropsychologia, 22,* 445–455.

——— 1989. Do visual field asymmetries intercorrelate? *Neuropsychologia, 27,* 697–704.

——— 1991. Factor analysis and the cerebral hemispheres: Pilot study and parietal functions. *Neuropsychologia, 29,* 59–92.

Borod, J. C., E. Koff, and B. White. 1983. Facial asymmetry in posed and spontaneous expressions of emotion. *Brain and Cognition, 2,* 165–175.

Borod, J. C., J. St. Clair, E. Koff, and M. Alpert. 1990. Perceiver and poser asymmetries in processing facial emotion. *Brain and Cognition, 13,* 167–177.

Bouma, A. 1990. *Lateral Asymmetries and Hemispheric Specialization: Theoretical Models and Research.* Amsterdam: Swets and Zeitlinger.

Bowers, D., R. M. Bauer, H. B. Coslett, and K. M. Heilman. 1985. Processing of face by patients with unilateral hemisphere lesions. I. Dissociations between judgements of facial affect and facial identity. *Brain and Cognition, 4,* 258–272.

Bracha, H. S. 1987. Asymmetric rotational (circling) behavior, a dopamine-related asymmetry: Preliminary findings in unmedicated and never-medicated schizophrenic patients. *Biological Psychiatry, 22,* 995–1003.

Bracha, H. S., D. J. Seitz, J. Otemaa, and S. D. Glick. 1987. Rotational movement (circling) in normal humans: Sex difference and relationship to hand, foot and eye preference. *Brain Research, 411,* 231–235.

Bradshaw, J. L. 1988. The evolution of human lateral asymmetries: New evidence and second thoughts. *Journal of Human Evolution, 17,* 615–637.

——— 1989. *Hemispheric Specialization and Psychological Function.* Chichester, England: Wiley.

Bradshaw, J. L., and N. C. Nettleton. 1981. The nature of hemispheric specialization in man. *Behavioral and Brain Sciences, 4,* 51–91.

——— 1983. *Human Cerebral Asymmetry.* Englewood Cliffs, NJ: Prentice-Hall.

Bridges, K. M. B. 1932. Emotional development in early infancy. *Child Development, 3,* 324–341.

Broca, P. 1861. Remarques sur le siege da la faculte du language articule, suive d'une observation d'aphemie. Trans. J. Kann (1950). *Journal of Speech and Hearing Disorders, 15,* 16–20.

Brownell, H. H., D. Michel, J. Powelson, and H. Gardner. 1983. Surprise but not coherence: Sensitivity to verbal humor in right-hemisphere patients. *Brain and Language, 18,* 20–27.

Brownell, H. H., T. L. Simpson, A. M. Bihrle, H. H. Potter, and H. Gardner. 1990. Appreciation of metaphoric alternative word meanings by left and right brain-damaged patients. *Neuropsychologia, 28,* 375–384.

Bruce, R., and M. Kinsbourne. 1974. Orientational model of perceptual asymmetry. Paper presented at the Annual Convention of the Psychonomic Society, Boston, MA.

Bryden, M. P. 1982. *Laterality: Functional Asymmetry in the Intact Brain.* New York: Academic Press.

——— 1987. Handedness and cerebral organization: Data from clinical and normal populations. In *Duality and Unity of the Brain: Unified Functioning and Specialisation of the Hemispheres,* ed. D. Ottoson, 55–70. London: Macmillan.

Bryden, M. P., and F. Allard. 1976. Visual hemifield differences depend upon typeface. *Brain and Language, 3,* 191–200.

Bryden, M. P., and L. MacRae. 1989. Dichotic laterality effects obtained with emotional words. *Neuropsychiatry, Neuropsychology, and Behavioral Neurology, 1,* 171–176.

Bryden, M. P., and R. E. Steenhuis. 1991. Issues in the assessment of handedness. In *Cerebral Laterality: Theory and Research,* ed. F. L. Kitterle, 35–52. Hillsdale, NJ: Erlbaum.

Byrd, M., and M. Moscovitch. 1984. Lateralization of peripherally and centrally masked words in young and elderly people. *Journal of Gerontology, 39,* 699–703.

Caramazza, A., and R. C. Martin. 1983. Theoretical and methodological issues in the study of aphasia. In *Cerebral Hemisphere Asymmetry: Method, Theory and Application,* ed. J. B. Hellige, 18–45. New York: Praeger.

Carlson, J. N., and S. D. Glick. 1989. Cerbral lateralization as a source of interindividual differences in behavior. *Experientia, 45,* 788–798.

Carson, R. G., D. Goodman, and D. Elliott. 1992. Asymmetries in the discrete and pseudocontinuous regulation of visually guided reaching. *Brain and Cognition, 18,* 169–191.

Chi, J. G, E. C. Dooling, and F. H. Gilles. 1977. Left-right asymmetries of the temporal speech areas of the human fetus. *Archives of Neurology, 34,* 346–348.

Chiarello, C. 1985. Hemisphere dynamics in lexical access: Automatic and controlled priming. *Brain and Language, 26,* 146–172.

——— 1988. Lateralization of lexical processes in the normal brain: A review of visual half-field research. In *Contemporary Reviews in Neuropsychology,* ed. H. A. Whitaker, 36–76. New York: Springer.

——— 1991. Interpretation of word meanings by the cerebral hemispheres: One is not enough. In *The Psychology of Word Meanings,* ed. P. Schwanenflugel, 251–278. Hillsdale, NJ: Erlbaum.

Chiarello, C., M. A. McMahon, and K. Schaefer. 1989. Visual cerebral lateralization over phases of the menstrual cycle: A preliminary investigation. *Brain and Cognition, 11,* 18–36.

Chiarello, C., J. Senehi, and M. Soulier. 1986. Viewing conditions and hemispheric asymmetry for the lexical decision. *Neuropsychologia, 24,* 521–530.

Christman, S. 1989. Perceptual characteristics in visual laterality research. *Brain and Cognition, 11,* 238–257.

——— 1990. Effects of luminance and blur on hemispheric asymmetries in temporal integration. *Neuropsychologia, 28,* 361–374.

Christman, S., F. L. Kitterle, and J. B. Hellige. 1991. Hemispheric asymmetry in the processing of absolute versus relative spatial frequency. *Brain and Cognition, 16,* 62–73.

Chomsky, N. 1980. Rules and representations. *Behavioral and Brain Sciences, 3,* 1–61.

Clarke, J. M. 1990. *Interhemispheric Functions in Humans: Relationships between Anatomical Measures of the Corpus Callosum, Behavioral Laterality Effects, and Cognitive Profiles.* Ph.D. dissertation, Department of Psychology, University of California at Los Angeles.

Cohen, H., and J. Levy. 1986. Cerebral and sex differences in the categorization of haptic information. *Cortex, 22,* 253–260.

Collins, R. L. 1985. On the inheritance of direction and degree of asymmetry. In *Cerebral Lateralization in Nonhuman Species,* ed. S. D. Glick, 41–72. New York: Academic Press.

Coltheart, M. 1983. The right hemisphere and disorders of reading. In

Functions of the Right Cerebral Hemisphere, ed. A. W. Young, 172–201. New York: Academic Press.

——— 1985. Right hemisphere reading revisited. *Behavioral and Brain Sciences, 8,* 363–365.

Cook, N. D. 1984. Homotopic callosal inhibition. *Brain and Language, 23,* 116–125.

——— 1986. *The Brain Code: Fundamental Mechanisms of Information Transfer in the Human Brain.* London: Routledge and Kegan Paul.

Corballis, M. C. 1991. *The Lopsided Ape: Evolution of the Generative Mind.* Oxford: Oxford University Press.

Corballis, M. C., and M. J. Morgan. 1978. On the biological basis of human laterality: I. Evidence for a maturational left-right gradient. *Behavioral and Brain Sciences, 1,* 261–269.

Corballis, M. C., and J. Sergent. 1988. Imagery in a commissurotomized patient. *Neuropsychologia, 26,* 13–26.

Coren, S., and D. F. Halpern. 1991. Left-handedness: A marker for decreased survival fitness. *Psychological Bulletin, 109,* 90–106.

Cowin, E. L., and J. B. Hellige. 1991. Effects of blurring on hemispheric asymmetry for processing spatial information. Paper presented at the Annual Meeting of the Psychonomic Society, San Francisco, CA.

Craft, S., L. Willerman, and E. D. Bigler. 1987. Callosal dysfunction in schizophrenia and schizo-affective disorder. *Journal of Abnormal Psychology, 96,* 205–213.

Cunningham, D. J. 1892. *Contribution to the Surface Anatomy of the Cerebral Hemispheres.* Dublin: Royal Irish Academy. Cited in Geschwind and Galaburda, 1987.

Dagenbach, D. 1986. Subject variable effects in correlations between auditory and visual language processing asymmetries. *Brain and Language, 28,* 169–177.

Damasio, A. R., and H. Damasio. 1986. The anatomical substrate of prosopagnosia. In *The Neuropsychology of Face Perception and Facial Expression,* ed. R. Bruyer, 31–38. Hillsdale, NJ: Erbaum.

Davidson, R. J. 1988. EEG measures of cerebral asymmetry: Conceptual and methodological issues. *International Journal of Neuroscience, 39,* 71–89.

——— 1992. Emotion and affective style: Hemispheric substrates. *Psychological Science, 3,* 39–43.

Davidson, R. J., and N. A. Fox. 1982. Asymmetrical brain activity discriminates between positive and negative affective stimuli in human infants. *Science, 218,* 1235–1237.

——— 1989. Frontal brain asymmetry predicts infants' response to maternal separation. *Journal of Abnormal Psychology, 98,* 127–131.

Davidson, R. J., G. E. Schwartz, C. Saron, J. Bennett, and D. J. Goleman. 1979. Frontal versus parietal EEG asymmetry during positive and negative affect. *Psychophysiology, 16,* 202–203.

de Lacoste, M-C., D. S. Horvath, and D. J. Woodward. 1991. Possible sex differences in the developing human fetal brain. *Journal of Clinical and Experimental Neuropsychology, 13,* 831–846.

Delis, D. C., L. C. Robertson, and R. Efron. 1986. Hemispheric specialization of memory for visual hierarchical stimuli. *Neuropsychologia, 24,* 205–214.

Denenberg, V. H., and D. A. Yutzey. 1985. Hemispheric laterality, behavioral asymmetry, and the effects of early experience in rats. In *Cerebral Lateralization in Nonhuman Species,* ed. S. D. Glick, 110–135. New York: Academic Press.

Deruelle, C., and S. de Schonen. 1991. Hemispheric asymmetries in visual pattern processing in infancy. *Brain and Cognition, 16,* 151–179.

de Schonen, S., and I. Bry. 1987. Interhemispheric communication of visual learning: A developmental study in 3- to 6-month old infants. *Neuropsychologia, 25,* 601–612.

de Schonen, S., and C. Deruelle. 1991. Visual field asymmetries for pattern processing are present in infancy: A comment on T. Hatta's study on children's performance. *Neuropsychologia, 29,* 335–338.

de Schonen, S., M. Gil de Diaz, and E. Mathivet. 1986. Hemispheric asymmetry in face processing in infancy. In *Aspects of Face Processing,* ed. H. D. Ellis, M. A. Jeeves, F. Newcombe, and A. Young. Dordrecht: Nijhoff.

de Schonen, S., and E. Mathivet. 1989. First come first served: A scenario about the development of hemispheric specialization in face recognition during infancy. *European Bulletin of Cognitive Psychology (CPC), 9,* 3–44.

——— 1990. Hemispheric asymmetry and interhemispheric communication of face processing in infancy. *Child Development, 61,* 1192–1205.

De Valois, R. L., and K. K. De Valois. 1980. Spatial vision. *Annual Review of Psychology, 31,* 117–153.

Diamond, M. C. 1985. Rat forebrain morphology: Right-left; male-female; young-old; enriched-impoverished. In *Cerebral Lateralization in Nonhuman Species,* ed. S. D. Glick, 73–88. New York: Academic Press.

Dimond, S. 1972. *The Double Brain.* Baltimore: Williams and Wilkins.

Druckman, D., and J. A. Swets, eds. 1988. *Enhancing Human Performance: Issues, Theories, and Techniques.* Washington, DC: National Academy Press.

Dunn, J. C., and K. Kirsner. 1988. Discovering functionally independent mental processes: The principle of reversed associations. *Psychological Review, 95,* 91–101.

Eberstaller, O. 1884. Zur oberflachten anatomie des grosshirn hemispharen. *Wien Med Blatter, 7,* 479+. Cited in Geschwind and Galaburda, 1987.

Efron, R. 1990. *The Decline and Fall of Hemispheric Specialization.* Hillsdale, NJ: Erlbaum.

Ehret, G. 1987. Left hemisphere advantage in the mouse brain for recognizing ultrasonic communication calls. *Nature, 325,* 249–251.

Ehrlichman, H., and J. Barrett. 1983. Right hemispheric specialization for mental imagery: A review of the evidence. *Brain and Cognition, 2,* 55–76.

Ehrlichman, H., and A. Weinberger. 1978. Lateral eye movements and hemispheric asymmetry: A critical review. *Psychological Bulletin, 85,* 1080–1101.

Eidelberg, D., and A. M. Galaburda. 1984. Inferior parietal lobule: Divergent architectonic asymmetries in the human brain. *Archives of Neurology, 41,* 843–852.

Ellis, A. W., and A. W. Young. 1988. *Human Cognitive Neuropsychology.* Hillsdale, NJ: Erlbaum.

Ellis, A. W., A. W. Young, and C. Anderson. 1988. Modes of word recognition in the left and right cerebral hemispheres. *Brain and Language, 35,* 254–273.

Ellis, R, J. 1987. *Cerebral Functional Asymmetries in Aging and Chronic Alcoholism.* Unpublished doctoral dissertation, Boston University, Boston, MA. Cited in Ellis and Oscar-Berman, 1989.

Ellis, R. J., and M. Oscar-Berman. 1989. Alcoholism, aging, and functional cerebral asymmetries. *Psychological Bulletin, 106,* 128–147.

Entus, A. K. 1977. Hemispheric asymmetry in processing of dichotically presented speech and nonspeech stimuli by infants. In *Language Development and Neurological Theory,* ed. S. J. Segalowitz and F. A. Gruber, 63–73. San Diego, CA: Academic Press.

Eviatar, Z., J. B. Hellige, and E. Zaidel. 1992. Differences in hemispheric division of labor for right and left handers. Paper presented at the Annual Convention of the American Psychological Society, San Diego, CA.

Faglioni, P., G. Scotti, and H. Spinnler. 1969. Impaired recognition of written letters following unilateral hemispheric damage. *Cortex, 5,* 120–133.

Fagot, J., and J. Vauclair. 1991. Manual laterality in nonhuman primates:

A distinction between handedness and manual specialization. *Psychological Bulletin, 109,* 76–89.

Farah, M. J. 1984. The neurological basis of mental imagery: A componential analysis. *Cognition, 18,* 245–272.

——— 1986. The laterality of mental image generation: A test with normal subjects. *Neuropsychologia, 24,* 541–551.

——— 1989. The neuropsychology of visual imagery. In *Neuropsychology of Perception,* ed. J. W. Brown, 183–201. Hillsdale, NJ: Erlbaum.

——— 1990. *Visual Agnosia: Disorders of Object Recognition and What They Tell Us about Normal Vision.* Cambridge: MIT Press.

Farah, M. J., J. L. Brunn, A. B. Wong, M. A. Wallace, and P. A. Carpenter. 1990. Frames of reference for allocating attention to space: evidence from the neglect syndrome. *Neuropsychologia, 28,* 335–348.

Fendrich, R., and M. Gazzaniga. 1990. Hemispheric processing of spatial frequencies in two commissurotomy patients. *Neuropsychologia, 28,* 657–664.

Fiorentini, A., and N. Berardi. 1984. Right-hemisphere superiority in the discrimination of spatial phase. *Perception, 13,* 695–708.

Fisk, J. D., and M. A. Goodale. 1988. The effects of unilateral brain damage on visually guided reaching: Hemispheric differences in the nature of the deficit. *Experimental Brain Research, 72,* 425–435.

Flor-Henry, P. 1987. Cerebral dynamics, laterality and psychopathology: A commentary. In *Cerebral dynamics, laterality and psychopathology,* ed. R. Takahashi, P. Flor-Henry, J. Gruzelier, and S. I. Niwa, 3–22. Amsterdam: Elsevier.

Fontes, V. 1944. *Morfologia do Cortex Cerebral (Desenvolvimentao).* Lisbon: Instituto Antonio Aurelio da Costa Ferreira. Cited in Geschwind and Galaburda, 1987.

Fox, N. A. 1991. If it's not left, it's right. *American Psychologist, 46,* 863–872.

Fox, N. A., and R. J. Davidson. 1986. Taste-elicited changes in facial signs of emotion and the asymmetry of brain electrical activity in human newborns. *Neuropsychologia, 24,* 417–422.

Friedman, A., and M. C. Polson. 1981. Hemispheres as independent resource systems: Limited-capacity processing and cerebral specialization. *Journal of Experimental Psychology: Human Perception and Performance, 7,* 1031–1058.

Friedman, A., M. C. Polson, and C. G. Dafoe. 1988. Dividing attention between the hands and the head: Performance trade-offs between rapid finger tapping and verbal memory. *Journal of Experimental Psychology: Human Perception and Performance, 14,* 60–68.

Friedman, A., M. C. Polson, C. G. Dafoe, and S. J. Gaskill. 1982. Dividing attention within and between hemispheres: Testing a multiple resources approach to limited capacity information processing. *Journal of Experimental Psychology: Human Perception and Performance, 8,* 625–650.

Gainotti, G. 1985. Constructive apraxia. In *Handbook of Clinical Neurology,* vol. 45, *Clinical Neuropsychology,* ed. P. J. Vinken, G. W. Bruyn, and H. J. Klawans. Amsterdam: Elsevier Science Publishers.

——— 1987. Disorders of emotional behaviour and of autonomic arousal resulting from unilateral brain damage. In *Duality and Unity of the Brain,* ed. D. Ottoson, 161–179. London: Macmillan Press.

Galaburda, A. M., G. D. Rosen, and G. F. Sherman. 1990. Individual variability in cortical organization: Its relationship to brain laterality and implications to function. *Neuropsychologia, 28,* 529–546.

Galaburda, A. M., F. Sanides, and N. Geschwind. 1978. Human brain: Cytoarchitectonic left-right asymmetries in the temporal speech region. *Archives of Neurology, 35,* 812–817.

Galin, D., J. Johnstone, L. Nakoff, and J. Herron. 1979. Development of the capacity for tactile information transfer between hemispheres in normal children. *Science, 204,* 1330–1332.

Galin, D., and R. Ornstein. 1972. Lateral specialization of cognitive mode: An EEG study. *Psychophysiology, 9,* 412–418.

Gardner, H., H. H. Brownell, W. Wapner, and D. Michelow. 1983. Missing the point: The role of the right hemisphere in the processing of complex linguistic materials. In *Cognitive Processing in the Right Hemisphere,* ed. E. Perecman, 169–192. New York: Academic Press.

Gazzaniga, M. S. 1970. *The Bisected Brain.* New York: Appleton-Century-Crofts.

——— 1983. Right hemisphere language following brain bisection: A 20-year perspective. *American Psychologist, 38,* 525–537.

——— 1985. *The Social Brain.* New York: Basic Books.

Gazzaniga, M. S., and J. E. LeDoux. 1978. *The Integrated Mind.* New York: Plenum Press.

Geffen, G., J. L. Bradshaw, and N. C. Nettleton. 1973. Attention and hemispheric differences in reaction time during simultaneous audiovisual tasks. *Quarterly Journal of Experimental Psychology, 25,* 404–412.

Geschwind, N., and P. Behan. 1982. Left-handedness: Association with immune disease, migraine, and developmental learning disorder. *Proceedings of the National Acadamy of Sciences, 79,* 5097–5100.

Geschwind, N., and A. M. Galaburda. 1987. *Cerebral Lateralization: Biological Mechanisms, Associations, and Pathology.* Cambridge, MA: MIT Press.

Geschwind, N., and W. Levitsky. 1968. Human brain: Left-right asymmetries in temporal speech region. *Science, 161,* 186–187.

Gevins, A. S. 1983. Brain potential (BP) evidence for lateralization of higher cognitive functions. In Hellige, 1983, 335–382.

Gilger, J. W., B. F. Pennington, P. Green, S. M. Smith, and S. D. Smith. 1992. Reading disability, immune disorders and non-right-handedness: Twin and family studies of their relations. *Neuropsychologia, 30,* 209–228.

Girotti, F., M. Casazza, M. Musicco, and G. Avanzini. 1983. Oculomotor disorders in cortical lesions: The role of unilateral neglect. *Neuropsychologia, 21,* 543–553.

Gladstone, M., and C. T. Best. 1985. Developmental dyslexia: The potential role of interhemispheric collaboration in reading acquisition. In *Hemispheric Function and Collaboration in the Child,* ed. C. T. Best, 87–118. New York: Academic Press.

Glick, S. D., J. N. Carlson, K. L. Drew, and R. M. Shapiro. 1987. Functional and neurochemical asymmetry in the corpus striatum. In *Duality and Unity of the Brain,* ed. D. Ottoson, 3–16. London: Macmillan.

Glick, S. D., D. A. Ross, and L. B. Hough. 1982. Lateral asymmetry of neurotransmitters in human brain. *Brain Research, 234,* 53–63.

Glick, S. D., and R. M. Shapiro. 1985. Functional and neurochemical mechanisms of cerebral lateralization in rats. In *Cerebral Lateralization in Nonhuman Species,* ed. S. D. Glick, 158–184. New York: Academic Press.

Glick, S. D., L. M. Weaver, and R. C. Meibach. 1981. Amphetamine-induced rotation in normal cats. *Brain Research, 208,* 227–229.

Goldman-Rakic, P. S. 1988. Topography of cognition: Parallel distributed networks in primate association cortex. *Annual Review of Neuroscience, 11,* 137–156.

Goldstein, G., and C. Shelly. 1981. Does the right hemisphere age more rapidly than the left? *Journal of Clinical Neuropsychology, 3,* 65–78.

Goldstein, S. G., and L. S. Braun. 1974. Reversal of expected transfer as a function of increased age. *Perceptual and Motor Skills, 38,* 1139–1145.

Goodall, J. 1970. Tool use in primates and other vertebrates. In *Advances in the Study of Behavior, vol 1,* ed. D. S. Lehrman, R. A. Hinde, and E. Shaw, 195–249. New York: Academic Press.

Gordon, H. W. 1980. Cognitive asymmetry in dyslexic families. *Neuropsychologia, 18,* 645–656.

——— 1986. The cognitive laterality battery: Tests of specialized cognitive function. *International Journal of Neuroscience, 29,* 223–244.

Gordon, H. W., and S. Kravetz. 1991. The influence of gender, handedness, and performance level on specialized cognitive functioning. *Brain and Cognition, 15,* 37–61.

Gospe, S. M., B. J. Mora, and S. D. Glick. 1990. Measurement of spontaneous rotational movement (circling) in normal children. *Journal of Child Neurology, 5,* 31–34.

Green, J. 1984. Effects of intrahemispheric interference on reaction times to lateral stimuli. *Journal of Experimental Psychology: Human Perception and Performance, 10,* 292–306.

Guiard, Y. 1987. Asymmetric division of labor in human skilled bimanual action: The kinematic chain as a model. *Journal of Motor Behavior, 19,* 486–517.

Gunturkun, O. 1985. Lateralization of visually controlled behavious in pigeons. *Physiology and Behavior, 34,* 575–577.

Haaland, K. Y., and D. L. Harrington. 1989a. Hemispheric control of the initial and corrective components of aiming movements. *Neuropsychologia, 27,* 961–969.

——— 1989b. The role of the hemispheres in closed loop movements. *Brain and Cognition, 9,* 158–180.

Haaland, K. Y., D. L. Harrington, and R. Yeo. 1987. The effects of task complexity on motor performance in left and right CVA patients. *Neuropsychologia, 25,* 783–794.

Habib, M., D. Gayraud, A. Oliva, J. Regis, G. Salamon, and R. Khalal. 1991. Effects of handedness and sex on the morphology of the corpus callosum: A study with brain magnetic resonance imaging. *Brain and Cognition, 16,* 41–61.

Hahn, W. K. 1987. Cerebral lateralization of function: From infancy through childhood. *Psychological Bulletin, 101,* 376–392.

Halpern, D. F. 1986. *Sex Differences in Cognitive Abilities.* Hillsdale, NJ: Erlbaum.

Hamilton, C. R., and B. A. Vermeire. 1982. Hemispheric differences in split-brain monkeys learning sequential comparisons. *Neuropsychologia, 20,* 691–698.

——— 1988. Complementary hemispheric specialization in monkeys. *Science, 242,* 1691–1694.

——— 1991. Functional lateralization in monkeys. In *Cerebral Laterality: Theory and Research,* ed. F. L. Kitterle, 19–34. Hillsdale, NJ: Erlbaum.

Hammond, G. R. 1990. Manual performance asymmetries. In *Cerebral Control of Speech and Limb Movements,* ed. G. R. Hammond, 59–78. Amsterdam: North-Holland.

Hampson, E., and D. Kimura. 1988. Reciprocal effects of hormonal fluctuations on human motor and perceptual-spatial skills. *Behavioral Neuroscience, 102,* 456–459.

Hardyck, C. 1991. Shadow and substance: Attentional irrelevancies and perceptual constraints in hemispheric processing of language stimuli. In *Cerebral Laterality: Theory and Research,* ed. F. L. Kitterle, 133–154. Hillsdale, NJ: Erlbaum.

Hardyck, C., C. Chiarello, N. F. Dronkers, and G. V. Simpson. 1985. Orienting attention within visual fields: How efficient is interhemispheric transfer? *Journal of Experimental Psychology: Human Perception and Performance, 11,* 650–666.

Hardyck, C., and L. F. Petrinovich. 1977. Left-handedness. *Psychological Bulletin, 84,* 385–404.

Harrington, D. L., and K. Y. Haaland. 1991. Hemispheric specialization for motor sequencing: Abnormalities in levels of programming. *Neuropsychologia, 29,* 147–163.

Harshman, R. A., and E. Hampson. 1987. Normal variation in human brain organization: Relation to handedness, sex and cognitive abilities. In *Duality and Unity of the Brain: Unified Functioning and Specialisation of the Hemispheres,* ed. D. Ottoson, 83–99. London: Macmillan.

Hecaen, H., M. DeAgostini, and A. Monzon-Montes. 1981. Cerebral organization in left-handers. *Brain and Language, 12,* 261–284.

Heffner, H. E., and R. S. Heffner. 1984. Temporal lobe lesions and perception of species-specific vocalizations by macaques. *Science, 226,* 75–76.

——— 1986. Effect of unilateral and bilateral auditory cortex lesions on the discrimination of vocalizations by Japanese macaques. *Journal of Neurophysiology, 56,* 683–701.

Heilman, K. M., R. T. Watson, and E. Valenstein. 1985. Neglect and related disorders. In *Clinical Neuropsychology,* ed. K. M. Heilman and E. Valenstein. Oxford: Oxford University Press.

Hellige, J. B. 1976. Changes in same-different laterality patterns as a function of practice and stimulus quality. *Perception and Psychophysics, 20,* 267–273.

——— 1980. Effects of perceptual quality and visual field of probe stimulus presentation on memory search for letters. *Journal of Experimental Psychology: Human Perception and Performance, 6,* 639–651.

——— 1983. *Cerebral Hemisphere Asymmetry: Method, Theory and Application.* New York: Praeger.

——— 1987. Interhemispheric interaction: Models, paradigms and recent findings. In *Duality and Unity of the Brain,* ed. D. Ottoson, 454–465. London: Macmillan.

——— 1989. Endogenous and experiential determinants of cerebral laterality: What develops? *European Bulletin of Cognitive Psychology (CPC), 9,* 85–89.

——— 1990. Hemispheric asymmetry. *Annual Review of Psychology, 41,* 55–80.

——— 1991. Cerebral laterality and metacontrol. In *Cerebral Laterality: Theory and Research,* ed. F. L. Kitterle, 117–132. Hillsdale, NJ: Erlbaum.

Hellige, J. B., M. I. Bloch, and T. L. Eng. 1992. Individual differences in hemispheric asymmetry and interhemispheric interaction. Manuscript in preparation.

Hellige, J. B., M. I. Bloch, and A. K. Taylor. 1988. Multitask investigation of individual differences in hemispheric asymmetry. *Journal of Experimental Psychology: Human Perception and Performance, 14,* 176–187.

Hellige, J. B., E. L. Cowin, and T. L. Eng. In press. Recognition of CVC syllables from LVF, RVF and central locations: Hemispheric differences and interhemispheric interaction. *Journal of Cognitive Neuroscience.*

Hellige, J. B., and P. J. Cox. 1976. Effects of concurrent verbal memory on recognition of stimuli from the left and right visual fields. *Journal of Experimental Psychology: Human Perception and Performance, 2,* 210–221.

Hellige, J. B., P. J. Cox, and L. Litvac. 1979. Information processing in the cerebral hemispheres: Selective hemispheric activation and capacity limitations. *Journal of Experimental Psychology: General, 108,* 251–279.

Hellige, J. B., J. E. Jonsson, and W. H. Corwin. 1984. Effects of perceptual quality on the processing of human faces presented to the left and right cerebral hemispheres. *Journal of Experimental Psychology: Human Perception and Performance, 10,* 90–107.

Hellige, J. B., J. E. Jonsson, and C. Michimata. 1988. Processing from LVF, RVF and BILATERAL presentations: Metacontrol and interhemispheric interaction. *Brain and Cognition, 7,* 39–53.

Hellige, J. B., and D. W. Kee. 1990. Asymmetric manual interference as an indicator of lateralized brain function. In *Cerebral Control of Speech and Limb Movements,* ed. G. R. Hammond, 635–660. Amsterdam: North-Holland.

Hellige, J. B., and L. E. Longstreth. 1981. Effects of concurrent hemisphere-specific activity on unimanual tapping rate. *Neurospychologia, 19,* 395–405.

Hellige, J. B., and C. Michimata. 1989a. Categorization versus distance: Hemispheric differences for processing spatial information. *Memory and Cognition, 17,* 770–776.

——— 1989b. Visual laterality for letter comparison: Effects of stimulus factors, response factors and metacontrol. *Bulletin of the Psychonomic Society, 27,* 441–444.

Hellige, J. B., and J. Sergent. 1986. Role of task factors in visual field asymmetries. *Brain and Cognition, 5,* 200–222.

Hellige, J. B., A. K. Taylor, and T. L. Eng. 1989. Interhemispheric interaction when both hemispheres have access to the same stimulus information. *Journal of Experimental Psychology: Human Perception and Performance, 15,* 711–722.

Hellige, J. B., and R. Webster. 1979. Right hemisphere superiority for initial stages of letter processing. *Neuropsychologia, 17,* 653–660.

Hellige, J. B., and T. M. Wong. 1983. Hemisphere-specific interference in dichotic listening: Task variables and individual differences. *Journal of Experimental Psychology: General, 112,* 218–239.

Hepper, P. G., S. Shahidullah, and R. White. 1991. Handedness in the human fetus. *Neuropsychologia, 29,* 1107–1112.

Herdman, C. M., and A. Friedman. 1985. Multiple resources in divided attention: A cross-modal test of the independence of hemispheric resources. *Journal of Experimental Psychology: Human Perception and Performance, 11,* 40–49.

Hines, T. 1987. Left brain / right brain mythology and implications for management and training. *Academy of Management Review, 12,* 600–606.

Hinton, G. E., and T. Shallice. 1991. Lesioning an attractor network: Investigations of acquired dyslexia. *Psychological Review, 98,* 74–95.

Hiscock, M., C. K. Hiscock, and R. Inch. 1991. Is there a sex difference in visual laterality? Paper presented at the Annual Meeting of the International Neuropsychological Society, San Antonio, TX.

——— 1992. Is there a sex difference in tactile laterality? Paper presented at the Annual Meeting of the International Neuropsychological Society, San Diego, CA.

Hiscock, M., C. K. Hiscock, and K. M. Kalil. 1990. Is there a sex difference in auditory laterality? Paper presented at the Annual Meeting of the International Neuropsychological Society, Kissimmee, FL.

Hiscock, M., and Kinsbourne, M. 1978. Ontogeny of cerebral dominance: Evidence from time-sharing asymmetry in children. *Developmental Psychology, 14,* 321–329.

——— 1980. Asymmetry of verbal-manual time sharing in children: A follow-up study. *Neuropsychologia, 18,* 151–162.

Hochberg, F. M., and M. LeMay. 1975. Arteriographic correlates of handedness. *Neurology, 25,* 218–222.

Holloway, R. L., and M. C. de la Coste-Lareymondie. 1982. Brain endocast asymmetry in pongids and hominids: Some preliminary findings on the paleontology of cerebral dominance. *American Journal of Physical Anthropology, 58,* 101–110.

Holtzman, J. D., and M. S. Gazzaniga. 1982. Dual task interactions due exclusively to limits in processing resources. *Science, 218,* 1325–1327.

Hopkins, W. D., and R. D. Morris. 1989. Laterality for visual-spatial processing in two language-trained chimpanzees (*Pan troglodytes*). *Behavioral Neuroscience, 103,* 227–234.

Hopkins, W. D., R. D. Morris, and E. S. Savage-Rumbaugh. 1991. Evidence for asymmetrical hemispheric priming using known and unknown warning stimuli in two language-trained chimpanzees (*Pan troglodytes*). *Journal of Experimental Psychology: General, 120,* 46–56.

Hopkins, W. D., D. A. Washburn, and D. M. Rumbaugh. 1990. Processing of form stimuli presented unilaterally in humans, chimpanzees (*Pan troglodytes*) and monkeys (*Macaca mulatta*). *Behavioral Neuroscience, 104,* 577–582.

Horn, J. L. 1988. Thinking about human abilities. In *Handbook of Multivariate Psychology,* ed. J. R. Nesselroade. New York: Academic Press.

Horn, J. L., and R. B. Cattell. 1972. Age differences in primary mental ability factors. *Journal of Gerontology, 176,* 539–541.

Hough, M. S. 1990. Narrative comprehension in adults with right and left hemisphere brain-damage: Theme organization. *Brain and Language, 38,* 253–277.

Hugdahl, K. 1988. *Handbook of Dichotic Listening: Theory, Methods and Research.* New York: Wiley.

Hugdahl, K., B. Synnevag, and P. Satz. 1990. Immune and autoimmune diseases in dyslexic children. *Neuropsychologia, 28,* 673–679.

Hynd, G. W., M. Semrud-Clikeman, A. R. Lorys, E. S. Novey, and D. Eliopulos. 1990. Brain morphology in developmental dyslexia and attention deficit disorder/hyperactivity. *Archives of Neurology, 47,* 919–926.

Ivry, R. B., and P. Lebby. 1993. Hemispheric differences in auditory perception are similar to those found in visual perception. *Psychological Science, 4.*

Jason, G. W., A. Cowey, and L. Weiskrantz. 1984. Hemispheric asymmetry for a visuospatial task in monkeys. *Neuropsychologia, 22,* 777–784.

Jaynes, J. 1977. *The Origin of Consciousness in the Breakdown of the Bicameral Mind.* Boston: Houghton Mifflin.

Jerison, H. J. 1982. The evolution of biological intelligence. In *Handbook of Intelligence,* ed. R. Sternberg, 723–791. Cambridge: Cambridge University Press.

Jonsson, J. E., and J. B. Hellige. 1986. Lateralized effects of blurring: A test of the visual spatial frequency model of cerebral hemisphere asymmetry. *Neuropsychologia, 24,* 351–362.

Juraska, J. M., and J. R. Kopcik. 1988. Sex and environmental influences on the size and ultrastructure of the rat corpus callosum. *Brain Research, 450,* 1–8.

Kaplan, J. A., H. H. Brownell, J. R. Jacobs, and H. Gardner. 1990. The effects of right hemisphere damage on the pragmatic interpretation of conversational remarks. *Brain and Language, 38,* 315–333.

Kee, D. W., J. B. Hellige, and K. Bathurst. 1983. Lateralized interference of repetitive finger tapping: Influence of family handedness, cognitive load, and verbal production. *Neuropsychologia, 21,* 617–625.

Kim, H., and S. C. Levine. 1991a. Sources of between-subjects variability in perceptual asymmetries: A meta-analytic review. *Neuropsychologia, 29,* 877–888.

——— 1991b. Inferring patterns of hemispheric specialization for individual subjects from laterality data: A two-task criterion. *Neuropsychologia, 29,* 93–106.

Kim, H., S. C. Levine, and S. Kertesz. 1990. Are variations among subjects in lateral asymmetry real individual differences or random error in measurement?: Putting variability in its place. *Brain and Cognition, 14,* 220–242.

Kimura, D. 1973. Manual activity during speaking. *Neuropsychologia, 11,* 45–50.

——— 1977. Acquisition of a motor skill after left-hemisphere damage. *Brain, 100,* 527–542.

——— 1987. Are men's and women's brains really different? *Canadian Psychologist, 28,* 133–147.

Kimura, D., and Y. Archibald. 1974. Motor functions of the left hemisphere. *Brain, 97,* 337–350.

Kinsbourne, M. 1970. The cerebral basis of lateral asymmetries in attention. *Acta Psychologica, 33,* 193–201.

——— 1972. Eye and head turning indicates cerebral lateralization. *Science, 176,* 539–541.

——— 1975. The mechanism of hemispheric control of the lateral gradient of attention. In *Attention and Performance,* vol. 5, ed. P. M. A. Rabbitt and S. Dornic. New York: Academic Press.

——— 1982. Hemispheric specialization and the growth of human understanding. *American Psychologist, 37,* 411–420.

Kinsbourne, M., and J. Cook. 1971. Generalized and lateralized effects of concurrent verbalization on a unimanual skill. *Quarterly Journal of Experimental Psychology, 23,* 341–345.

Kinsbourne, M., and M. Hiscock. 1983. Asymmetries of dual-task performance. In Hellige, 1983, 255–334.

Kitterle, F. L., and S. Christman. 1991. Hemispheric symmetries and asymmetries in the processing of spatial frequencies. In *Cerebral Laterality: Theory and Research,* ed. F. L. Kitterle, 201–224. Hillsdale, NJ: Erlbaum.

Kitterle, F. L., S. Christman, and J. B. Hellige. 1990. Hemispheric differences are found in the identification, but not the detection, of low vs. high spatial frequencies. *Perception and Psychophysics, 48,* 297–306.

Kitterle, F. L., J. B. Hellige, and S. Christman. 1992. Visual hemispheric asymmetries depend on which spatial frequencies are task relevant. *Brain and Cognition, 20.*

Kitterle, F. L., and L. M. Selig. 1991. Visual field effects in the discrimination of sine-wave gratings. *Perception and Psychophysics, 50,* 15–18.

Klisz, D. 1978. Neuropsychological evaluation in older persons. In *The Clinical Psychology of Aging,* ed. M. Storandt, I. C. Siegler, and M. F. Elias, 71–95. New York: Plenum Press.

Koenig, O., L. P. Reiss, and S. M. Kosslyn. 1990. The development of spatial relation representations: Evidence from studies of cerebral lateralization. *Journal of Experimental Child Psychology, 50,* 119–130.

Kolb, B., and B. Milner. 1981. Performance of complex arm and facial movements after focal brain lesions. *Neuropsychologia, 19,* 491–504.

Kolb, B., R. J. Sutherland, A. J. Nonneman, and I. Q. Whishaw. 1982. Asymmetry in the cerebral hemispheres of the rat, mouse, rabbit and cat: The right hemisphere is larger. *Experimental Neurology, 78,* 348–359.

Kosslyn, S. M. 1980. *Image and Mind.* Cambridge, MA: Harvard University Press.

——— 1987. Seeing and imagining in the cerebral hemispheres: A computational approach. *Psychological Review, 94,* 148–175.

——— 1988. Aspects of a cognitive neuroscience of mental imagery. *Science, 240,* 1621–1626.

Kosslyn, S. M., C. F. Chabris, C. J. Marsolek, and O. Koenig. 1992. Categorical versus coordinate spatial relations: Computational analyses and computer simulations. *Journal of Experimental Psychology: Human Perception and Performance, 18,* 562–577.

Kosslyn, S. M., R. A. Flynn, J. B. Amsterdam, and G. Wang. 1990. Components of high-level vision: A cognitive neuroscience analysis and accounts of neurological syndromes. *Cognition, 34,* 203–277.

Kosslyn, S. M., J. Holtzman, M. J. Farah, and M. S. Gazzaniga. 1985. A computational analysis of mental image generation: Evidence from functional dissociations in split-brain patients. *Journal of Experimental Psychology: General, 114,* 311–341.

Kosslyn, S. M., O. Koenig, A. Barrett, C. B. Cave, J. Tang, and J. D. E. Gabrieli. 1989. Evidence for two types of spatial representations: Hemispheric specialization for categorical and coordinate relations. *Journal of Experimental Psychology: Human Perception and Performance, 15,* 723–735.

Lamb, M. R., L. C. Robertson, and R. T. Knight. 1989. Attention and interference in the processing of hierarchical patterns: Inferences from patients with right and left temporal-parietal lesions. *Neuropsychologia, 27,* 471–483.

——— 1990. Component mechanisms underlying the processing of hierarchically organized patterns: Inferences from patients with unilateral cortical lesions. *Journal of Experimental Psychology: Learning, Memory, and Cognition, 16,* 471–483.

Lambert, A. J. 1991. Interhemispheric interaction in the split-brain. *Neuropsychologia, 29,* 941–948.

Landwehrmeyer, B., J. Gerling, and C. W. Wallesch. 1990. Patterns of task-related slow brain potentials in dyslexia. *Archives of Neurology, 47,* 791–797.

Larsen, J. P., T. Hoien, I. Lundberg, and H. Odegaard. 1990. MRI evaluation of the size and symmetry of the planum temporale in adolescents with developmental dyslexia. *Brain and Language, 39,* 289–301.

Lavergne, J., and D. Kimura. 1987. Hand movement asymmetry during speech: No effect of speaking topic. *Neuropsychologia, 25,* 689–694.

Lecours, A. R. 1975. Myelogenetic correlates of the development of speech and language. In *Foundations of Language Development: A Multidisciplinary Approach,* vol. 1, ed. E. H. Lenneberg and C. Lenneberg. New York: Academic Press.

LeMay, M. 1985. Asymmetries of the brains and skulls of nonhuman primates. In *Cerebral Lateralization in Nonhuman Species,* ed. S. D. Glick, 234–246. New York: Academic Press.

LeMay, M., and A. Culebras. 1972. Human brain: Morphologic differences in the hemispheres demonstrable by carotid arteriography. *New England Journal of Medicine, 287,* 168–170.

Lenneberg, E. H. 1967. *Biological Foundations of Language.* New York: Wiley.

Levine, S. C. 1985. Developmental changes in right-hemisphere involvement in face recognition. In *Hemispheric Function and Collaboration in the Child,* ed. C. T. Best, 157–192. New York: Academic Press.

——— 1989. The question of faces: Special in the brain of the beholder.

In *Handbook of Research in Face Processing,* ed. A. W. Young and H. D. Ellis, 37–40. Hillsdale, NJ: Erlbaum.

Levine, S. C., M. T. Banich, and H. Kim. 1987. Variations in arousal asymmetry: Implications for face processing. In *Duality and Unity of the Brain: Unified Functioning and Specialisation of the Hemispheres,* ed. D. Ottoson, 207–222. London: Macmillan.

Levine, S. C., and J. Levy. 1986. Perceptual asymmetry for chimeric faces across the life span. *Brain and Cognition, 5,* 291–306.

Levy, J. 1969. Possible basis for the evolution of lateral specialization of the human brain. *Nature, 224,* 614–615.

——— 1982. Handwriting posture and cerbral organization: How are they related? *Psychological Bulletin, 91,* 589–608.

——— 1983. Language, cognition, and the right hemisphere: A response to Gazzaniga. *American Psychologist, 38,* 538–541.

Levy, J., W. Heller, M. T. Banich, and L. A. Burton. 1983a. Are variations among right-handed individuals in perceptual asymmetries caused by characteristic arousal differences between hemispheres? *Journal of Experimental Psychology: Human Perception and Performance, 9,* 329–359.

——— 1983b. Asymmetry of perception in free viewing of faces. *Brain and Cognition, 2,* 404–419.

Levy, J., and L. Kueck. 1986. A right hemispatial advantage on a verbal free-vision task. *Brain and Language, 27,* 24–37.

Levy, J., and M. Reid. 1978. Variations in cerebral organization as a function of handedness, hand posture in writing, and sex. *Journal of Experimental Psychology: General, 107,* 119–144.

Levy, J., and C. Trevarthen. 1976. Metacontrol of hemispheric function in human split-brain patients. *Journal of Experimental Psychology: Human Perception and Performance, 2,* 299–312.

Levy, J., C. Trevarthen, and R. W. Sperry. 1972. Perception of bilateral chimeric figures following hemisphere deconnexion. *Brain, 95,* 61–78.

Ley, R. G., and M. P. Bryden. 1982. A dissociation of right and left hemispheric effects for recognizing emotional tone and verbal content. *Brain and Cognition, 1,* 3–9.

Liederman, J. 1986. Interhemispheric interference during word naming. *International Journal of Neuroscience, 30,* 43–56.

Liederman, J., J. Merola, and C. Hoffman. 1986. Longitudinal data indicate that hemispheric independence increases during early adolescence. *Developmental Neuropsychology, 2,* 183–201.

Liederman, J., J. Merola, and S. Martinez. 1985. Interhemispheric collaboration in response to simultaneous bilateral input. *Neuropsychologia, 23,* 673–683.

Liotti, M., and D. M. Tucker. 1992. Right hemisphere sensitivity to arousal and depression. *Brain and Cognition, 18,* 138–151.

Lovejoy, C. V. 1981. The origin of man. *Science, 211,* 341–350.

Luh, K. E., L. M. Rueckert, and J. Levy. 1991. Perceptual asymmetries for free viewing of several types of chimeric stimuli. *Brain and Cognition, 16,* 83–103.

MacNeilage, P. F., M. G. Studdert-Kennedy, and B. Lindblom. 1987. Primate handedness reconsidered. *Behavioral and Brain Sciences, 10,* 247–303.

Marr, D. 1982. *Vision.* San Francisco: Freeman.

Martin, M. 1979. Hemispheric specialization for local and global processing. *Neuropsychologia, 17,* 33–40.

Mateer, C., and D. Kimura. 1977. Impairment of nonverbal oral movements in aphasia. *Brain and Language, 4,* 262–276.

McGlone, J. 1980. Sex differences in human brain organization: A critical survey. *Behavioral and Brain Sciences, 3,* 215–227.

——— 1986. The neuropsychology of sex differences in human brain organization. In *Advances in Clinical Neuropsychology,* vol. 3, ed. G. Goldstein and R. E. Tarter, 1–30. New York: Plenum Press.

McKeever, W. F. 1991. Handedness, language laterality, and spatial ability. In *Cerebral Laterality: Theory and Research,* ed. F. L. Kitterle, 53–70. Hillsdale, NJ: Erlbaum.

McRoberts, G. W., and B. Sanders. 1992. Sex differences in performance and hemispheric organization in a nonverbal auditory task. *Perception and Psychophysics, 51,* 118–122.

Mehta, Z., F. Newcombe, and H. Damasio. 1987. A left hemisphere contribution to visuospatial processing. *Cortex, 23,* 447–461.

Michimata, C. 1988. *Lateralized Effects of Semantic Priming in Lexical Decision Tasks: Examinations of the Time Course of Semantic Activation Build-up in the Left and Right Cerebral Hemispheres.* Unpublished doctoral dissertation, University of Southern California, Los Angeles, CA.

Michimata, C., and J. B. Hellige. 1987. Effects of blurring and stimulus size on the lateralized processing of nonverbal stimuli. *Neuropsychologia, 25,* 397–407.

Milberg, W., and S. E. Blumstein. 1981. Lexical decision and aphasia: Evidence for semantic processing. *Brain and Language, 14,* 371–385.

Milner, B. 1968. Visual recognition and recall after temporal lobe excision in man. *Neuropsychologia, 6,* 191–209.

Mittenberg, W., M. Seidenberg, D. S. O'Leary, and D. V. DiGiulio. 1989. Changes in cerebral functioning associated with normal aging. *Journal of Clinical and Experimental Neuropsychology, 11,* 918–932.

Molfese, D. L., and L. M. Burger-Judisch. 1991. Dynamic temporal-spatial allocation of resources in the human brain: An alternative to the static view of hemisphere differences. In *Cerebral Laterality: Theory and Research,* ed. F. L. Kitterle, 71–102. Hillsdale, NJ: Erlbaum.

Molfese, D. L., and V. J. Molfese. 1979. Hemisphere and stimulus differences as reflected in the cortical responses of newborn infants to speech stimuli. *Developmental Psychology, 15,* 505–511.

——— 1980. Cortical responses of preterm infants to phonetic and nonphonetic speech stimuli. *Developmental Psychology, 16,* 574–581.

Molfese, V. J., D. L. Molfese, and C. Parsons. 1983. Hemispheric processing of phonological information. In *Language Functions and Brain Organization,* ed. S. J. Segalowitz, 29–49. New York: Academic Press.

Morgan, E. 1984. The aquatic hypothesis. *New Scientist,* April 12, 11–13.

Morrel-Samuels, P., L. Herman, and T. Bever. 1989. A left hemisphere advantage for gesture-language signs in the dolphin. Paper presented at the meeting of the Psychonomic Society, Atlanta, GA.

Morrel-Samuels, P., L. Herman, and A. Pack. 1990. Cerebral asymmetry during picture recognition: Preliminary evidence from dolphins. Paper presented to the Psychonomic Society, New Orleans, LA.

Morse, P., D. L. Molfese, N. K. Laughlin, S. L. Linnville, and F. Wetzel. 1987. Categorical perception for voicing contrasts in normal and lead-treated rhesus macaques: Electrophysiological indices. *Brain and Language, 30,* 63–88.

Morton, L. L., and L. S. Siegel. 1991. Left ear dichotic listening performance on consonant-vowel combinations and digits in subtypes of reading-disabled children. *Brain and Language, 40,* 162–180.

Moscovitch, M. 1987. Lateralization of language in children with developmental dyslexia: A critical review of visual half-field studies. In *Duality and Unity of the Brain,* ed. D. Ottoson, 324–346. London: Macmillan.

Moscovitch, M., and D. Klein. 1980. Material-specific perceptual interference for visual words and faces. *Journal of Experimental Psychology: Human Perception and Performance, 6,* 590–604.

Moscovitch, M., and J. Olds. 1982. Asymmetries in spontaneous facial expression and their possible relation to hemispheric specialization. *Neuropsychologia, 20,* 71–82.

Myers, J. J., and R. W. Sperry. 1985. Interhemispheric communication after section of the forebrain commissures. *Cortex, 21,* 249–260.

Navon, D. 1984. Resources—A theoretical stone soup? *Psychological Review, 91,* 216–234.

Navon, D., and D. Gopher. 1979. On the economy of the human information processing system. *Psychological Review, 86,* 214–225.

Nebes, R. D. 1978. Direct examination of cognitive function in the right and left hemispheres. In *Asymmetrical Function of the Brain,* ed. M. Kinsbourne. London: Cambridge University Press.

Neely, J. H. 1977. Semantic priming over retrieval from lexical memory:

Roles of inhibitionless spreading activation and limited-capacity attention. *Journal of Experimental Psychology: General, 106,* 226–254.

Nestor, P. G., and M. A. Safer. 1990. A multi-method investigation of individual differences in hemisphericity. *Cortex, 26,* 409–421.

Nottebohm, F. 1970. Ontogeny of bird song. *Science, 167,* 950–956.

——— 1979. Origins and mechanisms in the establishment of cerebral dominance. In *Handbook of Behavioral Neurobiology,* vol. 2, *Neuropsychology,* ed. M. S. Gazzaniga, 295–344. New York: Plenum.

Obler, L. K., S. Woodward, and M. L. Albert. 1984. Changes in cerebral lateralization with aging? *Neuropsychologia, 22,* 235–240.

O'Boyle, M. W. 1986. Hemispheric laterality as a basis of learning: What we know and don't know. In *Cognitive Classroom Learning: Understanding, Thinking and Problem Solving,* ed. T. Andre and G. Phye, 21–48. New York: Academic Press.

O'Boyle, M. W., J. E. Alexander, and C. P. Benbow. 1991. Enhanced right hemisphere activation in the mathematically precocious: A preliminary EEG investigation. *Brain and Cognition, 17,* 138–153.

O'Boyle, M. W., and C. P. Benbow. 1990a. Handedness and its relationship to ability and talent. In *Left-handedness: Behavioral Implications and Anomalies,* ed. S. Coren, 343–372. Amsterdam: Elsevier.

——— 1990b. Enhanced right hemisphere involvement during cognitive processing may relate to intellectual precocity. *Neuropsychologia, 28,* 211–216.

O'Boyle, M. W., and J. B. Hellige. 1989. Cerebral hemisphere asymmetry and individual differences in cognition. *Learning and Individual Differences, 1,* 7–35.

O'Boyle, M. W., F. van Wyhe-Lawler, and D. A. Miller. 1987. Recognition of letters traced in the right and left palms: Evidence for a process-oriented tactile asymmetry. *Brain and Cognition, 6,* 474–494.

Ojemann, G. A. 1983. Brain organization for language from the perspective of electrical stimulation mapping. *Behavioral and Brain Sciences, 6,* 189–230.

Ojemann, G., and C. Mateer. 1979. Human language cortex: Localization of memory, syntax, and sequential motor-phoneme identification systems. *Science, 205,* 1401–1403.

Oke, A., R. Keller, I. Mefford, and R. Adams. 1978. Lateralization of norepinephrine in human thalamus. *Science, 200,* 1411–1413.

Orsini, D. L., and P. Satz. 1986. A syndrome of pathological left-handedness: Correlates of early left hemisphere injury. *Archives of Neurology, 43,* 333–337.

Orton, S. T. 1937. *Reading, Writing and Speech Problems in Children.* New York: Norton.

Palmer, S. E. 1977. Hierarchical structure in perceptual representation. *Cognitive Psychology, 9,* 441–474.

Pardo, J. V., P. T. Fox, and M. E. Raichle. 1991. Localization of a human system for sustained attention by positron emission tomography. *Nature, 349,* 61–64.

Patterson, K., and D. Besner. 1984. Is the right hemisphere literate? *Cognitive Neuropsychology, 1,* 315–341.

Penfield, W. 1958. *The Excitable Cortex in Conscious Man.* Springfield, IL: Charles C. Thomas.

Peterson, M. R., M. D. Beecher, S. R. Zoloth, D. B. Moody, and W. C. Stebbins. 1978. Neural lateralization of species-specific vocalizations by Japanese macaques. *Science, 202,* 324–326.

Peterzell, D. H. 1991. On the nonrelation between spatial frequency and cerebral hemispheric competence. *Brain and Cognition, 15,* 62–68.

Peterzell, D. H., L. O. Harvey, and C. Hardyck. 1989. Spatial frequencies and the cerebral hemispheres: Contrast sensitivity, visible persistence, and letter classification. *Perception and Psychophysics, 46,* 443–455.

Pfeiffer, J. E. 1985. *The Emergence of Humankind.* New York: Harper and Row.

Poizner, H., E. Klima, and U. Bellugi. 1987. *What the Hands Reveal about the Brain.* Cambridge, MA: MIT Press.

Posner, M. I. 1992. Attention as a cognitive and neural system. *Current Directions in Psychological Science, 1,* 11–14.

Posner, M. I., T. S. Early, E. Reiman, P. J. Pardo, and M. Dhawan. 1988. Asymmetries in hemispheric control of attention in schizophrenia. *Archives of General Psychiatry, 45,* 814–821.

Posner, M. I., S. E. Petersen, P. T. Fox, and M. E. Raichle. 1988. Localization of cognitive operations in the human brain. *Science, 240,* 1627–1631.

Posner, M. I., and C. R. R. Snyder. 1975. Attention and cognitive control. In *Information Processing and Cognition: The Loyola Symposium,* ed. R. L. Solso, 55–86. New York: Halsted Press.

Posner, M. I., J. A. Walker, F. A. Friedrich, and R. D. Rafal. 1984. Effects of parietal injury on covert orienting of attention. *Journal of Neuroscience, 4,* 1863–1874.

——— 1987. How do the parietal lobes direct covert attention? *Neuropsychologia, 25,* 135–146.

Potter, S. M., and R. E. Graves. 1988. Is interhemispheric transfer related to handedness and gender? *Neuropsychologia, 26,* 319–326.

Previc, F. H. 1991. A general theory concerning the prenatal origins of cerebral lateralization in humans. *Psychological Review, 98,* 299–334.

Pribram, K. H., and D. McGuinness. 1975. Arousal, activation, and effort in the control of attention. *Psychological Review, 82,* 116–149.

Raine, A., G. Harrison, G. Reynolds, C. Sheard, J. E. Cooper, and I. Medley. 1990. Structural and functional characteristics of the corpus callosum in schizophrenics, psychiatric controls, and normal controls: An MRI and neuropsychological evaluation. *Archives of General Psychiatry, 47,* 1061–1064.

Raine, A., M. O'Brien, N. Smiley, A. Scerbo, and C. J. Chan. 1990. Reduced lateralization in verbal dichotic listening in adolescent psychopaths. *Journal of Abnormal Psychology, 99,* 272–277.

Rasmussen, T., and B. Milner. 1977. The role of early left-brain injury in determining lateralization of cerebral speech functions. *Annals of the New York Academy of Sciences, 299,* 355–369.

Rebai, M., L. Mecacci, J. Bagot, and C. Bonnet. 1986. Hemispheric asymmetries in the visual evoked potentials to temporal frequency: Preliminary evidence. *Perception, 15,* 589–594.

——— 1989. Influence of spatial frequency and handedness on hemispheric asymmetry in visually steady-state evoked potentials. *Neuropsychologia, 27,* 315–324.

Reuter-Lorenz, P. A., and K. Baynes. 1992. Modes of lexical access in the callosotomized brain. *Journal of Cognitive Neuroscience, 4,* 155–164.

Reuter-Lorenz, P. A., and M. I. Posner. 1990. Components of neglect from right-hemisphere damage: An analysis of line bisection. *Neuropsychologia, 28,* 327–333.

Risberg, J., J. H. Halsey, E. L. Wills, and B. M. Wilson. 1975. Hemispheric specialization in normal man studied by bilateral measurements of the regional cerebral blood flow. *Brain, 98,* 511–524.

Robertson, L. C. 1986. From gestalt to neo-gestalt. In *Approaches to Cognition: Contrasts and Controversies,* ed. T. J. Knapp and L. C. Robertson, 159–188. Hillsdale, NJ: Erlbaum.

Robertson, L. C., M. R. Lamb, and R. T. Knight. 1988. Effects of lesions of temporal-parietal junction on perceptual and attentional processing in humans. *Journal of Neuroscience, 8,* 3757–3769.

——— 1991. Normal global-local analysis in patients with dorsolateral frontal lobe lesions. *Neuropsychologia, 29,* 959–968.

Rogers, L. J. 1980. Lateralization in the avian brain. *Bird Behaviour, 2,* 1–12.

——— 1986. Lateralization of learning in chicks. *Advances in the Study of Behavior, 16,* 147–189.

Rosenbaum, D. A. 1990. *Human Motor Control.* New York: Academic Press.

Ross, E. D. 1985. Modulation of affect and nonverbal communication by the right hemisphere. In *Principles of Behavioral Neurology,* ed. M. Mesulam, 239–258. Philadelphia, PA: F. A. Davis.

Rossor, M., N. Garret, and L. Iversen. 1980. No evidence for lateral asymmetry of neurotransmitters in post-mortem human brain. *Journal of Neurochemistry, 35,* 743–745.

Rueckl, J. G., K. R. Cave, and S. M. Kosslyn. 1989. Why are "what" and "where" processed by separate cortical visual systems? A computational investigation. *Journal of Cognitive Neuroscience, 1,* 171–186.

Rybash, J. M., and W. J. Hoyer. 1992. Hemispheric specialization for categorical and coordinate spatial representations: A reappraisal. *Memory and Cognition, 20,* 271–276.

Schaie, K. W., and J. P. Schaie. 1977. Clinical assessment and aging. In *Handbook of the Psychology of Aging,* ed. J. E. Birren and K. W. Schaie, 692–723. New York: Van Nostrand Reinhold.

Scheibel, A. M. 1984. A dendritic correlate of human speech. In *Cerebral Dominance: The Biological Foundations,* ed. N. Geschwind and A. M. Galaburda, 43–52. Cambridge, MA: Harvard University Press.

Scheibel, A. M., I. Fried, L. Paul, A. Forsythe, U. Tomiyasu, A. Wechsler, A. Kao, and J. Slotnick. 1985. Differentiating characteristics of the human speech cortex: A quantitative Golgi study. In *The Dual Brain,* ed. D. F. Benson and E. Zaidel, 65–74. New York: Guilford Press.

Segalowitz, S. J., and M. P. Bryden. 1983. Individual differences in hemispheric representation of language. In *Language Functions and Brain Organization,* ed. S. J. Segalowitz, 341–372. New York: Academic Press.

Semmes, J. 1968. Hemispheric specialization: A possible clue to mechanism. *Neuropsychologia, 6,* 11–26.

Seidenberg, M. S., and J. L. McClelland. 1989. A distributed, developmental model of word recognition and naming. *Psychological Review, 96,* 523–568.

Sereno, A. B., and S. M. Kosslyn. 1991. Discrimination within and between hemifields: A new constraint on theories of attention. *Neuropsychologia, 29,* 659–676.

Sergent, J. 1982a. The cerebral balance of power: Confrontation or cooperation? *Journal of Experimental Psychology: Human Perception and Performance, 8,* 253–272.

——— 1982b. About face: Left-hemisphere involvement in processing of physiognomies. *Journal of Experimental Psychology: Human Perception and Performance, 8,* 253–272.

——— 1983a. The role of the input in visual hemispheric asymmetries. *Psychological Bulletin, 93,* 481–514.

——— 1983b. Unified response to bilateral hemispheric stimulation by a split-brain patient. *Nature, 305,* 800–802.

——— 1984. Configural processing of faces in the left and right cerebral hemispheres. *Journal of Experimental Psychology: Human Perception and Performance, 10,* 554–572.

——— 1985. Influence of task and input factors on hemispheric involvement in face processing. *Journal of Experimental Psychology: Human Perception and Performance, 11,* 846–861.

——— 1986. Subcortical coordination of hemisphere activity in commissurotomized patients. *Brain, 109,* 357–369.

——— 1987a. Information processing and laterality effects for object and face perception. In *Visual Object Processing: A Cognitive Neuropsychological Approach,* ed. G. W. Humphreys and M. J. Riddoch, 145–173. Hillsdale, NJ: Erlbaum.

——— 1987b. Failures to confirm the spatial-frequency hypothesis: Fatal blow or healthy complication? *Canadian Journal of Psychology, 41,* 412–428.

——— 1989. Image generation and processing of generated images in the cerebral hemispheres. *Journal of Experimental Psychology: Human Perception and Performance, 15,* 170–178.

——— 1990. Furtive incursions into bicameral minds: Integrative and coordinating role of subcortical structures. *Brain, 113,* 537–568.

——— 1991. Judgments of relative position and distance on representations of spatial relations. *Journal of Experimental Psychology: Human Perception and Performance, 17,* 762–780.

——— In press. Processing of spatial relations within and between the disconnected cerebral hemispheres. *Brain.*

Sergent, J. and D. Bindra. 1981. Differential hemispheric processing of faces: Methodological considerations and reinterpretation. *Psychological Bulletin, 89,* 541–554.

Sergent, J., and M. C. Corballis. 1990. Generation of multipart images in the disconnected cerebral hemispheres. *Bulletin of the Psychonomic Society, 28,* 309–311.

Sergent, J., and J. B. Hellige. 1986. Role of input factors in visual-field asymmetries. *Brain and Cognition, 5,* 174–199.

Sergent, J., S. Ohta, and B. MacDonald. 1992. Functional neuroanatomy of face and object processing: A positron emission tomography study. *Brain, 115,* 15–36.

Shapiro, R. M., N. A. Camarota, and S. D. Glick. 1987. Nocturnal rotational behavior in rats: Further neurochemical support for a two-population model. *Brain Research Bulletin, 19,* 421–427.

Sidtis, J. J., and M. S. Gazzaniga. 1983. Competence versus performance after callosal section: Looks can be deceiving. In Hellige, 1983, 152–176.

Simonds, R. J., and A. B. Scheibel. 1989. The postnatal development of the motor speech area: A preliminary study. *Brain and Language, 37,* 42–58.

Sperry, R. W., M. S. Gazzaniga, and J. E. Bogen. 1969. Interhemispheric relationships: The neocortical commissures, syndromes of hemispheric disconnection. In *Handbook of Clinical Neurology,* vol. 4, *Disorders of Speech, Perception and Symbolic Behavior,* ed. P. J. Vinken and G. W. Bruyn, 145–153. Amsterdam: Elsevier/North Holland Biomedical Press.

Springer, S. P., and G. Deutsch. 1989. *Left Brain, Right Brain.* New York: W. H. Freeman.

Sroufe, A. L. 1979. Socioemotional development. In *Handbook of Infant Development,* ed. J. D. Osofsky, 462–516. New York: Wiley.

Steenhuis, R. E., and M. P. Bryden. 1989. Different dimensions of hand preference that relate to skilled and unskilled activities. *Cortex, 25,* 289–304.

Stern, J. A., and A. C. Baldinger. 1983. Hemispheric differences in preferred modes of information processing and the aging process. *International Journal of Neuroscience, 18,* 97–106.

Sternberg, S. 1969. Memory scanning: Mental processes revealed by reaction-time experiments. *American Scientist, 57,* 421–457.

——— 1975. Memory scanning: New findings and current controversies. *Quarterly Journal of Experimental Psychology, 27,* 1–32.

Stewart, J., and B. Kolb. 1988. The effects of neonatal gonadectomy and prenatal stress on cortical thickness and asymmetry in rats. *Behavioral and Neural Biology, 49,* 144–160.

Stillings, N. A., M. H. Feinstein, J. Garfield, E. L. Risslandy, D. A. Rosenbaum, S. E. Weisler, and L. Baker-Ward. 1987. *Cognitive Science: An Introduction.* Cambridge: MIT Press.

Summers, J. J., and C. A. Sharp. 1979. Bilateral effects of concurrent verbal and spatial rehearsal on complex motor sequencing. *Neuropsychologia, 17,* 331–343.

Sussman, H. M. 1979. Evidence for left hemisphere superiority in processing movement-related tonal signals. *Journal of Speech and Hearing Research, 22,* 224–235.

Teuber, H. L. 1955. Physiological psychology. *Annual Review of Psychology, 6,* 267–296.

Thatcher, R. W., R. A. Walder, and S. Giudice. 1987. Human cerebral hemispheres develop at different rates and ages. *Science, 236,* 1110–1113.

Thomas, J. 1986. Seeing spatial patterns. In *Handbook of Perception and Human Performance,* vol. 1, ed. K. R. Boff, L. Kaufman, and J. P. Thomas. New York: Wiley.

Thompson, R. A. 1984. Language, the brain and the question of dichotomies. *American Anthropologist, 86,* 98–105.

Tobias, P. V. 1987. The brain of *Homo habilis:* A new level of organization in cerebral evolution. *Journal of Human Evolution, 16,* 741–761.

Toth, N. 1985. Archaeological evidence for preferential right handedness in the lower and middle Pleistocene and its possible implications. *Journal of Human Evolution, 14,* 607–614.

Trope, I., B. Fishman, R. C. Gur, N. M. Sussman, and R. E. Gur. 1987. Contralateral and ipsilateral control of fingers following callosotomy. *Neuropsychologia, 25,* 287–292.

Tucker, D. M. 1987. Hemisphere specialization: A mechanism for unifying anterior and posterior brain regions. In *Duality and Unity of the Brain,* ed. D. Ottoson, 180–193. London: Macmillan Press.

Tucker, D. M., and P. A. Williamson. 1984. Asymmetric neural control systems in human self-regulation. *Psychological Review, 91,* 185–215.

Turkewitz, G. 1988. A prenatal source for the development of hemispheric specialization. In *Brain Lateralization in Children: Developmental Implications,* ed. D. L. Molfese and S. J. Segalowitz, 73–81. New York: Guilford Press.

Ungerleider, L. G., and M. Mishkin. 1982. Two cortical visual systems. In *Analysis of visual behavior,* ed. D. J. Ingle, M. A. Goodale, and R. J. W. Mansfield. Cambridge, MA: MIT Press.

Van Kleeck, M. H. 1989. Hemispheric differences in global versus local processing of hierarchical visual stimuli by normal subjects: New data and a meta-analysis of previous data. *Neuropsychologia, 27,* 1165–1178.

Van Lancker, D., and J. J. Sidtis. In press. The identification of affective-prosodic stimuli by left and right hemisphere damaged subjects: All errors are not created equal. *Journal of Speech and Hearing Research.*

Vargha-Khadem, F., and M. C. Corballis. 1979. Cerebral asymmetry in infants. *Brain and Language, 8,* 1–9.

Verrillo, R. T. 1983. Vibrotactile subjective magnitude as a function of hand preference. *Neuropsychologia, 21,* 383–396.

von Fersen, L., and O. Gunturkun. 1990. Visual memory lateralization in pigeons. *Neuropsychologia, 28,* 1–7.

Wagner, H. N., D. H. Burns, R. F. Dannals, D. F. Wong, B. Langstrom,

T. Duelfer, J. J. Frost, H. T. Ravert, J. M. Links, S. B. Rosenbloom, S. E. Lucas, A. V. Kramer, and M. J. Kuhlar. 1983. Imaging dopamine receptors in the human brain by positron tomography. *Science, 221,* 1264–1266.

Walker, E., and M. McGuire. 1982. Intra- and interhemispheric information processing in schizophrenia. *Psychological Bulletin, 92,* 701–725.

Ward, J. P. 1991. Prosimians as animal models in the study of neural lateralization. In *Cerebral Laterality: Theory and Research,* ed. F. L. Kitterle, 1–18. Hillsdale, NJ: Erlbaum.

Warrington, E. K. 1985. Agnosia: The impairment of object recognition. In *Handbook of Clinical Neurology,* vol. 45, *Clinical Neuropsychology,* ed. P. J. Vinken, G. W. Bruyn, and H. L. Klawans. Amsterdam: Elsevier Science Publishers.

Warrington, E. K., and A. M. Taylor. 1973. The contribution of the right parietal lobe to object recognition. *Cortex, 9,* 152–164.

Weber, A. M., and J. L. Bradshaw. 1981. Levy and Reid's model in relation to writing hand/posture: An evaluation. *Psychological Bulletin, 90,* 74–88.

Wechsler, D. 1958. *The Measurement and Appraisal of Adult Intelligence.* Baltimore, MD: Williams and Wilkins.

Wernicke, C. 1874. The symptom of complex aphasia. Translated and republished in *Disorders of the Nervous System,* ed. A. Church, New York: Appleton-Century-Crofts.

Westergaard, G. C. 1991. Hand preference in the use and manufacture of tools by tufted capuchin (*Cebus apella*) and loin-tailed macaque (*Macaca silenus*) monkeys. *Journal of Comparative Psychology, 105,* 172–176.

Wexler, B. E., E. L. Giller, and S. Southwick. 1991. Cerebral laterality, symptoms, and diagnosis in psychotic patients. *Biological Psychiatry, 29,* 103–116.

Whitaker, H. A., and G. A. Ojemann. 1977. Lateralization of higher cortical functions: A critique. *Annals of the New York Academy of Science, 299,* 459–473.

Whitehead, R. 1991. Right hemisphere superiority during sustained visual attention. *Journal of Cognitive Neuroscience, 3–4,* 329–331.

Witelson, S. F. 1983. Bumps on the brain: Right-left anatomic asymmetry as a key to functional asymmetry. In *Language Functions and Brain Organization,* ed. S. J. Segalowitz, 117–144. New York: Academic Press.

——— 1977. Developmental dyslexia: Two right hemispheres and none left. *Science, 195,* 309–311.

——— 1985. The brain connection: The corpus callosum is larger in left handers. *Science, 229,* 665–668.

——— 1987. Neurobiological aspects of language. *Child Development, 58,* 653–688.

——— 1989. Hand and sex differences in the isthmus and genu of the human corpus callosum. *Brain, 112,* 799–835.

Witelson, S. F., and D. L. Kigar. 1987. Neuroanatomical aspects of hemisphere specialization in humans. In *Duality and Unity of the Brain,* ed. D. Ottoson, 466–495. London: MacMillan.

Wolcott, C. L., R. E. Saul, J. B. Hellige, and S. Kumar. 1990. Effects of stimulus degradation on letter-matching performance of left and right hemispheric stroke patients. *Journal of Clinical and Experimental Neuropsychology, 2,* 222–234.

Wolf, M. E., and M. A. Goodale. 1987. Oral asymmetries during verbal and non-verbal movements of the mouth. *Neuropsychologia, 25,* 375–396.

Wood, F. 1983. Laterality of cerebral function: Its investigation by measurement of localized brain activity. In *Cerebral Hemisphere Asymmetry: Method, Theory, and Application,* ed. J. B. Hellige, 383–410. New York: Praeger.

Wood, F. B., D. L. Flowers, and C. E. Naylor. 1991. Cerebral laterality in functional neuroimaging. In *Cerebral Laterality: Theory and Research,* ed. F. L. Kitterle, 103–116. Hillsdale, NJ: Erlbaum.

Woods, B. T., and H. L. Teuber. 1978. Changing patterns of childhood aphasia. *Annals of Neurology, 3,* 273–280.

Wyke, M. 1971. The effects of brain lesions on the performance of bilateral arm movements. *Neuropsychologia, 9,* 33–42.

Yeni-Komshian, G. H., and D. H. Benson. 1976. Anatomical study of cerebral asymmetry in the temporal lobe of humans, chimpanzees, and rhesus monkeys. *Science, 192,* 387–389.

Young, G., S. J. Segalowitz, P. Misek, I. E. Alp, and R. Boulet. 1983. Is early reaching left-handed? Review of manual specialization research. In *Manual Specialization and the Developing Brain,* ed. G. Young, S. J. Segalowitz, C. Corter, and S. E. Trehaub, 13–32. New York: Academic Press.

Zaidel, E. 1983a. Disconnection syndrome as a model for laterality effects in the normal brain. In *Cerebral Hemisphere Asymmetry: Method, Theory, and Application,* ed. J. B. Hellige, 95–151. New York: Praeger.

——— 1983b. A response to Gazzaniga: Language in the right hemisphere, convergent perspectives. *American Psychologist, 38,* 542–546.

——— 1985. Language in the right brain. In *The Dual Brain: Hemispheric*

Specialization in Humans, ed. D. F. Benson and E. Zaidel, 205–231. New York: Guilford Press.

Zaidel, E., and A. M. Peters. 1981. Phonological encoding and ideographic reading by the disconnected right hemisphere: Two case studies. *Brain and Language, 14,* 205–234.

Index of Authors Cited

General Index

Bradford College Library
3 6660 00107 8418